Reactivity and Structure
Concepts in Organic Chemistry

Volume 3

Editors:

Klaus Hafner Jean-Marie Lehn
Charles W. Rees P. von Ragué Schleyer
Barry M. Trost Rudolf Zahradník

H. Kwart   K. King

d-Orbitals in the Chemistry of Silicon, Phosphorus and Sulfur

Springer-Verlag
Berlin Heidelberg New York 1977

Harold Kwart

Professor at the Department of Chemistry, University of Delaware, Newark, Delaware, USA

Kenneth King

Franklin Avenue, Massepequa, USA

ISBN-3-540-07953-X Springer-Verlag Berlin Heidelberg New York
ISBN-0-387-07953-X Springer-Verlag New York Heidelberg Berlin

Library of Congress Cataloging in Publication Data. Kwart, H. 1916– D-orbital involvement in the organo-chemistry of silicon, phosphorus, and sulfur. (Reactivity and structure ; v. 3) Bibliography: p. Includes index. 1. Chemical bonds. 2. Organosilicon compounds. 3. Organophosphorus compounds. 4. Organosulphur compounds. I. King, Kenneth, 1941– joint author. II. Title. III. Series. QD461.K9 547'.1'224 77-1555

This work is subject to copyright. All rights are reserved, whether the whole or part of the material is concerned, specifically those of translation, reprinting, re-use of illustrations, broadcasting, reproduction by photocopying machine or similar means, and storage in data banks. Under § 54 of the German Copyright Law where copies are made for other than private use, a fee is payable to the publisher, the amount of the fee to be determined by agreement with the publisher.

© by Springer-Verlag Berlin Heidelberg 1977.
Printed in Germany.

Typesetting: R. u. J. Blank, München
Bookbinding: Konrad Triltsch, Würzburg
2152/3120 – 543210

Explanatory Preface

This book was undertaken for the purpose of bringing together the widely diverse lines of experimental work and thinking which has been expressed but has often been unheard on the title question. It will be clear to the reader that a critical viewpoint has been maintained in assembling the material of this rapidly expanding area of concern to organic chemists. It should be clear, too, that the authors are not purveying a singular viewpoint and do not regard the discussions presented as the ultimate word on the subject. In fact, it should be anticipated that many of the viewpoints presented may have to be altered in the light of new developments. In recognition of this and to show the way an appendix of recent results and interpretation has been included where an alteration in viewpoint on some of the material treated in the text has been necessitated by developments in the most recent literature. This appendix should be regarded as the reader's opportunity to maintain currency in all aspects of this subject if it is kept abreast of the literature.

The bibliography, from which most of the material of discussion has been drawn, is organized in a somewhat unusual manner which deserves some explanation here. A reference citation can consist of (as much as) a six space combination of letters and numerals. The first three spaces can correspond to the first letters of the first three author's (if there are that many) last names, or, if there are more than three authors, the first letter of each of the first two author's last names followed by the symbol "&". If there are only two authors the reference is a five space combination and, in the case of a sole author, a four space combination. Since there has been no need to reference the literature before 1920, the last two spaces of the combination comprise the last numerals of the year of the twentieth century in which the publication was issued. Finally, a hyphen occupies a space separating the authors initials from the year number. Thus, for example, (WRC-69) can be uniquely located in the bibliography as an article by Wolfe, Rauk, and Csizmadia published in 1969, and (UM&-70) represents an article by Ugi, Marquarding, and others which appeared in 1970. On the other hand, in the purely alphabetical and chronological listing of these references in the bibliography, one will find that (U-69) corresponds to an article by Usher which issued in 1969.

An index of authors names has also been compiled which lists every author cited in the bibliography alongside of all the references in which his or her name appears as an author. Supplementing this is a list of reference-page citations in which are compiled the numbers of every page on which the reference of interest has appeared in the text.

Explanatory Preface

A subject index, the material for which has been gathered in an unexceptional way for the purpose of assisting the reader to locate discussions in the text through key words, has also been included. This index seems to find its greatest use in connection with the table of contents listing the various categories in which the general subject has been treated.

We are pleased to acknowledge here our debt to several sources of inspiration and unremitting aid in carrying out our mission; to Dr. William A. Sheppard of the Central Research Department of the E. I. Dupont Experimental Station who provided a catalytic file of references to initiate the project, to Helen W. Kwart for invaluable stenographic and bibliographic services which held the program on course, and to Dr. Robert Louw and the Gorleus Lab of the University of Leiden where one of us (H.K.) was provided with the opportunities of a stimulating sabbatical leave during 1975.

Newark, Delaware,
June 27, 1977

Harold Kwart
Kenneth King

Contents

	Explanatory Preface	V
I.	Introduction – dπ vs. dσ Bonding	1
II.	Theoretical Basis for d-Orbital Involvement	2
	A. Geometry of Bonding Orbitals	2
	B. Size of d-Orbitals	3
	C. Effect of Electronegative Ligands	4
	D. Energies of d-Orbitals	6
	E. Alternate Explanations for dσ and dπ Bonding	7
	F. dπ Delocalized Bonding	13
III.	Physical Properties Related to dp-π Bonding	14
	A. Bond Lengths and Angles	14
	1. Survey of Relevant Articles	22
	B. Infrared Spectra	25
	1. Survey of Relevant Articles	30
	C. NMR and Related Indices	32
	1. Survey of Relevant Articles	44
	D. UV, ESR and Dipole Moments	52
	1. Survey of Relevant Articles	66
IV.	The Effects of dp-π Bonding on Chemical Properties and Reactivity	71
	A. Acidity-Basicity	71
	B. Electrophilic Aromatic Substitution	73
	C. Alpha-Sulfonyl and -Sulfinyl Carbanions	74
	D. Alpha-Thio Carbanions	80
	E. Thiazolium Ylides	84
	F. Addition to Conjugated Carbon-Carbon Double Bonds	88
	G. Survey of Relevant Articles Sections IV A-IV F	90
V.	Pentacovalency	96
	A. The Pentavalent State in Organic Chemistry	96
	1. Survey of Relevant Articles	111
	B. dp-π Bonding in Phosphate Hydrolysis	116
	1. Survey of Relevant Articles	119
	C. Pentavalent Intermediates in Phosphate Hydrolysis	121
	1. Survey of Relevant Articles	133

Contents

      D. Pentacoordinate Sulfur Intermediates 137
          1. Survey of Relevant Articles 146
      E. Pentacoordinate Silicon Intermediates 149
          1. Survey of Relevant Articles 159
      F. Synthetic Applications of Pentacoordinate Intermediates . . . 164

VI. References 179

VII. Appendix of Recent Results and Interpretations 197

VIII. Index of Authors and References 200

IX. Index of Reference – Page Citations 213

X. Subject Index 218

I. Introduction

Ever since the formulation of the Lewis octet theory in 1916, valence shell expansion in organic compounds of third and higher period elements has been a controversial question. Much research has been directed to this problem, particularly with sulfur compounds but also with phosphorus, silicon, and the halogens, but definitive data are lacking. Valence shell expansion, almost invariably by the use of d-orbitals, has commonly been invoked to explain all sorts of anomalous and unexpected effects in the spectra, bond lengths and angles, and the chemical activity of organic compounds of sulfur, phosphorus, silicon, and the halogens. Often alternative explanations such as lesser electronegativity and greater polarizability of these third row elements with respect to their second period analogs were never considered as explanations for apparent anomalies.

In this article we critically review evidence cited for d-orbital participation in order to assess the contribution of d-orbitals in bonding of silicon, phosphorus and sulfur, particularly in organic molecules. In effect we have two problems to consider:
1) $d-\pi$ bonding where d-orbitals interact to accept charge density from a π system or unshared p electrons;
2) $d-\sigma$ bonds, where the valence shell of the third row element is expanded, presumably by hybridization with d-orbitals, to form a higher than normal number of σ bonds, for example in PCl_5 or SF_6.

However, in actual bonding situations, this separation is artificial since $d\pi$ and $d\sigma$ are not independent and contributions from each not easily separated.

II. Theoretical Basis for d-Orbitals

A. Geometry of Bonding Orbitals

Since d-orbitals are central to the problem of valence shell expansion, we should first discuss the nature of d-orbitals. In the free atom, there are five d-orbitals, each of equal energy and having the electron distributions shown in Fig. 1. Although the degenerate energy level will be split by the field of the ligands, the angular parts of the orbital wave functions will remain unchanged. On the basis of angularity (geometry alone), d-orbitals should have considerable overlap with suitably oriented $2s$ or $2p$ orbitals on the ligand.

Depending on the type of overlap, either σ, π or δ bonds may result (Fig. 2.) The δ bonds have two nodes and result only from overlap of two appropriately directed d-orbitals; we will not be concerned with these. The most important type of d-orbital bonding is the π bond, which has a node along the bond axis and which may result from overlap with a p or d-orbital. In most cases of interest to us, the overlap is with a $2p$ orbital on carbon, nitrogen or oxygen. Since π bonds have a node along the bond-

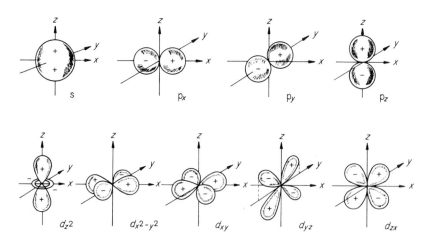

Fig. 1. Ballon pictures of s, p and d atomic orbitals

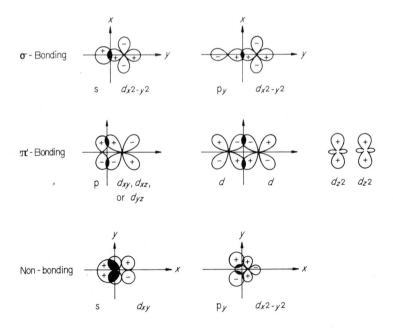

Fig. 2. Examples of σ, π and δ Bonding

ing axis, they always occur in conjunction with a σ bond, thus forming the familiar double bond. However, double bonds involving $d\pi$ orbitals appear to have very different properties from ordinary $p\pi$ double bonds, as will be discussed later. Of lesser importance, at least in, organic chemistry, are the σ bonds which may be formed from overlap of a d-orbital with an s, p, or another d-orbital. In this connection we will discuss later a number of single-bonded organic derivatives of pentavalent phosphorus (*HF-68*). In addition, $S_N 2$ reactions at phosphorus or silicon are generally thought to involve pentavalent intermediates with $d\sigma$ bonds (*E-60, H-65a*).

B. Size of d-Orbitals

The angular electron distribution is only half the story. In order to compute overlap integrals, we must know the radial wave function (the size of the d-orbital), a much more difficult problem. The one-electron radial wave function is generally taken to be a distribution function of the form in equation (1) where Z*, the effective nuclear charge, depends on the screening of the nucleus by the other electrons. Since this screening is difficult to evaluate quantiatively, Z* is often computed by means of Slater's rules (*S-30*).

Theoretical Basis for d-Orbitals

The electron distribution functions

$$R(r) = r^a \exp.(-Z^*/ar) \qquad \text{eq. (1)}$$

calculated by this means for $3d$ orbitals of second-row elements are found to be extremely diffuse with radial maxima far in excess of normal bond lengths. Clearly, an orbital this large could never have any effective overlap.

One way out of this problem would be to assume that Slater's rules are inadequate for $3d$ orbitals and to calculate Hartree-Fock SCF wave functions, which are much more exact. Such calculations generally indicate that singly occupied d-orbitals (e.g. the sp^3d configuration of sulfur) are of high energy and, consequently, large size, but that doubly occupied d-orbitals may be compact enough for bonding, particularly for spectroscopic terms of low energy, e.g. the ^7F term of the sp^3d^2 configuration of sulfur, which was found to have a mean $3d$ orbital radius of 1.18 Å (*CWM-64*). However, for the sp^3d^2 valence state of sulfur, which has a random spin distribution and is thus a combination of spectroscopic terms, two separate calculations gave mean $3d$-orbital radii of 2.72 Å and 1.98 Å respectively (*BZ-65, CT-66*).

C. Effect of Electronegative Ligands

Since d-orbitals in the free atom appear far too diffuse for effective overlap, we must introduce some means of radial contraction. The electrostatic field of the ligands can promote contraction as construed by a quantitative model due to *Craig* and *Magnusson* (*CM-56*). The ligands are considered as point charges q in, let us say, octahedral array at a distance from the central nucleus equal to the bond length. Since an electron is most likely to be found where the positive potential is highest, the position of maximum electron density should lie closest to the ligands. Thus, an orbital whose radial maximum lies beyond the ligands will be contracted until the radial maximum, the position of highest electron density, lies near the ligands. One would expect this contraction to be anisotropic; that is, orbitals which point in the direction of the ligands should be contracted more than those which do not. However, the calculations of *Craig* and *Magnusson* show that the departure of the ligand potential field from spherical symmetry has a negligible effect (at least) upon the electron density distribution.

If this point charge model is a true representation of the radial contraction, we would expect that only highly electronegative ligands could cause valence shell expansion in third row elements.

While most compounds involving valence shell expansion are fluorides, chlorides, or oxides, there are many cases where the ligand is carbon or nitrogen, weakly electronegative elements. Some examples are the penta(aryl) phosphoranes and the phosphazenes. Carbon must also be the ligand in any case where conjugation in organic compounds of third and higher period elements involves $d-p$ bonding. To consider the ligands more explicitly, *Craig* and *Zauli* (*CZ-62*) calculated the potential due to the ligands from the best available SCF wave functions for those ligands. In order to simplify the mathematics all exchange between bonding and nonbonding orbitals was neglected, thus minimizing only the electrostatic part of the total energy. The energy so calculated was minimized with respect to the exponent in a Slater type orbital. For compounds of the type AB_n, large radial contractions were found for $n = 4$ and $n = 6$. The largest contractions, of course, occurred with B = Fluorine, but sizable contractions were also found for chlorine and for carbon in the sp^3 configuration.

A major difference between the results of the point charge model and those of *Craig* and *Zauli* is the radial contraction of d-orbitals engaged in π bonding. In the latter case the contraction was found to be quite small and very anisotropic. However, radial contraction is of much lesser importance in π than in σ bonding. Whereas σ bonds and $pp-\pi$ orbitals have regions of maximum overlap equidistant between the two nuclei, the maximum overlap for the $dp-\pi$ bond lies much closer to the ligand; in fact, the lobes of the d-orbital should lie almost above and below the ligand (*S-68*). Thus, calculations using Slater $3d$-orbitals indicate fairly large $dp-\pi$ overlap integrals, generally above 0.2 (*CM&-54, J-54*). It should be noted however that such bonds would be expected to be highly polar and fairly weak since the maximum overlap occurs in a region of weak nuclear field.

Mitchell (*M-68*) has carried out electrostatic calculations similar to those of *Craig* and *Zauli* (*CZ-62*) on X_3PO, where X is F, Cl, sp^3 C, or H. Considerable increases in the $3d$-orbital exponent, which is inversely proportional to the radial maximum, were found for $d_\sigma(d_{z^2})$, d_π (the degenerate d_{xz} and d_{yz}), and $d_\delta(d_{x^2-y^2}$ and $d_{xy})$ orbitals, these increases being in the order

$$F > C > Cl > H \quad \text{and} \quad \sigma > \pi > \delta.$$

However, when interatomic exchange was included, using a simple perfect-pairing approach in the calculations on F_3PO, repulsive interactions with inner-shell and lone pair electrons of the ligands prevented any contraction of the d_σ and d_δ-orbitals, but the contraction of the d_π-orbitals was even greater than that without the interatomic exchange.

In summary then, the perturbing power of ligands on d-orbitals decreases in the order: $F > C > Cl > H$ where fluorine, chlorine and carbon are all very effective relative to hydrogen. These results are in accord with experimental observations that PF_5, PCl_5 and $P(C_6H_5)_5$ are all stable compounds, but that PH_5 is unknown. In the case of sulfur, SF_6 is the only stable hexavalent derivative of the series, and SF_4 is stable to normal handling, but SCl_4 decomposes at $-31\,°C$, (*CW-66*), and $S(C_6H_5)_4$ and related pentacoordinate sulfur compounds decompose at $-80\,°C$, (*TLA-69*). Thus, these data clearly show that the stabilization of d-orbitals for effective $d\sigma$ (or $d\pi$) bonding is not derived simply from the electronegativity of the ligand, but is related to both the orbital exponent and bond length.

D. Energies of d-Orbitals

The third factor to be considered in d-orbital bonding is the energy of the d-orbital. This may be expressed in two ways depending on how the bond is formed: as the promotion energy needed to excite an electron into the higher orbital, or as the electron affinity related to the energy gain when the higher orbital accepts an electron from some outside source. Since most $dp-\pi$ bonds are dative, the electron affinity is more relevant; however a high promotion energy generally means a low electron affinity. For example, the difference between the first ionization potential of sulfur (10.3 eV) and the promotion energy needed to reach the s^2p^3d state (8.4 eV) as determined by atomic spectroscopy (*CWM-64*) will be equal to the electron affinity of the empty d-orbital of the $s^2 p^3$ state of the singly charged cation. The d-orbitals of a cation will of course have a larger electron affinity than those of a neutral atom. In fact, in the case of silicon the electron affinity for the process $s^2p^2 \rightarrow s^2p^2\ 4s$ is only 0.15 eV (*PZ-67*), which would mean an even lower electron affinity for the $3d$ orbitals; atomic spectroscopy as well as the positions of potassium and calcium in the periodic table both show that the $3 d$-orbitals are of higher energy than the $4 s$.

Quantum-mechanical calculations confirm the high energy of the $3 d$-orbitals; for example, *Bendazzoli* and *Zauli* (*BZ-65*) calculated a promotion energy of 31.3 eV for the $sp^3 d^2$ valence state of sulfur (*CZ-62*). For comparison the energy of the process

$$S(s^2 p^4) \rightarrow S^{++}(sp^3)$$

is 34.1 eV (*H-60, M-68*). In addition, the point charge model indicates even larger promotion energies than those in the free atom, since the increase in binding energy due to the positive electric field of the ligands will be less for orbitals whose radial maxima lie outside the bonding radius than for those whose radial maxima lie inside it.

Electrostatic SCF calculations by *Mitchell* (*M-68*), similar to those of *Craig* and *Zauli*, showed that the promotion energy of the sp^3d^2 state was reduced from 24.3 eV in the free atom to 7.9 eV in SF_6, a value 4.6 eV less than the promotion energy of the $s^2p^3\ 4s$ state of SF_6. The promotion energy must be recovered from the energy difference between molecular SF_6 and the free atoms in their ground states, a difference calculated to be 36.7 eV. When interatomic exchange was taken into account, this atom-molecule energy difference increased to as much as 190 eV, depending on how much mixing of the fluorine $2s$ and $2p$ atomic orbitals was assumed. On the other hand the $s^2p^3d \rightarrow sp^3d^2$ promotion energy was 16.8 eV compared to an atom-molecule energy difference of only 5.4 eV in SH_6, thus explaining the nonexistence of this compound. The measured heat of formation of SF_6 is 11.4 eV (*PZ-67*), but this comparison is not meaningful since the thermodynamic standard states for sulfur and fluorine are not the free atoms.

E. Alternate Explanations for $d\sigma$ and $d\pi$ Bonding

Compounds such as SF_6 and $(C_6H_5)_5P$ of third and higher period elements appear to contain ten (and sometimes twelve and fourteen) valence electrons. This valence shell expansion is most easily explained by use of d-orbitals in sigma bonding. The most characteristic feature of these "valence expanded" compounds that must be explained by any theory of bonding is stereochemical configurations. The stereochemistry of such compounds is similar to that of the transition metal complexes and has been commonly accepted as a result of contributions of properly shaped d-orbital hybrids. Recently, alternative explanations have been presented.

One of the most interesting explanations of the stereochemistry that does not involve the question of chemical bonding has been proposed by *Gillespie* (*G-60*). This model considers a number of like-charged particles, i.e. electron pairs, on the surface of a sphere and asks what configurations will minimize the repulsions between the particles. For two, three or four particles the configurations with minimum repulsions are respectively a straight line, an equilateral triangle, and a tetrahedron, all of which are very familiar to chemists, especially the tetrahedron. For the case of five particles we must consider more explicitly the repulsions, which will be proportional to r^{-n}; for purely electrostatic repulsions, $n = 2$. However, the Pauli exclusion principle requires that particles of like spin will have a repulsion due to their spins superimposed on the electrostatic repulsion; thus, electron pairs will repel each other with a force proportional to r^{-n}, $n > 2$. For very large n, five particles have two configurations of minimum energy, a trigonal bipyramid and a square pyramid. Using d-orbitals an sp^3d hybrid may have either of these configurations depending on whether the d_{z^2} and $d_{x^2-y^2}$ orbital is used (*G-63*), but calculations do not define which is more energetically favorable. However, *Gillespie* (*G-63*) has shown that for $8 \leqslant n \leqslant 12$ the square pyramid becomes distorted and is about 8% less stable than the trigonal bipyramid. As a matter of fact, almost all compounds with five single bonds and ten valence electrons appear to assume the trigonal bipyramid configuration. However, certain bicyclic phosphoranes do assume a modified square pyramidal configuration, as will be discussed in Section V.A.

One of the most important features of Gillespie's model is that repulsions involving nonbonding electron pairs should be somewhat larger than those involving

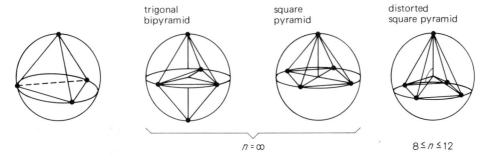

Fig. 3. Gillespie's Model of Stereochemistry

only bonding pairs. For example, ammonia is best considered as a tetrahedron with one of the four vertices occupied by a nonbonding pair. Similarly SF$_4$ is best considered as a trigonal bipyramid with one of the basal positions occupied by the nonbonding electron pair. The preference of the nonbonding pair for the equatorial position becomes obvious when it is realized that, for third row elements, electron pairs experience repulsion whenever the angle between them is less than or equal to 90° (*G-63*). An apical position in the trigonal bipyramid forms 90° angles to the three basal positions, while an basal position forms 90° angles only to the two apical positions, thus giving less repulsive strain overall at the basal position. This difference in repulsive strain between the two positions also accounts for the longer bond lengths to the apical position, a characteristic of all compounds in the trigonal bipyramid configuration. The structure of SF$_4$ is shown below (*CW-66*). The resemblance of this structure to a trigonal bipyramid is clearly apparent; the bond angles of any real compound are not expected to be those of a perfect trigonal bipyramid.

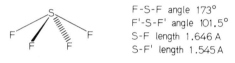

F-S-F angle 173°
F'-S-F' angle 101.5°
S-F length 1.646 A
S-F' length 1.545 A

Structure of sulfur tetrafluoride

A similar situation is found in the phosphoranyl radical (*SJB-75*). The first of these to be observed was the (PF$_4$) radical (*M-63a*), formed by γ-irradiation of solid ammonium hexafluorophosphate. ESR spectroscopy (*MS-74*) shows this radical to have a trigonal bipyramidal structure with the lone pair basal and two nonequivalent types of fluorines, i.e. no indication of a permutational isomerism or polytopal rearrangement of its bond structure. Similar structures have been observed in solution for the radicals $R_n\dot{P}(OR)_{4-n}$, $Me_3CO(R)\dot{P}Cl_{3-n}$, and $Me_3CO(R)\dot{P}(NR_2)_{3-n}$ where R = alkyl. The same preference of electronegative substituents for the apical positions prevails here as in ordinary phosphoranes. Studies have been carried out on the formation of various tetraalkoxyphosphoranyl radicals by attack of an alkoxy radical on a phosphite ester, and on the decomposition of the radical to a phosphate ester and an alkyl radical. These data showed both that phosphoranyl radical formation and decomposition occur preferentially at the apical position, although reaction at the basal position is possible, and that polytopal bond rearrangement is considerably slower than decomposition, implying a rearrangement barrier of at least 10 kcal/mole (*BM-76, BH&-76*).

Ab-initio Hartree-Fock calculations on the PH$_4$ · radical (*HO-76*) were in agreement with these results. The optimum structure, a slightly distorted trigonal bipyramid with the unpaired electron basal, was 18 kcal more stable than a trigonal bipyramid with unpaired electron apical, 57.5 kcal more stable than a tetrahedron, but only 3.0 kcal more stable than a square pyramid with the unpaired electron basal. Similar calculations for SH$_4$ showed an even smaller difference between the trigonal bipyramid and the square pyramid (*SS-75*). Analysis of possible ligand scrambling mechanisms showed the lowest energy transition state to be ~ 15 kcal for a turnstile rotation. Whether the unpaired electron is in the duet or trio of ligands, the transition state resembles a distorted trigonal bipyramid with the unpaired electron apical, thus explaining the high energy requirement. On the other hand ligand scrambling involving cleavage and reformation of an apical P–H bond might be of low energy; calculations of the potential energy surface for the reaction $PH_3 + H\cdot$ showed that the incoming hydrogen is preferably located at the apical position of the resulting trigonal bipyramid and that cleavage of this apical P–H bond is a very low energy process.

However, the presence of ligands with π electrons results in a tetrahedral configuration, with the spin delocalized over the π electron system, as observed by *Buck* for the biphenylenephosphonium system, with or without substituents on the benzene rings (*RLB-72, RFB-73*). Calculations (*RF &-74*) for the system

Biphenylene phosphoranyl Radical

indicate a spin density of 0.85 in the biphenylene system and 0.2 on phosphorus, apparently delocalized over all five $3d$ orbitals. The biphenylene system is not necessary; $Me_3C(Ph)\dot{P}(OMe)_2$ has been found to be tetrahedral (*BJB-74*). CNDO calculations (*DPB-75*) show that the preference for a tetrahedral over a trigonal-bipyramidal configuration depends on the extent of spin delocalization in the π ligands and on the electronegativity of the other ligands. To illustrate, $Ph\dot{P}(OMe)_2 OCF_3$ is a trigonal bipyramid, but $Ph\dot{P}(OMe)_3$ prefers the tetrahedral configuration. This can be explained by the fact that spin delocalization would require formation of positive charge on phosphorus, which would become untenable under the influence of strongly electron-withdrawing substituents. However, it should be noted that recent MO calculations on $Cl_3\dot{P}O\cdot$ (*GW-74*) suggest that the spin in this trigonal bipyramidal system is not localized in a nonbonding, basal hybrid orbital but is delocalized into p orbitals of the apical ligands *via* a three-center nonbonding orbital composed of a symmetry-allowed combination of the phosphorus $3s$ and $3p$ orbitals in antibonding combination. In this case the spin densities were 0.29 each on the apical chlorines and 0.38 on the phosphorus $3s$. ESR data (*KMK-72*) on $Me_nH_{3-n}\dot{P}OCMe_3$ can also be interpreted in terms of delocalization into the apical ligands. Similar results were found in extended *Hückel* (*CH-76*) and ab-initio SCF calculations (*SS-75*) on SH_4 with optimized geometry. They demonstrated the severe mixing of the basal lone pair and the nonbonding orbital of the three-center apical bond; the lone pair orbital and the nonbonding orbital are of the same symmetry, and either one may be the HOMO depending on the electronegativity of the ligands around sulfur. The ab-initio calculations (*SS-75*) also showed negligible d-orbital contributions to the bonding in this hypothetical sulfurane, which is predicted to be thermodynamically unstable with respect to the $H_2S + H_2$ system.

Theoretical Basis for d-Orbitals

For compounds with twelve valence electrons, Gillespie's model predicts that the configuration of minimum energy will be, as expected, the octahedron. In fact, all hexavalent compounds assume an octahedral configuration, although in many cases the bond angles and lengths are distorted. The bonding in such compounds may, but need not necessarily, involve $sp^3d_{z^2}d_{x^2-y^2}$ hybrid orbitals. If one of the vertices of an octahedron is occupied by a nonbonding electron pair, the result is a square pyramid. This is the structure of BrF$_5$ (*PP-68*) and the fact that the angle between the base and the vertex is only 85° instead of 90° indicates that the larger steric requirement of the lone pair is forcing the four basal fluorine atoms out of coplanarity with the bromine. Where there are two nonbonding pairs, a square-planar configuration is assumed; this is the structure of XeF$_4$ and BrF$_4$ (*PP-68*). This configuration is quite common in transition metal complexes but is rather rare outside the transition series.

The apparent advantage of *d*-orbital hybrids in discussing the bonding of compounds involving valence shell expansion is a straightforward explanation of the stereochemistry; but Gillespie's repulsive model also explains the stereochemistry. Two major problems in the use of *d*-orbital hybridization are promotion energies and radial contraction, already discussed. Also, in certain cases, notably the polyhalide anions, the use of *d*-orbital hybridization gives completely erroneous results. For example, the linear structure of I$_3^-$ may be explained in terms of a trigonal bipyramid sp^3d hybrid with the three basal positions occupied by nonbonding electron pairs. Then, by analogy, I$_5^-$ should be an octahedral sp^3d^2 hybrid with two apical positions occupied by nonbonding electron pairs, giving a square planar structure. This is indeed the structure of ICl$_4^-$, but X-ray diffraction analysis of the $(CH_3)_4N^+I_5^-$ crystal gives a structure which appears impossible to interpret in terms of *d*-orbital hybridization (*HR-51*).

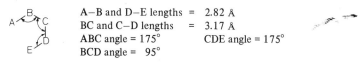

A—B and D—E lengths = 2.82 Å
BC and C—D lengths = 3.17 Å
ABC angle = 175°   CDE angle = 175°
BCD angle = 95°

Structure of $(CH_3)_4N^+I_5^-$: A = B = C = E = I

Havinga and *Wiebinga* (*HW-59*) explain the structure of such anions in terms of a molecular orbital theory using 3-center-4-electron bonds formed from only the *p*-orbitals of the atoms involved. When three p_x (or three p_y or three p_z) orbitals on three adjacent atoms (arranged linearly) overlap, three 3-center molecular orbitals result, where one will be strongly bonding, one just as strongly antibonding, and the third nonbonding. That such a bond is energetically favorable in spite of the occupied nonbonding orbital can be proven qualitatively by simple *Hückel* type calculations. For I$_3^-$ the secular determinant can be expressed as

$$\begin{vmatrix} \alpha - E & \beta & 0 \\ \beta & \alpha - E & 0 \\ 0 & \beta & \alpha - E \end{vmatrix} = 0$$

and its solution gives the orbital energies of $E = \alpha + \beta\sqrt{2}$, α, and $\alpha - \beta\sqrt{2}$. The total energy of this 16 electron system (six p electrons from the iodide ion and five each from the two iodine atoms) is thus $16\alpha + 2\sqrt{2}\beta$. For comparison the $I_2 + I^-$ system has a total energy of $(10\alpha + 2\beta) + 6\alpha$, giving a resonance energy of $2\sqrt{2}\beta - 2\beta = .83\beta$. Similar calculations may be made for different configurations of various polyhalide anions to give the configuration of highest resonance energy. For chlorine and bromine, *Havinga* and *Wiebinga* (*HW-59*) express the Coulomb integrals as $\alpha + .4\beta$ and $\alpha + .2\beta$ respectively, where the coefficient of β is based on two-thirds of the electronegativity difference between iodine and chlorine or bromine. Since the factor 2/3 is arbitrary, the results are qualitative only. Such bonding has also been discussed by *Musher* (*M-69a*).

The extension of this bonding scheme to other systems is evident. Using for example the p_x and p_y orbitals in two mutually perpendicular 3-center bonds gives a square planar structure, while using all three p-orbitals gives an octahedron. A trigonal bipyramid would be formed from one 3-center bond and an sp^2 hybrid to form the three basal bonds. The bonding in I_5^- must again involve two perpendicular 3-center bonds but with different central atoms for the two bonds.

The nonbonding orbital of a three-center bond has a node on the central atom, consequently shifting most of the electron density to the two outer atoms and making the 3-center bond rather ionic in character. In valence bond terms this involves resonance contributions from ionic structures such as $[PCl_4]^+Cl^-$ for PCl_5 (*P-50*). Nuclear quadrapole resonance studies on $CsICl_4$ and $RbICl_2$ show the Cl–I bond to be about 50% ionic in character (*CY-57*). This explains why all compounds involving valence shell expansion have fairly electronegative ligands. In almost all cases where carbon serves as a ligand atom it is part of an aromatic system which can delocalize negative charge. More important than the electronegativity of the ligands, however, is the difference in electronegativity between the ligands and the central atom; this explains why, for example, AsI_5 exists but not PI_5. Similarly, tetravalent selenium and tellurium both form bis(biphenylyl) derivatives, but the selenium compound is less stable in spite of the decreased ring strain in the 5-membered heterocyclic spiro rings, due to the smaller atomic radius of selenium (*HF-68*). Tellurium but not selenium or sulfur forms a tetraphenyl derivative by reaction of the triphenyl "onium" salt with phenyllithium at $-70°$ (*WF-52*), although tetraphenylsulfurane is apparently involved as a transient intermediate in the formation of diphenyl from triphenylsulfonium tetrafluoroborate and carbon-14 labelled phenyllithium (*HW&-72*, see p. 167). The electronegativity difference also explains the failure of first row elements to show valence shell expansion. Here, however, there is also a steric factor in that the small atomic radii of carbon, nitrogen, and oxygen makes it difficult for these elements to accommodate large numbers of ligands in bulky trigonal bipyramid or octahedral structures around them.

Recent MO calculations confirm the instability of 3-center-4-electron bonding for first row elements. *Gleiter* and *Hoffmann* (*GH-68*) considered the possible stabilization of singlet nitrene, NH, with respect to the triplet ground state by means of a 3-center bond with the two nonbonding electron pairs on two ammonia molecules. This system is indeed more stable than an isolated nitrene, but it is unstable with respect to dissociation into one ammonia molecule and a 2-center bond:

$HN=NH_3 \leftrightarrow H\overset{\oplus}{N}-\overset{\ominus}{N}H_3 \leftrightarrow H\overset{\ominus}{N}-\overset{\oplus}{N}H_3$.

This instability probably results from the fact that the nonbonding, 3-center orbital becomes strongly antibonding below a certain internuclear separation distance. On the other hand the system $H_2S-S-SH_2$ appears to be a true minimum in the energy level diagram (*GH-68*) because it contains empty d-orbitals which can accommodate an electron pair when the nonbonding orbital becomes antibonding. Similarly, pentavalent carbon occurs only as a transition state in the $S_N 2$ reaction while pentavalent silicon appears to be an intermediate in some substitution reactions (*E-60*), and a few pentavalent silicon compounds have actually been isolated (*WC-63*).

The theory discussed above is as much an oversimplification as the use of simple $sp^3 d^n$ hybrids. Any valid MO theory must take explicitly into account all relevant atomic orbitals including d-orbitals, and should include all the p-electrons on the ligands, not just those formally engaged in bonding to the central atom. *Santry* and *Segal* (*SS-67, S-68*), have performed CNDO calculations for a number of molecules including SF_6 which show more clearly the role of d-orbitals. The CNDO method is an SCF scheme simplified by the complete neglect of differential overlap and by empirical calculation of all one-electron matrix elements. Calculations were performed both with and without the inclusion of d-orbitals, and the bond angles, bond lengths, and dipole moments obtained were compared with experimental data. In most cases, including SF_6, the bond angles could be satisfactorily approximated without the use of d-orbitals, but d-orbitals were required to reproduce the experimental bond lengths and dipole moments. In other words, d-orbitals affect stability and electronic charge distribution but not the molecular geometry. Most of the increase in stability is due to $dp-\pi$ bonding, which decreases the positive charge on the central atom, but for SF_6 some $sp^3 d^2$ hybridization was involved in lowering the energy of the e_{lg} orbitals. In short, then, while most compounds involving valence shell expansion probably utilize 3-center-4-electron bonding, this does not preclude the use of d-orbitals.

Recent self-consistent, *ab-initio* MO calculations (*CS&-76*) of the geometry and binding energies for 20 different compounds of third-row elements show that satisfactory agreement of bond angles and lengths for compounds of chlorine, sulfur, and phosphorus generally requires inclusion of d-orbitals in the basis set. This is particularly true for the hypervalent molecules molecules ClF_3, SO_2, and POF_3, but it also applies to Cl_2 and HCl. It does not apply to hydrogen sulfide, methanethiol, and silanes. Inclusion of d-orbitals in the basis set increased the binding energies by an average of 42 kcal for normal-valent molecules and 235 kcal for hypervalent compounds.

Coulson has pointed out that any calculations based on a full set of atomic orbitals and using a variational principle to determine the best energy will inevitably show some d-orbital participation, but this is chemically meaningless unless the $3d$ atomic coefficients are large (*C-69*). For example, extended *Hückel* calculations on IF_7 indicated the hybridization at iodine (*OD-67*) to be $s^{1.323} p^{2.723} d^{0.204}$, clearly indicating negligible contribution from the $5d$ orbitals of iodine to the three-center bonding based on the $5s$ and $5p$ orbitals of iodine. However, since these calculations were based on Slater orbitals, the results obtained are perhaps not surprising. Similarly, p-orbitals may be polarized by an external electric field to a shape which may be described as a *pd–π* or *pd–σ* hybrid; this, too, is chemically meaningless since it can even involve 2*p*-orbitals. When ab-initio SCF–MO calculations were carried out on phosphane and trimethylphosphane, the results showed very small 3*d* populations, and a negligible *d*-orbital effect on the total energy, but the calculated dipole moments were in much better agreement with experimental data when *d*-orbitals were included in the basis set (*HS-70a*). However, the same calculations also indicated that *dp–π* bonding does exist in PF_3, the highest occupied molecular orbital containing about 3% *d* character.

Similar SCF calculations showed the total 3*d* orbital population to be 0.49e in PCl_3, 0.66e in PF_3, and 1.07e in $POCl_3$ (*HS-70*). The major contributions to the total P–O overlap population in $POCl_3$ were 31.5% from $(3p-2p)\pi$, 24.3% from $(3p-2p)\sigma$, and 34.5% from $(3d-2p)\pi$. In POF_3, SO_2F, and ClO_3F the *dp–π* contributions to the bonds to oxygen were about double those of the fluorine bonds (*HS-70*). In the ClO_2^-, ClO_3^-, and ClO_4^- anions, the differences between the total energies calculated with and without *d*-orbitals were 8.4, 13.0, and 21.8 eV respectively; this is to be expected since increasing the number of electronegative ligands around the central atom should increase the *d*-orbital radial contractions. In an even more extreme example, the same sort of calculations (*HS-69*) showed a reduction in the total energy upon including *d*-orbitals in the basis set of only 1.36 eV for hydrogen sulfide, but 22.8 eV for sulfate ion. For phosphorus trichloride and oxychloride (*HS-70*) calculated and experimental ionization potentials were in good agreement, indicating that this SCF method does satisfactorily describe the bonding.

F. *dπ* Delocalized Bonding

Calculations on *dπ* bonding often are directed to cyclic delocalizations of molecules such as phosphazenes. The reviews by *Mitchell* (*M-69*) and by *Paddock* (*P-64*) discuss this important subject in sufficient detail so that only a summary of the conclusions need be given here. The problem is analyzed for $(AB)_n$ molecules where atoms A contribute *p* electrons and B empty *d*-orbitals. In order to compare these systems to benzenoid hydrocarbons planarity is assumed. Using classical Hückel type calculations, delocalization energies for molecules such as $(NPX_2)_z$ are calculated to be less than for benzene, but still significant. When the d_{xz} and d_{yz} orbitals contribute equally, nodes result at each phosphorus giving a three-center or "island" (*DLW-60*) type of delocalization. This heteromorphic type of bonding does not require ring planarity and does not show the alternating dependence on ring size characteristic of both a chromomorphic system based on d_{yz} alone, and of nonbenzenoid aromaticity. Experimental results, including the nonplanarity of rings with $z > 3$, are in better agreement with heteromorphic than homomorphic bonding. The other three *d*-orbitals contribute to exocyclic P–X, *dp–π* bonding.

III. Physical Properties Relative to dp-π Bonding

The question of d–π bonding has long been an area of controversy in organic compounds containing sulfur, phosphorus or silicon. Although the involvement of d-orbitals in valence-shell-expanded compounds (d–σ bonding) is commonly accepted (although no longer required — see theoretical discussion in preceding section), dp–π bonding is on a much more questionable experimental basis and has usually been invoked to explain the "unexplainable". Most cases where dp–π bonding has been invoked have been concerned with physical and spectroscopic properties, since the effect on chemical properties is generally smaller and complicated by other factors.

A. Bond Length and Angles

d–π Bonding is usually claimed when bond lengths are shorter than predicted by the *Schomaker-Stevenson* equation (*SS-41*) in which the bond length is taken to be the sum of the atomic radii, less a correction for electronegativity. *Cruickshank* (*C-61*) tried to correlate bond lengths with bond orders calculated by a simple MO theory, according to which only the $3 d_{x^2-y^2}$ and $3 d_{z^2}$ orbitals are capable of π overlap in tetrahedral molecules, e.g. orthosilicate, phosphate, and sulfate; this theory resulted in a linear correlation between bond length and order.

Unfortunately, Cruickshank's theory neglects the inductive effect of ligands on d-orbital exponents. *Gillespie* and *Robinson* (*GR-63*), based on published results for large numbers of compounds, empirically correlated S–O and P–O bond lengths and orders by the nonlinear Pauling equation as proposed by *Robinson* (*R-63*).

$$r_n = r_1 - \frac{r_1 - r_2}{1 + C\left(\frac{2-n}{n-1}\right)}.$$

Here n is the total bond order and C an empirical constant. Thus, P–O bond lengths ranged from 1.45 Å in POF$_3$ ($n = 2.22$), through 1.50 Å for dimethylphosphinate ($n = 1.64$), and 1.54 Å for orthophosphate ($n = 1.50$) ions, to 1.71 Å for a hypothetical P—O single bond as calculated by the *Schomaker-Stevenson* equation (*SS-41, R-63*). The bond orders were determined from a linear correlation between bond order and infrared force constant calculated on the assumption of no coupling between the P—O frequency and other vibrations, so that the P—O bond acts as a simple diatomic molecule. The force constant also correlated linearly with bond length on a log-log plot.

P—O and P—S bond lengths have been determined in trimethylphosphine oxide and sulfide (*WH&-75*). At 1.94 Å, the P—S bond length was about equal to the sum of the double-bond radii, whereas the P—O bond length, 1.476 Å, was even shorter than the sum of the double-bond radii. Since there are two $2p$ oxygen orbitals and at least two $3d$ phosphorus orbitals capable of π bonding, there is really nothing unexpected in the idea of a bond order > 2.00. Indeed, by proper adjustment of the constants in the Pauling equation P—S and P—O bond orders of 2.00 and 2.68 have been calculated (*WH&-75*). On the other hand *Robinson's* correlation (*R-63*) gives a P—O bond order of 1.75 for Me_3PO, and the *Siebert* equation (*S-53*) gives a bond order of 1.68 based on a force constant of 8.03 mdyne/Å (*M-75a*). Since it seems unlikely that all the parameters in the Pauling equation would have the same values for all P—O and P—S bonds, the high bond orders calculated by *Hedberg* et. al. (*WH&-75*) may be an artifact. While insufficient data are available for correlating P—S bond order and force constants, the fact that the force constant in Me_3PS is only 1.0 mdyne/Å above that for an ideal single bond (*RNW-75*) makes a bond order of 2.0 seem somewhat unlikely for this compound.

Both correlations fail for silicon since almost all Si—O bonds are in the range $1.58 - 1.67$ Å, as compared to $1.40 - 1.64$ for S—O bonds (*MC-67, PC-68*). Although the bond lengths are considerably less than the 1.76 Å calculated for an Si—O "single bond" by the Schomaker-Stevenson equation, the lack of variation of bond length with bond order is surprising; the bond length is 1.64 Å in tetramethyl silicate (*C-61*) and 1.63 Å in disiloxane (*AB&-63*). This, of course, may not be a good example since the two bonding d-orbitals in disiloxane (one from each Si) may compete for the single nonbonding electron pair in a p orbital on oxygen (i.e. cross-conjugation), and because the lack of electronegative substituents on silicon may make the d-orbitals too diffuse for effective overlap. But other evidence indicates greater $dp-\pi$ bonding in disiloxane; thus the Si—O—C angle in tetramethyl silicate is 113° (*C-61*), as compared to 111° for dimethyl ether (*GR&-70*), 144° in disiloxane (*AB&-63*), and 130° for hexamethyldisiloxane, (*E-69*). An increase in bond angle at oxygen must involve an increase in $dp-\pi$ bonding, since, as the σ bonding assumes more sp hybrid character, the $p\pi$ (or $p\pi'$) orbital must become more available for π bonding, thus enabling more participation by the $3d_{x^2-x^2}$ (or d_{z^2}) orbital. Therefore, in the series $X_3SiOSiX_3$, the Si—O bond length decreases and the SiOSi angle and Si—O force constant and bond order all increase with increasing electronegativity of X, as shown below:

Physical Properties Relative to $dp-\pi$ Bonding

X	Si–O Length	SiOSi angle	Ref.	Si–O force constant mdyne/Å	Ref.
H	1.63 Å	144°	(AB&-63)	5.05	(TN-60)
Me	1.63	130	(Y-57)	4.27	(BGS-70)
Cl	1.592	146	(AG&-71)	5.38	(BGS-68) (BB&-69a)
F	1.580	156	(AG&-70)	5.48	(DKF-75)

Due to the small variation in bond lengths, bond orders are calculated from the force constants by the Siebert equation and found to range from 1.12 for X = H to 1.38 for X = F; clearly, Si–O, $dp-\pi$ bonding is apparently quite weak in all cases.

Electron diffraction is less useful in analyzing $dp-\pi$ bonding in halosilanes than in siloxanes since in the halosilanes there is no information to be derived from bond angles. All known silicon-halogen bonds are shorter than the sum of the bond radii (E-63), but only Si–F bonds were shorter than those predicted by the Schomaker-Stevenson equation. This is good evidence for $dp-\pi$ bonding to the other halogens. Other evidence, e.g. photoelectron spectra (CE-71), dipole moments (BM-70b) and force constants (B-68c), does indicate that chlorine and even bromine are capable of $dp-\pi$ bonding to silicon, but not all of it is considered to be reliable. The question has been discussed in detail by Ebsworth (E-63, E-69).

Probably the most obvious example of $dp-\pi$ bonding is trisilylamine, which was found to be planar rather than pyramidal as most amines, with Si–N–Si bond angles of 120° and an Si–N bond length of 1.74 Å, 0.06 Å shorter than predicted by the Schomaker-Stevenson equation (H–55). The related compound $(Me_3Si)_3N$ has no basic properties (AAB-66), indicating that the nonbonding electron pair on nitrogen is completely delocalized. Ebsworth (E-69) has calculated that the $dp-\pi$ bond energy must be 10–15 kcal/Si–N bond in trisilylamine in order to permit this compound to assume a planar form; a recent MO calculation gives a π bond energy of 16 kcal/bond, (P-67). Similar results have been found for related compounds. The bond angles in $(H_3Si)_2NH$ and $(Me_3Si)_2NH$ are 127° and 125° respectively and the Si–N bond lengths are 1.724 Å and 1.735 Å, respectively (RR&-69, RS&-68). Preliminary results on $(H_3Si)_2NMe$ indicate that all the atoms other than hydrogen are coplanar and that the Si–N–Si bond angle is 126°. A bond angle of 116° has been found for trigermylamine, indicating that Ge–N, $dp-\pi$ bonding is also possible (R–69a) but $dp-\pi$ bonding to phosphorus appears impossible, the bond angles being 98.6° in trimethylphosphane, 96.5° in trisilylphosphane, and 95.4° in trigermylphosphane (R-69a). Lastly, $(Me_2N)_3SiCl$ has been found (VT-69) to have a planar configuration around each nitrogen, the Si–N–C bond angle being 120.5° and the C–N–C angle 118.5°. The Si–N bond length is 1.715 Å and the Si–Cl length 2.082 Å, which is considerably longer than that in chlorosilane and would seem to indicate that no $dp-\pi$ bonding to chlorine is involved. Since we should expect a π bond order of only 1/3 in tris-(dimethylamino) chlorosilane, the planarity around nitrogen may be due to steric factors rather than to $dp-\pi$ bonding. The reason for the assignment of π bond order of 1/3 is that each nitrogen used the same p orbital to hold its nonbonding electron pair; thus, only one d orbital will overlap with the three nonbonding pairs.

Sulfones and sulfonamides show a conspicuous lack of variation of S—O bond lengths, these being 1.42–1.44 Å in Me_2SO_2, $(p-ClC_6H_4)_2SO_2$, $PhSO_2NHC_6H_4Br-p$, sulfanilamide, the tosylate ester I and the sulfonylhydrazines II and III (*GR-63, SA-60, OM-67, SSK-68, TS-68, MMS-66*). In all cases the OSO bond angles were about 120°. *Koch* and *Moffitt* (*KM-51*) have shown that conjugation of X with the sulfone group in X_2SO_2 depends on whether the free valence orbital (a $pp-\pi$ MO or a lone pair in a p AO) of X lies in (Case I) or perpendicular to (Case II) the OSO plane. Only for Case I can overlap occur with the S—O, $dp-\pi$ MO, thereby weakening the S—O bond. The steric requirements of the ortho substituents in diaryl sulfones preclude this type of conjugation. The benzene rings lie in approximately parallel planes perpendicular to the CSC plane; in this conformation (Case II) they may conjugate with vacant d-orbitals on sulfur but cannot affect the S—O bonds. A theoretical study (*G-67*) indicated that Case I conjugation is to be expected when the sulfone group is bonded to a benzene ring and to a nitrogen; here again, however, the steric requirements of the ortho substituents prevent the necessary coplanarity of the benzene ring with the CSN plane. Thus, the dihedral angle between the XSX plane and the benzene ring attached to sulfur was 68° in III (*G-70*), 84.7° in di (*p*-chlorophenyl) sulfone (*G-67*) and 88° in both sulfanilamide and $PhSO_2NHC_6H_4Br-p$ (*OM-67, RR&-67*). The dihedral angle

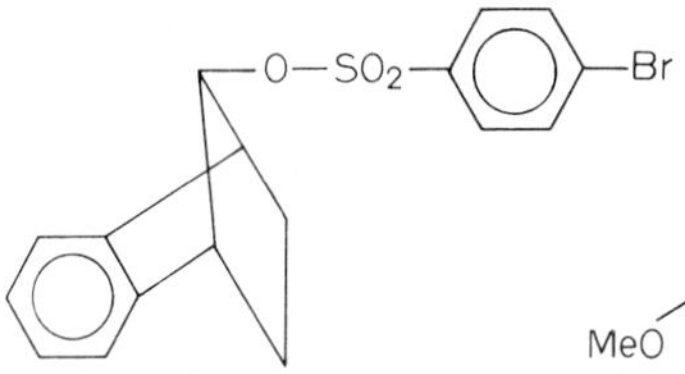

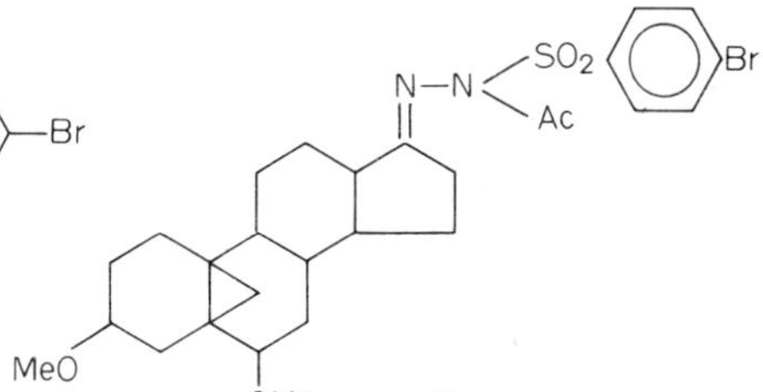

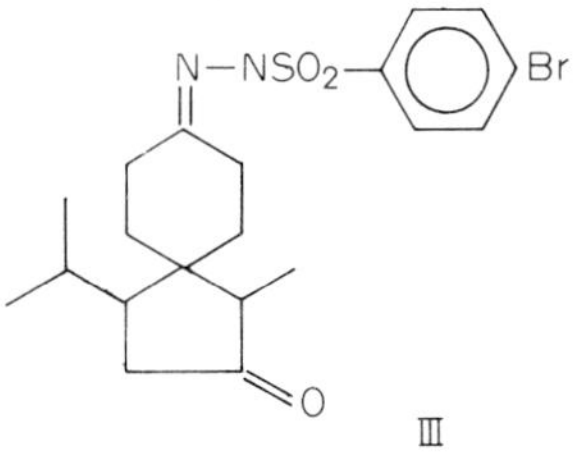

between the CSN plane and the benzene ring attached to nitrogen was 72° in $PhSO_2NHC_6H_4Br-p$ and 80° in IV (*RR&-67*), thus indicating a compromise between conjugation and steric hindrance. However, the geometry of $(Me_2N)_2SO_2$ (*JS&-63*) was clearly Case II and not Case I, the dihedral angle between the CNC and NSN planes being 89° and the S—O bond lengths 1.44–1.45 Å.

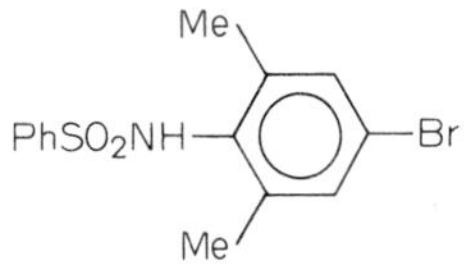

Twisting out of the plane does not hinder $dp-\pi$ conjugation of the benzene ring or of nitrogen with the sulfur atom. Thus, S—N bond lengths were 1.62 Å in sulfanilamide (*OM-67*), (Me$_2$N)$_2$SO$_2$ (*JS&-63*), and *p*–ClC$_6$H$_4$SO$_2$NHC$_6$H$_4$Br, and 1.56 Å in PhSO$_2$NHC$_6$H$_4$Br (*RR&-67*), all considerably shorter than the 1.67 Å S–N single bond in sulfamic acid. Similarly the sulfur/alkoxy oxygen bond length was 1.57 Å in III (*SSK-68*), corresponding to a total bond order of 1.256 by Robinson's equation (*R-63*). Similarly C–S bond lengths were 1.71 Å in PhSO$_2$NHC$_6$H$_4$Br, 1.75 Å in sulfanilamide and 1.76 Å in (*p*–ClC$_6$H$_4$)$_2$SO$_2$ and in *p*–ClC$_6$H$_4$SO$_2$NHC$_6$H$_4$Br, and 1.77 Å in II, all slightly shorter than the 1.80 Å C–S bond length in dimethyl sulfone and indicating fairly weak phenyl-sulfur dp–π bonding. The configuration at N in (Me$_2$N)$_2$–SO$_2$ corresponded to $sp^{2.23}$ hybridization, affording the lone pair on nitrogen 93% *p* character; high *p* character is of course essential for large d–π overlap. This indicates that dp–π bonding should be even stronger in α-sulfonyl carbanions since there the lone pair is in a larger orbital of 100% *p* character. (*JS&-63*).

Westheimer (*DW-66*) has suggested that the extremely rapid hydrolysis of cyclic five-membered phosphate and sulfate esters is due, at least in part, to a decrease in dp–π bonding caused by ring strain. This is easily confirmed by X-ray diffraction. Methyl ethylene phosphate (MEP) and methyl pinacol phosphate (MPP) both had bond lengths of 1.44 Å to the phosphoryl oxygen; the bonds to the other three oxygens were all about 1.57 Å (*NCB-66*). For comparison the P–O bond lengths in triphenyl phosphate (*DS-62*) were 1.43 Å and 1.63 Å, thus showing that ring strain, as indicated by the POC angles, which were 112° in the ring and 119° and 123° for the methoxy groups in MEP and MPP, respectively, does not affect dp–π bonding. This occurs in spite of the fact that the three ester oxygens in MEP and MPP are oriented so as to permit overlap with only two of the five $3d$ orbitals on phosphorus, whereas in acyclic phosphates, three of the *d*-orbitals should be capable of overlap (*NCB-66*) given sufficient radial contraction. Ring strain is even more obvious in the vinylene phosphate V (*SC&-67*) and in catechol sulfate VI which hydrolyzes two million times faster than diphenyl sulfate (*KKW-65*), the endocyclic OPO and OSO angles being 98.5° and 97.1°, respectively. In catechol sulfate VI the endocyclic OCC angles were 112.5° and 110.7°; thus, an endocyclic pp–π system greatly increases strain in the five-membered ring (*BF-69*). The endocyclic S–O bond lengths were 1.585 Å and 1.601 Å, quite long compared to the 1.533 Å in ethylene sulfate, (*BF-69*). Using the Pauling equation we get 1.23 for the average S–O bond order in the catechol sulfate ring and 1.41 for the ethylene sulfate ring. In the vinylene phosphate the P–O bond lengths were 1.38 Å for the phosphoryl oxygen, 1.59 Å in the ring, and 1.53 Å for the methoxyl oxygen (*SC&-67*); the corresponding bond orders as calculated by the Pauling equation (*R-63*) were 1.32 in the ring and 1.55 for the methoxyl oxygen. Clearly, the major factor affecting $dp-\pi$ bonding in the five-membered rings appears to be not ring strain but the presence of adjacent $pp-\pi$ systems which can compete for the lone pairs on the oxygens. Probably the principal reason for the rapid hydrolysis of the cyclic esters is the low endocyclic OPO and OSO angles, which permits formation of the trigonal bipyramidal intermediate or transition state with only very small deformation of the bond angles around phosphorus or sulfur.

On the other hand, recent MO calculations on phosphoric acid (*LW-75*) prove that steric constraints generated by a bicyclic ring can affect the strength of the P—O bonds in phosphate esters as a result of overlap of antiperiplanar oxygen lone pairs with P—O, σ^* orbitals. Compared to an acyclic phosphate, this can result in a destabilization of 21.9 kcal for a bridgehead bicyclic phosphate in which all three ester oxygens have two antiperiplanar lone pairs each. This effect is not dependent on the presence of d-orbitals in the basis set, but d-orbitals do magnify the effect, increasing the destabilization from 21.9 to 24.2 kcal. Inclusion of d-orbitals strengthened all the P—O bonds but did so to a much greater extent for those to the ester oxygens than for the phosphoryl "double" bond. In fact, the ratio of phosphoryl to ester P—O overlap populations decreased from 3.92 (6.53 in the presence of antiperiplanar overlap, which affects primarily the ester P—O bonds) to 1.95 (2.2 in the presence of antiperiplanar overlap) upon inclusion of d-orbitals in the basis set.

Reduced $dp-\pi$ bonding due to ring strain has recently been observed experimentally, but the effect is small. In VII the P=O, P—O', and P—O'' lenths were 1.445, 1.577, and 1.594 Å, respectively (*MS&-76*). The P—O' bond lengths are about those expected for a phosphate ester, but the P—O'' length is about the same as in the endocyclic bonds of V. The O'PO'' angle in VII is 97.4°, similar to the endocyclic OPO angle in vinylene phosphate, suggesting that reduced $dp-\pi$ bonding originates from the same source in both cases; the O'PO angle was 105°, about the same as in unstrained phosphates. However, ring strain does exist at O' as indicated by the PO'C and PO''C angles of 106° and 95.3° compared to 117—118° for the POC angle in the completely unstrained phosphite VIII (*MS&-76*). Thus, it would appear that only fairly severe ring strain can affect $dp-\pi$ bonding, as measured by P—O bond lengths and OPO angles.

Another interesting application of X-ray diffraction involves "no bond resonance" in the thiothiophthenes. Thus, in IX, the reaction product of acetylacetone and P_2S_5, all the sulfur atoms are on a straight line with equal S—S distances of 2.36 Å, considerably longer than the normal disulfide bond of 2.04 Å. Such "no bond resonance" actually involves 3-center-4-electron bonding, as explained by *Gleiter* and *Hoffmann* (*GH-68*). The stability of the 3-center bond is increased somewhat by the formation of a ten π electron system, but primarily by the use of d-orbitals. SCF calculations without d-orbitals on X gave a slightly smaller S—S bond energy than would be expected on the basis of a proportionality between bond length and energy, assuming 3.0 Å as the minimum distance between nonbonded sulfur atoms (*K-69*). For X these same calculations showed that the S—O bond energy was less than that of a normal hydrogen bond, i.e. essentially nonexistent, in spite of the 2.41 Å distance which is

Physical Properties Relative to $dp-\pi$ Bonding

only 75% of the sum of the van der Waals radii; the S—S bond was a normal single bond with a length of 2.12 Å (*K-69*).

An interesting characteristic of 3-center-4-electron bonding is that increasing the electronegativity of an atom increases the lengths of the bonds to that atom. In XI the S_1-S_2 distance is 2.22 Å while the S_2-S_3 distance is 2.50 Å; S_3 is more electronegative than S_1 because it is closer to the electron-withdrawing Ph_3 than is S_1 to Ph_1 (*HSS-66*). However, in XII the S—S distances are equal to 2.35 Å since the dimethylamino group reduces the electron-withdrawing effect of Ph_3 until it is no stronger than that of Ph_1 (*HSS-66*). Since thiothiophthenes have been recently reviewed (*K-69*) there seems to be no need for further discussion of this subject here.

Some X-ray diffraction data is available on phosphorus ylides. Most of these are of the stabilized type, but the P=C bond length in triphenylphosphoranylidenemethane, $Ph_3P=CH_2$, has been measured as 1.661 Å (*B-69*); for comparison the sum of the Pauling double bond radii is 1.665 Å, which would seem to indicate considerable $dp-\pi$ bonding if one follows the usual assumption that decreased bond length is associated with increased double bond character. Trost (*TM-75*) has recently questioned this assumption, and his doubts are supported by a recent MO study (*WWB-75*), which showed that the C—X bond lengths in the CH_3X (X=O or S) radicals, anions, and cations could not be correlated with overlap populations but instead correlated linearly with ionic bond order. Although extended Hückel calculations on $H_3P=CH_2$ (*HBG-70*) did show a considerable increase in the P—C overlap population upon addition of d-orbitals to the basis set, later *ab-initio* calculations (*AW-72*), while agreeing with this finding, showed that the conventional picture of separate $dp-\pi$ and ionic contributions to the bonding is an oversimplification. The P—C rotational barrier was extremely small, 3—16 cal/mole with or without d-orbitals, which is hardly characteristic of a double bond, and even with d-orbitals the ionic bond order was far higher in $H_3P=CH_2$ than in H_3PCH_3 (*AW-72a*). Only minor changes were observed in the electron density contour maps of $H_3P=CH_2$ upon addition of d-orbitals to the basis set; the only observable effect was a stabilization of those orbitals having π character.

Stabilized ylides showed much longer P—C bond lengths, e.g. 1.73 Å in $Ph_3P=CClCOPh$ (*S-65b*). The C—O and C—C bond lengths were 1.30 and 1.36 Å, respectively. The C–O length is much longer than expected for a carbonyl (1.23 Å), but the C–C bond is almost a double bond, the normal C=C bond length being 1.33 Å (*S-65b*). Evidently, this ylide is primarily stabilized by the enolic system, $dp-\pi$ bonding playing only a secondary role. In $Ph_3P=CHSO_2C_6H_4Me$ the P=C length is 1.71 Å, indicating the lesser stabilization (i.e., negative charge dispersal) available by $dp-\pi$ bonding with the sulfonyl group (*SR-65*). The P=C bond order here was 1.40 as calculated by the Pauling equation, compared to 1.30 Å in $Ph_3P=CClCOPh$. In the triphenylphosphoranylideneacetophenone the PCCO system was planar as required by resonance; the same applied to the PCSO system in the sulfonyl-ylide, the second oxygen being noncoplanar (*SR-65*). Since one of the sulfonyl oxygens is thus conjugated, the two have very different lengths, 1.469 Å and 1.444 Å. The S–C bond length here was 1.686 Å, considerably shorter than the sulfur-tolyl bond length of 1.76 Å and clearly indicating extensive S–C, $dp-\pi$ bonding (*SR-65*), although the S–C bond is probably not as strong as the P–C bond. The PCC and PCS bond angles in the carbonyl and sulfonyl-ylides respectively were 120°, thus proving the ylide carbon to be sp^2 hybridized; the same probably applies in $Ph_3P=CH_2$ since this gives the largest $dp-\pi$ overlap.

The reaction of an ylide with a carbonyl compound goes through a betaine intermediate with a four-membered ring structure. One such intermediate, XIII, has been isolated from the reaction of $Ph_3P=C=PPh_3$ with hexafluoroacetone (*CD-68*). The four membered ring was planar with a basal P–C bond of 1.76 Å, which was also the length of the exocyclic P–C bond, probably indicating a continuous π system with a Pauling bond order of ~ 1.2. The C–O bond was 1.39 Å, almost a single bond. The apical P–O bond was 2.01 Å, considerably longer than the apical P–O bonds in the oxyphosphorane derived from phenanthrenequinone and triisopropyl phosphite XIV, (*SHL-67*), suggesting some significant ionic character.

The excellent reviews (*M-69, P-64*) on phosphonitriles include structural data which need not be reproduced here. It should be noted, however, that a linear correlation has been drawn (*CW-72*) between the endocyclic P–N and S–N bond lengths in phosphazenes, thiazyl halides, and sulfuranic halides, all lengths ranging between 1.52 and 1.60 Å, and the total π and σ overlaps calculated by an LCAO-MO method with d-orbital exponents being optimized in terms of energy by a valence-bond perfect-pairing scheme. Deviations from this line were all less than 1.5%.

Survey of relevant articles on physical properties related to dp–π bonding – bond lengths and angles Table (III A)

System and Results	Conclusions	References
X-ray diffraction of $Ph_2FP{=}NMe$ P=N 1.64 Å, P–C 1.81 Å, P–F 1.49 Å, N–C 1.45 Å, FPN 118.7°, P=N–C 119.1° The molecule is monomeric.	P–F distance is very short compared to 1.52 Å in $(F_2PN)_3$. P=N indicates a double bond, but longer than in phosphonitriles (1.51–1.61 Å). P is sp^3 hybridized. N is sp^2 hybridized, with 3rd position a lone pair not involved in dp–$π$ bonding, since this would increase PNC angle. Thus, d-overlap is to an unhybridized orbital on N.	(AB-70)
Electron diffraction study of $(F_3Si)_2O$ Si–F 1.554 Å, F–Si–F 108.8° ± 0.5° Si–O 1.580 Å, Si–O–Si 155.7° ± 2.0°	Increased Si–F and Si–O, dp–$π$ compared to $SiMe_3F$ and $(SiH_3)_2O$.	(AG&-70)
Extended Hückel calculation with d-orbitals on A and B. 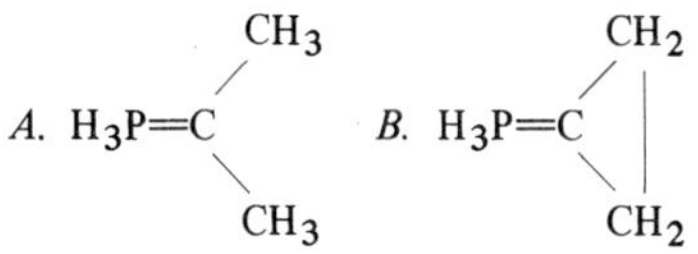 A. $H_3P{=}C{\begin{smallmatrix}CH_3\\CH_3\end{smallmatrix}}$ B. $H_3P{=}C{\begin{smallmatrix}CH_2\\|\\CH_2\end{smallmatrix}}$ P=C overlap population is 1.43 for A and 1.53 for B.	Stronger bonding in B attributed to a σ overlap between d_{yz} on P and the bent Walsh-Type σ orbitals on the cyclopropane ring.	(BH-71)
X-ray diffraction study of 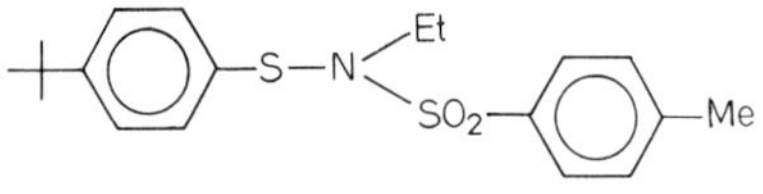 [Ar]–S–N(Et)(SO_2–[Ar]–Me) The nitrogen is planar with an SNS angle of 114° and SNC angles of 119° and 121°. The S–N bond lengths are 1.644 Å to the ring sulfur and 1.68 Å to the sulfonyl sulfur.	sp^2 hybridization at N with strong dp–$π$ bonding. The longer sulfonyl-N bond length suggests that the same d-orbitals are involved in both S–N and S–O, dp-bonding, thus weakening both (see TP&-70 and text Sec. III.B., Type I conjugation). The ring configuration indicates that the sulfur and nitrogen lone pairs are gauche to each other (W-72).	(CG&-71)
Electron diffraction analysis of $MeN(PF_2)_2$ showed the CNP_2 system to be planar with C–N–P and P–N–P angles of 122° and 116° respectively. The P–N bond length was 1.68 Å compared to 1.65 Å in Me_2NPF_2.	The nitrogen is essentially sp^2 hybridized, which indicates that the lone pair is fully involved in dp–$π$ bonding to both phosphorus atoms. However, since there are two dp–$π$ bonds in $MeN(PF_2)_2$ compared to one in Me_2NPF_2, each bond is weaker, thus explaining the longer bond length in the former.	(HHH-74)

System and Results	Conclusions	References
Microwave spectrum of H$_3$SiNCS showed all heavy atoms to be linear. Si–N 1.714 Å vs 1.73 Å in (H$_3$Si)$_3$N, N–C 1.211 Å, same as in HCNS and CH$_3$NCS.	The nitrogen lone pair must be involved in dp–π bonding.	(JKS-62)
X-ray diffraction Me$_2$S–N–SO$_2$Me S(IV)–N S–O 1.44, 1.45 Å S(VI)–N 1.58 Å SNS 116.2°, OSO 116.4°, NSO 106°	SNS is a continuous d–π system involving both sulfur atoms with both S–N bond orders equal. This is entirely independent of the OSO d–π systems. The SNS plane is approximately perpendicular to the OSO plane.	(KKK-67)
X-ray diffraction 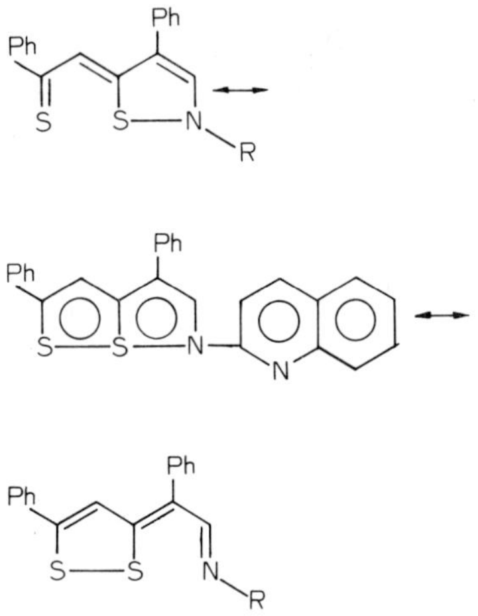 S–S = 2.36 Å S–N = 1.89 Å The ratio of the S–N bond length to the sum of the van der Waals radii is much less than the ratio of the S–O bond length in X to the sum of the S and O van der Waals radii. The S–S bond here is much longer than that in disulfides or in X.	Unlike the SSO system of X, the SSN system here forms a true three-center bond. Possibly the lower electronegativity of N compared to O reduces the energy difference between the two resonance hybrids compared to the corresponding hybrids for X.	(LN-71)
X-ray diffraction of a 1,1-dimethylphosphorin 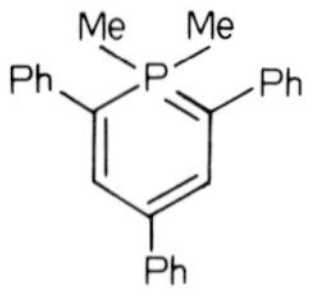 Ring is planar all P–C ring 1.75 Å all ring C–C 1.39 Å ring C–P–C 105°,	A planar aromatic ring, as with pyridine.	(D-70)

Physical Properties Relative to $dp-\pi$ Bonding

System and Results	Conclusions	References
102° on the demethylated P(III) phosphorin. The dihedral angle between the ring and the Me–P–Me plane is 84°.		
Based on published X-ray diffraction, dipole moment, and IR data, sulfoxides, sulfones, sulfonamides, sulfonates, etc., are classified into four types depending on the presence of π electrons or nonbonding pairs in p orbitals in the two substituents.	All compounds of the type X_2SO_2 show separate OSO and XSX dp-systems even though steric hindrance may prevent the p orbitals on X from forming exactly a 90° angle with the OSO plane. Sulfonamides, $ArSONR_2$, prefer Type I conjugation (classification of *KM-51*), but steric hindrance prevents the necessary coplanarity of the benzene ring and CSN systems; thus, these too fall into Type II conjugation.	(*G–70*)
X-ray diffraction tetrasilylhydrazine N–N 1.45 Å, similar to that in hydrazine; Si–N 1.73 Å, vs 1.74 Å in trisilylamine, Si–N–Si 129.5°. Each Si_2N_2 group forms a plane with an 82° dihedral angle between them.	Two independent Si–N–Si $dp-\pi$ bonding systems.	(*GR&-70*)
Extended Hückel calculations on $H_3P\!=\!CH_2$ with and without d-orbitals show an increase in the P=C overlap population from 0.88 to 1.36, and a decrease in the charge on the methylene carbon from -1.07 esu to -0.51 esu upon inclusion of d-orbitals in the basis set.	Unstabilized ylides show extensive $dp-\pi$ bonding.	(*HBG-70*)
X-ray diffraction of sodium metasilicate Na_2SiO_3 shows an infinite chain of SiO_4 tetrahedra with bridging and nonbridging Si–O bond lengths of 1.67 Å and 1.59 Å, respectively.	The lack of covalent bonding between Na and O emphasizes the difference in $dp-\pi$ bonding between bridging and nonbridging bonds, as expected by Cruickshank's theory (*C-61*). The Shomaker-Stevenson Si–O bond length is 1.67 Å; thus, the bridging bond shows no $dp-\pi$ overlap. The nonbridging Si–O bond is about as short as any known; apparently there are no really strong Si–$dp-\pi$ bonds.	(*MC-67*)
Microwave spectrum of Me_2SO_2 C–S = 1.777 ± .006 S–O = 1.431 ± .004 angle C–S–C = 103° 17' angle O–S–O = 121° 1'	Good agreement with X-ray diffraction data of *D.E. Sands*, (*S-63*), except for O–S–O. Sands gave 117.9° for this.	(*SM-72*)

System and Results	Conclusions	References
X-ray diffraction of A shows sp^3 N with S—N bond length of 1.68 Å, longer than the analogous bond in B (*CG&-71*), but much shorter than the Schomaker-Stevenson bond length, 1.74 Å.	The pyramidal configuration at N weakens $dp-\pi$ bonding but does not destroy it, in agreement with (*P-67*).	(*GD-76*)

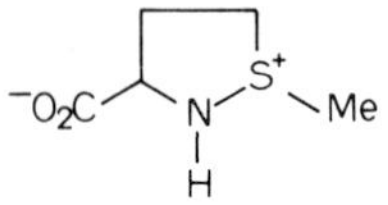

A

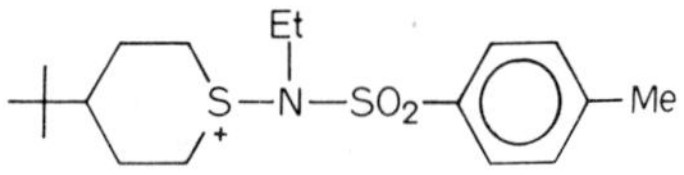

B

B. Infrared-Spectra

In theory, force constants calculated from infrared and Raman spectra should be an excellent indicator of the strength of $dp-\pi$ bonding; in fact, many studies have used force constants to calculate bond orders by means of the Siebert formula (*S-53*) which is a linear correlation between the total bond order and the ratio of the actual force constant to that for an ideal single bond A—B, this being defined as 7.2 ($Z_A Z_B/n_A^3 n_B^3$) where Z is the atomic number and n the principal quantum number. In practice, however, the difficulty of calculating force constants, particularly for complex molecules, renders their use rather limited. Moreover, *Ebsworth* (*E-63, E-69*) has pointed out that comparison of force constants for different bonds implies an assumption that the potential energy functions for the two bonds are similar in form, an assumption which may not always be justified.

Even where force constants are available, the wide variations observed in published values, e.g. 5.32 and 5.95 mdyne/Å for the Si—F bond in H_3SiF, 2.48 and 2.67 mdyne/Å for the Si—Cl bond in Me_3SiCl (*K-59, BGS-70*) render any conclusions reached from them exceedingly suspect. The assumptions made by *Robinson* (*R-63*), make his tabulated force constants clearly unsatisfactory. For instance, a larger Si—O force constant is assigned to hexamethyldisiloxane than to disiloxane, in violation of the relative electron-donating (inductive) effects of these groups. For comparison *Bürger* (*BGS-70*) assigns values of 5.05 mdyne/Å for disiloxane and 4.27 mdyne/Å for hexamethyldisiloxane; the latter is a surprisingly low value (Siebert bond order 1.12) and argues that $dp-\pi$ bonding in hexamethyldisiloxane is actually quite weak, as might be expected from the strong electron-donating (inductive) effect of methyl groups. Similarly, the force constants of trisilylamine and tris(trimethylsilyl)amine were 4.19 and 3.70 mdyne/Å, respectively, corresponding to Siebert bond orders of 1.25 and 1.15, *(BGS-70)*. The occurrence of a symmetric Si_3N skeletal vibration which is infrared but not Raman active and the existence of a dipole moment argue that $(Me_3Si)_3N$ is pyramidal rather than planar; the force constant was calculated on this assumption *(BGS-70)*, with the Si—N—Si angle assumed to be 115°. The failure of this compound to show basic properties may well be a steric effect of the bulky trimethylsilyl groups. This effect has also been invoked to account for the low rate of hydrolysis *(B-69a)*.

One case where force constants can be correlated to π bond order and to bond length involves the P—O *(MNG-67)*, and P—F *(MNG-66)*, bonds of $OPFX_2$, where X is halogen. In this case, however, the stretching frequencies and therefore the force constants varied linearly with the sum of the halogen electronegativities, indicating that there is no coupling between the stretching frequencies of adjacent bonds. Furthermore, there must be no P—X, $dp-\pi$ bonding since this would reduce the effective electronegativity (i.e. electron-withdrawing effect) of X. Coupling between the P=S and P—F stretching frequencies prevented any correlation for $SPFX_2$ *(MK&-67)* but stretching frequencies and force constants could be correlated linearly with halogen electronegativity in I, but with a different slope and intercept than for $OPFX_2$ *(R-68a)*.

$$\left[X_2P \underset{O}{\overset{S}{<}} \right]^{\ominus} Ph_4P^{\oplus} \qquad \qquad$$

I   II

These correlations have been extended to all other compounds of the type O=PXYZ; each compound fitted one of four parallel straight lines on a plot *(GL-71)* of force constant vs the sum of the electronegativities of X, Y, and Z, depending on the number of substituents (0 → 3) capable of $dp-\pi$ bonding.

Since the sum of electronegativities is the same, this correlation cannot apply to phosphate esters, for which phosphoryl stretching frequencies ranged from 1268 cm^{-1} *(M-57)* for $(EtO)_3PO$ to 1315 cm^{-1} *(HK-70)* for the vinylene phosphate *(SC&-67)*. The known weakness of endocyclic dp-bonding in vinylene phosphate V (p. 18) suggests that as the strength of $dp-\pi$ bonding to the ester oxygens decreases, that in the phosphoryl bond increases; however, a general correlation can be established. Phosphoryl bond force constants have been calculated as 10.3 mdyn/Å for II *(HK-70)* and 9.98 mdyn/Å for trimethyl phosphate *(GL-71)*, corresponding to stretching frequencies of 1308 and 1287 cm^{-1}, respectively, and Siebert bond orders of 2.22 and 2.07, respectively. However, a 1.91 bond order is calculated for II by the Pauling-Robinson equation from the observed 1.46 Å bond length *(NFV-68)*. The same P=O bond length, 1.46 Å, has also been reported in methyl diisopropyl phosphate *(NFV-68)* in spite of the variation in stretching frequencies.

In spite of the discrepancies between bond orders calculated from bond lengths and force constants, and between force constants from different sources, e.g. 12.5 mdyn/Å (*MNG-67*) and 11.9 mdyn/Å (*GL-71*) for F_3PO, all phosphoryl bonds show high bond orders. The minimum value found, 1.7, (calculated from the 8.3 mdyn/Å force constant for Me_3PO (*GL-71*)), supports the idea (*NCB-66*) that there are two π bonds using the d_{xy} and d_{yz} orbitals on the phosphoryl group.

Correlations between force constant, bond order, and vibration frequency have also been published for S—O (*GR-63*) and S—N (*BMP-67, GM&-68a*) bonds, but, as with the correlations for P—O and P—F discussed above, deviations are fairly high.

Using these correlations a Siebert bond order of 1.50 was found for the 960 cm^{-1} asymmetric SNS stretching frequency in $Ph_2S=NSO_2C_6H_4Me$, as compared to 1.40 for $Me_2S=NSO_2Me$. Substituents on the sulfonyl sulfur had a conjugative effect on the SNS frequency, but only a small inductive effect on the OSO frequencies, indicative of a continuous π orbital over the SNS system and Type II (*KM-51*) conjugation between the sulfonyl and tolyl groups.

However, the antisymmetric OSO stretching frequency was 1335 cm^{-1} in $(Me_2N)_2SO_2$ (*TP&-70*) but only 1265 cm^{-1} in $Me_2S=NSO_2C_6H_4Me$ (*TFO-70*). This is not simply an inductive effect as can be seen by comparison with a sulfonyl-stabilized ylide; the average of the symmetric and antisymmetric OSO stretching frequencies is 1200 cm^{-1} in $Ph_3P=CHSO_2C_6H_4Me$ and 1247 cm^{-1} in the corresponding hydrobromide (*SR-65*). The S—O bond lengths are 1.47 and 1.44 Å in this ylide (*SR-65*); this is about 0.025 Å longer than in sulfones, approximately the amount expected by *Robinson's* (*GR-63*) correlation of stretching frequency and bond length, suggesting that stabilization of ylides and imines by adjacent sulfonyl groups does involve the $dp-\pi$ bonding orbitals of the sulfonyl group.

Similar correlations of bond length with force constant apparently do not hold for P—N bonds. On a straight line drawn between the coordinate points representing $NaH(PO_3NH_2)$ and octafluorotetracyclophosphazene on a plot of the logarithms of the bond lengths versus stretching frequencies hexachlorotricyclophosphazene fell at 1150 cm^{-1} as compared to the observed frequency of 1218 cm^{-1} (*P-64*). In spite of this, PN stretching frequencies do generally increase with increasing π bonding, ranging from 697 cm^{-1} for $NaH(PO_3NH_2)$ and 1003 cm^{-1} for aminotrimethylphosphonium hexachloroantimonate to 1170 cm^{-1} for $Ph_3P=NC_6H_4Cl$ and 1435 cm^{-1} for $(PNF_2)_4$ (*P-64, HO-59*). For $(PNCl_2)_3$ a force constant of 8.12 mdyne/Å, corresponding to a Siebert bond order of 1.98, was obtained (*CC-63*).

Force constants have been determined for two unstabilized ylides. For trimethylphosphoranylidenemethane the force constant was 5.59 mdyne/Å (*S-69*) corresponding to a Siebert bond order of 1.65, compared to 1.70 for Me_3PO (*GL-71*) and 1.56 for aminotrimethylphosphonium hexachloroantimonate (*S-68a*). Since we should expect even more $dp-\pi$ bonding in triphenylphosphoranylidenemethane due to the electron-withdrawing phenyl groups, the published force constant of 4.90 mdyne/Å (*LW-65*) corresponding to a Siebert bond order of 1.50, seems rather anomalous, especially in view of the extremely short 1.661 Å P=C bond length (*B-69*).

Infrared studies on carbonyl stabilized ylides confirm and expand the X-ray diffraction results discussed earlier. In $Me_2S=CHCOOMe$ the carbonyl stretching frequency was 1620 cm^{-1} compared to 1735 cm^{-1} in the conjugate acid (*JA-69*). Almost the same results were obtained in a related imine, the carbonyl stretching frequency being 1620 cm^{-1} in Me$_2$S=NCOOEt and 1740 cm^{-1} in the conjugate acid; apparently both the ylide and the imine are primarily stabilized by an enolic resonance system. On the other hand, carbonyl stretching frequencies were 1508 cm^{-1} in Me$_2$S=CHCOPh and 1665 cm^{-1} in the conjugate acid (*WB&-70*); the stronger enolic character of this ylide is in line with the relative chemical stabilities of the phenacyl- and carbomethoxy-substituted ylides. The lesser stability of the carbomethyoxy-substituted ylide is probably due to competing resonance of the carbonyl carbon with the alkoxy and ylide groups. Evidently an inductive effect is also involved, since electron-withdrawing substituents interfere with the enolic resonance by removing electrons from the conjugate C=C=O system.

Analogies between the *dp*–π bonding in cyclic phosphonitriles and that in compounds with alternating S—N, Si—N, and Si—O rings have been discussed (*M-69, P-64*). Thus, ring stretching frequencies generally increase with increasing ring size in both phosphazenes and silazanes, (*P-64*). The antisymmetric Si—N—Si vibration frequency increased from 912 cm^{-1} to 940 cm^{-1} upon going from (Me$_2$SiNMe)$_2$ to (Me$_2$SiNMe)$_3$, and from 928 cm^{-1} to 939 cm^{-1} upon going from (Me$_2$SiNH)$_3$ to (Me$_2$SiNH)$_4$; this is clearly attributable to the greater flexibility of the larger rings which permits larger angles at nitrogen and thus greater *dp*–π bonding (*KD&-69*). Thus, for the acyclic case in which complete flexibility exists, a Siebert bond order of 1.18 has been calculated (*K-57*) from the IR spectral force constants for the Si—N bonds in hexamethyldisilazane; see also (*R-63* and *BGS-70*). Similarly, polymerization of (Me$_2$SiO)$_3$ is exothermic by 3.5 kcal/mole, whereas polymerization of (Me$_2$SiO)$_4$ involves no heat change; the base strength of the oxygen also decreased on going from trimer to the tetramer (*P-64*).

Exocyclic *dp*–π bonding is illustrated by the 1590–1593 cm^{-1} C=C stretching frequency in III and IV and vinylpentamethylcyclotrisilazane characteristic of conjugated double bonds; by comparison the C=C stretching frequency was 1632 cm^{-1} in V (*AKK-63*). The ultraviolet spectra of III, IV and of vinylpentamethylcyclotrisilazane were all very similar, indicating that the *dp*–π bonding is of the three-center (*M-69a, P-64*) type, with nodes at each silicon atom so that there is no conjugation between vinyl groups. The asymmetric Si—N—Si vibration frequency was 917 cm^{-1} in VI (*AK-67*); this fairly small increase over the 912 cm^{-1} in (Me$_2$SiMeN)$_2$ in spite of the much greater electronegativity of nitrogen compared to methyl is attributed to exocyclic *dp*–π bonding increasing the electron density on silicon.

In 1959 Smith and Angelotti (*SA-59*) found that the Si–H stretching frequency could be expressed as the sum of constants for each substituent on the hydrogen-containing silicon atom. This implies that ν(Si–H) in $\text{Me}_n X_{3-n}\text{SiH}$ depends only on the inductive effect of X, a supposition which was strengthened by the finding of a linear correlation between ν(Si–H) and the sum of the *Taft σ** constants (*T-60*). However, this correlation showed considerable scatter, and later work indicated that there is a separate linear correlation for each X, i.e., that ν(Si–H) should actually be correlated with $\Sigma\sigma'$, where $\sigma' = \sigma^* - \delta$, where δ is an empirical constant (*A-68*). The δ values were generally in line with the ability of X to donate electrons via dp–π bonding, being zero for hydrogen and methyl, 1.75 for fluorine, 0.70 for methoxy, 0.65 for dimethylamino, and 0.55 for phenyl substituents. The high δ value for chlorine, 1.70, is not in agreement with the relative dp–π bonding abilities of chlorine and oxygen as determined from bond lengths (*C-61, E-63*), and force constants (*BGS-70*), but does agree with a recent NMR Study (*B-64a*). Chvalovsky (*C-66*) found that ν(Si–H) varied linearly with n in the series $\text{Cl}_n\text{Me}_{3-n}\text{SiH}$, $\text{Ph}_n\text{Me}_{3-n}\text{SiH}$ and $(\text{CH}_2{=}\text{CH})_n\text{Me}_{3-n}\text{SiH}$. This is, of course, a basic requirement if the Smith-Angelotti correlation (*SA-59*) is to be valid. Assuming the validity of the Cruickshank formulation (*C-61*) or something similar, this also implies that there is no dp-bonding between silicon and chloro, phenyl or vinyl groups. As shown below, however, a non-linear variation was found for compounds containing Si–O bonds. The lack of importance of steric hindrance here is shown by the values for IX and for X, where the ethoxyl and propyl groups should have the same steric factors. These curves indicate that dp–π bonding is extensive to one oxygen, but that the extent of π-bonding per Si–O bond drops drastically as the number of such bonds increases, a result in accord with the Cruickshank model.

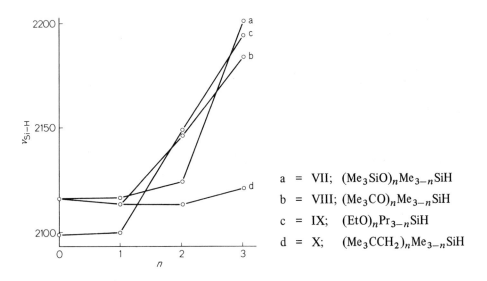

a = VII; $(\text{Me}_3\text{SiO})_n\text{Me}_{3-n}\text{SiH}$
b = VIII; $(\text{Me}_3\text{CO})_n\text{Me}_{3-n}\text{SiH}$
c = IX; $(\text{EtO})_n\text{Pr}_{3-n}\text{SiH}$
d = X; $(\text{Me}_3\text{CCH}_2)_n\text{Me}_{3-n}\text{SiH}$

Dependence of Si–H on Silicon Substitution from ref. (*C-66*).

Physical Properties Relative to dp–π Bonding

Survey of Relevant Articles on Physical Properties Related to dp–π Bonding – Infrared
Table (III B)

System and Results	Conclusions	References
Calculation of Si–X force constants from IR spectra for H_3SiX, and Me_3SiX (X = F, Cl, or Br)	Siebert bond orders range from 1.16 for H_3SiBr and Me_3SiBr to 1.35 for SiH_3F and 1.25 for Me_3SiF, indicating that although dp–π bonding decreases with increasing atomic number of the donor atom, it is still detectable even for bromine.	(B-68c)
Calculation of force constants from IR spectra for S–O, S–N and S–X bonds of Me_2NSO_2X (X = F, Cl or Br).	Siebert S–O bond orders ranged from 2.04 for X = Br to 2.15 for X = F, indicating strong dp–π bonding. Comparison of the S–N, S–F, and S–Cl force constants with those expected for a single bond indicated negligible dp–π contributions to these bonds.	(BB&-71)
Same as for (B-68c)	Si–X force constants are up to 10.6% lower than those of (B-68c). This demonstrates the unreliability of force constants as a means of studying dp–π bonding.	(D-64)
IR and Raman spectra of $(Cl_3Si)_2O$ could be interpreted by D_{3d} or D_{3H} structure with large Si–O–Si angle (but < 180°). Symm Si–O–Si str. 332 cm^{-1}, IR & Raman active, antisym Si–O–Si str. 1115 cm^{-1}, IR active only.	Strong coupling of symmetric SiOSi and SiCl$_3$ stretching vibrations, explaining low symm. SiOSi stretch.	(DH-69)
IR and Raman spectra show $N[P(CF_3)_2]_3$ to be pyramidal with very weak P–N bonds, in agreement with the easy cleavage of the P–N bond by HCl or ammonia. However, the P–N bond of $HN[P(CF_3)_2]$ is stable.	The lone pairs on P provide strong hindrance to planarity. The reduced dp–π overlap in the pyramidal conformation allows the σ bond weakening effect of electron withdrawal by the trifluoromethyl groups to override d-orbital contraction by trifluoromethyl.	(HJ&-70)
Si–Si stretching frequencies and force constants in $Si_2(NMe_2)_nMe_{6-n}$ increase as n increases from 0 to 6.	d-Orbital contraction (and Si–N, dp–π bonding) increase with increasing n, thus increasing $(d$–$d)$ or $(d$–$p)$ σ–π interactions.	(HPH-70)

Spectra-Infrared

System and Results	Conclusions	References
SiX_4(Si–X) Si_2X_6(Si–X) Si_2X_6(Si–Si) Cl 3.11 2.90 2.4 mdyn/Å Br 2.45 2.30 2.1 I 1.79 1.70 1.9 H – – 1.78	1) Decrease in f(Si–X) on going from SiX_4 to Si_2X_6 is due to smaller inductive effect of $-SiX_3$ compared to X. The p character and strength of the bond increases with electronegativity of substituents. Note similar variation of ν(Si–H). 2) Variation of f(Si–Si) attributed to $dp-\pi$ and $dd-\pi$ bonding with nonbonding pairs on Cl. Compare to results of Brune (B-64a, B-64b, B-65)	(HSH-70)

$$\begin{array}{c} X \\ \diagdown \\ P \\ \diagup \\ O \end{array} \begin{array}{c} O-CH_2 \\ \diagdown \\ \diagup \\ O-CH_2 \end{array} Z + BF_3 \rightarrow \text{acid-base complex}$$

Determination of ΔH for reaction

X	Z	$\Delta H \frac{\text{kcal}}{\text{mole}}$	νP=O eq.	νP–O apical
OPh	O	–18.22	1303 cm^{-1}	–
Cl	O	–19.57	1312 cm^{-1}	1288
OMe	O	–20.59	1298 cm^{-1}	1274
NMe$_2$	O	–23.15	–	1244
Cl	CPh$_2$	–16.74	1314 cm^{-1}	–
Cl	CMe$_2$	–19.57	1312	1288

IR and ΔH affected mainly by inductive effect of X, but ΔH also affected by electrostatic-field effect of Z.   (ME&-71)

Calculation of force const. for X = P(NMe$_2$)$_3$
ν = stretch freq., f = force constant

X=P(NMe$_2$)$_3$	O	S	Se	Te	
ν(P=X)	–	12.0	565	530	519 cm^{-1}
f(P–N)	3.54	3.95	402	3.75	3.68 m dyn/Å
f(P–X)	–	8.77	3.81	5.06	5.89 m dyn/Å
P=X bond order		1.88	1.46	2.02	2.83
P–N bond order	1.10	1.12	1.06	1.04	

In general, P–N force constants and bond orders decrease with decreasing electronegativity of X, i.e. decreasing positive charge on P. The cases of X = P and X = S are exceptions due to coupling of S–P and P–N vibrations.   (RP&-71a)

IR S–N stretching frequencies (symmetric and asymmetric) increase in order (Me$_2$N)$_2$SO < (Me$_2$N)$_2$S < (Me$_2$N)$_2$SO$_2$, whereas N-basicity, decreases in order (Me$_2$N)$_2$S > (Me$_2$N)$_2$SO > (Me$_2$N)$_2$SO$_2$, which is the expected order of increasing $dp-\pi$ bonding strength. | IR stretching frequencies are a poor measure of bond strength due to coupling with vibrations of adjacent bonds; here the basicity is a far better measure of $dp-\pi$ bonding.   (TP&-70)

NMR and Related Indices

The relationships (such as they are) between force constants and $dp-\pi$ bonding have been accepted by some scientists as a theoretically valid means of using infrared spectroscopy to study $dp-\pi$ bonding. Unfortunately, no theoretically valid approach is known for NMR chemical shifts because the molecular structural factors affecting chemical shifts are still not understood (*E-69*). Although inductive and resonance effects undoubtedly play a role in determining proton chemical shifts, magnetic anisotropy of adjacent bonds and possibly dispersion forces (of *van der Waals* origin) (*SR-64*) can often mask the former. With nuclei heavier than hydrogen (e.g., ^{19}F and ^{31}P) the situation is even more complicated since we must include a paramagnetic term involving interactions between the ground state and excited states having unpaired electrons (*E-69*). *Ebsworth*, in particular (*E-69*), has pointed out that these and other unknown factors create uncertainty with regard to many of the conclusions reached on the basis of empirical correlations involving chemical shifts. Thus, the increase in proton chemical shifts (*S-63a*) of Me_3SiX with increasing electronegativity of X, which might be attributed to the expected increase in Si–X, $dp-\pi$ bonding with increasing electronegativity of X, is also observed for Me_3CX, (*EF-63*), and must therefore be due to anisotropy. Only when all the members of a series of compounds have the same anisotropy can changes in chemical shift be reasonably attributed to $dp-\pi$ bonding. The very small increases in methyl proton chemical shifts with increasing n in the series $Me_{3-n}SiX_n$ for X = methoxy or dimethylamino, and perhaps the somewhat larger increase for X = Cl, compared to the corresponding decrease with increasing n in the analogous carbon compounds (*VN-68*) can reasonably be attributed to $dp-\pi$ bonding if we assume that the anisotropy contributions do not vary with n.

However convincing this evidence is for $dp-\pi$ bonding, the increases in chemical shifts were in all cases less than 21% of that observed between the α and β protons of fluoroethylene. Consequently, the magnitude of $dp-\pi$ bonding is considerably less than that of $pp-\pi$ bonding to the vinyl group, a result which is not exactly surprising. The $J(^{13}C-H)$ coupling constants for the X–Me protons increase with increasing n in these series, but the values for Me_3SiOMe and $Me_2Si(OMe)_2$ are larger than those for the analogous carbon compounds, while the values for $MeSi(OMe)_3$ and $Si(OMe)_4$ are smaller than those for the analogous carbon compounds. This is inferred (*CM&-66*) to mean that an increase in the amount of electron delocalization to silicon occurs with increasing n, but along with a decrease in the π contribution to each individual Si–O bond. This inference is in complete accord with the Cruickshank model but disagrees with conclusions reached by comparing the basicities of tetraethyl orthocarbonate and silicate, (*WWL-61*).

Few attempts have been made to quantitatively determine the anisotropy contribution to the chemical shift. *Brune* (*B-64b*, *B-65*) calculated the electronegativities of oxygen in the series $Cl_nMe_{3-n}SiOR$ by assuming that the anisotropy contribution to the chemical shift difference between the α and β protons is the same for R = ethyl and R = isopropyl. The Pauling electronegativities so calculated decreased from 3.22 for Me_3SiOR to 3.10 for Cl_3SiOR, whereas the opposite would be expected from the decrease in inductive electron donation by silicon with increasing chlorine substitution. This is proposed (*B-64a*) to signify strong Si–Cl, $dp-\pi$ bonding, which weakens Si–O, $dp-\pi$ bonding with increasing chlorine substitution, thereby increasing the electron density on oxygen. Since the decrease in oxygen electronegativity on going from Me_3SiOR to $Si(OR)_4$ was less than half that on going to Cl_3SiOR (*B-65*), Si–Cl, $dp-\pi$ bonding must be considerably stronger than Si–O, $dp-\pi$ bonding.

This conclusion conflicts with electron diffraction and IR spectroscopic evidence but is in agreement with other NMR work. The chemical shift of the *para* proton in ArX, where Ar is 2,5-dideuteriophenyl, decreased (greater deshielding) in the order $X = CMe_3 > MEt_3 > MCl_3$, indicating Ar–M, $dp-\pi$-bonding increasing as the number of electron-withdrawing substituents on M increases (*WS&-70*). However, the difference in *para* proton chemical shift between $ArMEt_3$ and $ArMCl_3$ decreased in the order $M = Sn > Ge > Si$, which is attributed to partial negation of the electron-withdrawing inductive effect of chlorine in the order $Si > Ge > Sn$. Due to the spatial distance between X and the *para* proton, this method is not likely to suffer from interference by anisotropic effects.

Egorochkin (*EV&-70*), who has recently reviewed the NMR and IR evidence for $dp-\pi$ bonding to silicon (*EVK-72*), compared published chemical shifts for the methyl protons in Me_3SiX with those calculated as the sum of inductive, anisotropic, and electric field contributions. The inductive contribution was determined from a linear correlation with the Taft σ^* values, and the anisotropic contribution was calculated directly from the anisotropic magnetic susceptibility and the molecular geometry of the Si–X bond, assuming free rotation around the C–Si bonds; a similar method was used to calculate the effect of the electric field created by the Si–X dipole. The difference between the experimental and calculated chemical shifts, assumed to depend only on $dp-\pi$ bonding, decreased in the order $X = F > OMe \sim Cl > Ph > Br$, which is essentially the expected order of $dp-\pi$ bonding ability.

Since an increase in $J(^{13}CH)$ in the XCH system implies an increase in the s character of the C–H bond, and therefore an increase in the p character of the X–C bond, it seems reasonable to correlate these coupling constants with the electronegativity of X (*HMT-64*). Thus, J increases from 131 cps in trimethylamine to 145 cps in tetramethylammonium chloride to 148 cps in nitromethane, which is the expected order of increase in nitrogen electronegativity (*MP-59a*). Similarly, the small decrease in the coupling constant from 118 cps in tetramethylsilane to 115 cps in Me_3SiONa may be attributed to $dp-\pi$ bonding or to inductive electron donation by the anionic oxygen (which is, of course, of very low electronegativity). Either of these effects would increase the electron density on silicon and thereby decrease its electronegativity (*S-63a, T-76*). In hexamethyldisiloxane $dp-\pi$ bonding is somewhat weaker, just cancelling out the electron-withdrawing, inductive effect of oxygen. However, the $J(^{13}CH)$ values for the protons of the alkyl groups of $Cl_nMe_{3-n}SiOR$ increased with increasing n (*B-64b*) in spite of the corresponding decrease in oxygen electronegativity (*B-64a*), proving that $J(^{13}CH)$ values are affected by anisotropy. On a *Goldstein-Reddy* plot (*GR-62*) of coupling constant vs. chemical shift, in which compounds of identical anisotropy form a straight line, the results show that the anisotropy increases with increasing n, as was found from the difference in chemical shifts between α and β protons in different alkyl groups. Thus, the lack of increase in $J(^{13}CH)$ in going from trimethylphosphane to its oxide and from dimethyl sulfide to the corresponding sulfoxide and sulfone (*HMT-64*) could be due to anisotropy, and need not necessarily imply strong P–O and S–O, $dp-\pi$ bonding.

A direct measure of hybridization at nitrogen may be obtained from $J(^{15}NH)$ coupling constants in ^{15}N-labelled amines; these coupling constants were almost identical in the ammonium ion, aniline, and N-trimethylsilylaniline (*RZ-68*). Similarly, $J(^{15}NH)$ for $(Me_3Si)_2NH$ was close to that for ammonia and less than that for ammonium ion, clearly indicating nonplanarity in all cases. From this *Randall and Zuckerman* (*RZ-66*) took "an agnostic view of supposed large $(p \rightarrow d)$ π contributions to the Si-, Ge-, and Sn–N bonds". This generalization made on the basis of a few compounds, may be somewhat extreme, since the electron donating methyl groups would be expected on theoretical grounds to greatly weaken $dp-\pi$ bonding to the trimethylsilyl group. In particular, $dp-\pi$ bonding in trimethylsilylaniline should be almost negligible. *Perkins* (*P-67*) showed that $dp-\pi$ bonding in trisilylamine, although favored by planarity, does not require it.

Proton chemical shifts have been used to study $dp-\pi$ conjugation between sulfur and a carbon-carbon double bond. The chemical shifts of the β-*trans* protons of phenyl vinyl ether and phenyl vinyl sulfide were at higher and lower field, respectively, than that in allylbenzene. Consequently, assuming anisotropy can be neglected in this system, the conjugation between sulfur and the olefinic electrons is $dp-\pi$ rather than $pp-\pi$ as with oxygen (*KKF-73*). In Hammett plots of the β-*trans* proton chemical shift in $p-XC_6H_4YCH=CH_2$, ρ was greater for Y = S than for Y = O, which in turn was greater than for the corresponding styrene (no Y). Similar results were found in Hammett plots based on ^{13}C chemical shifts and on the pKa values of $p-XC_6H_4$-YCH=CHCOOH (*FK&-73*). This has been attributed (*KKF-73*) to through conjugation at sulfur (but not at oxygen) involving d orbitals. Since the $dp-\pi$ bond is known to be nodal at the central atom, this explanation is unlikely; a more likely possibility is that the electron-donating X changes the vinyl-sulfur, but not, of course, the vinyl-oxygen conjugation from $dp-\pi$ to $pp-\pi$.

Price (*PH&-63*) found that all ring protons in I through V appeared as a single broad peak in the NMR, at 2.4–3.0 τ, relative to TMS at 10.0 τ, suggesting aromaticity. On the other hand, the equivalent carbon-2 and carbon-6 protons in VI fell at -5.75 *ppm* relative to TMS at 0, while the carbon-4 proton fell at -6.1 *ppm* (*HHM-74*). Similarly in VII and VIII, the carbon-2 protons appeared at -6.16 and -6.59 *ppm* respectively, and the carbon-4 protons appeared at -5.63 *ppm* in both cases (*TSK-68*). *Price* attributes the aromaticity of I through V to $pp-\pi$ bonding, the sulfur being sp^2 hybridized with the lone pair occupying a nonbonding 3d orbital; this is, of course, impossible in VI, VII and VIII. On a *Goldstein-Reddy* plot (*GR-62*), the values for VI lie well below the line for substitued thiophenes, indicating that there is no aromatic ring current, and the chemical shifts for the carbon-2 and carbon-6 protons are well upfield of those expected for olefinic protons α to sulfur, (*HHM-74*). This is evidence for negative charge on carbon-2 and carbon-6, i.e. ylide character. For $Ph_3P=CH_2$ δ is $+0.61$ *ppm* compared to -3.61 *ppm* for $Ph_3P=CHCOMe$ in which the negative charge on the α carbon is reduced by $pp-\pi$ resonance with the acetyl group. By comparison, then, VI, VII, and VIII must have fairly low negative charge on their ylide carbons. Proton exchange with solvent (*HHM-74*) occurs at carbon-2 and carbon-6 in VI, but it is slow on the NMR time scale except in strong acid (e.g. CDCL$_3$/CF$_3$COOH), and the equilibrium between ylide and onium salt in IX is much in favor of the former.

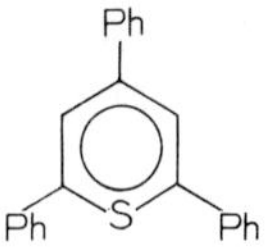

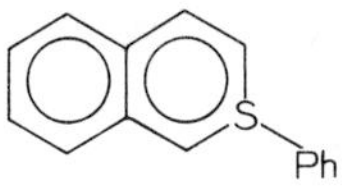

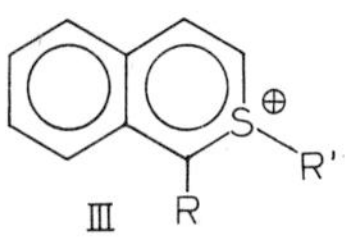

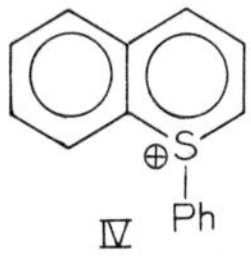

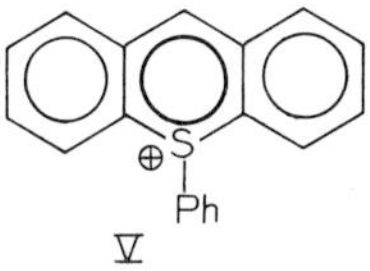

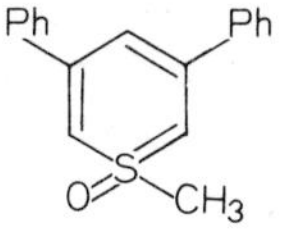

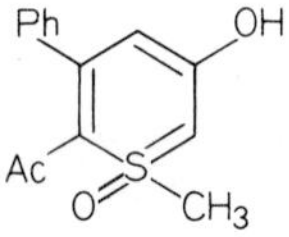

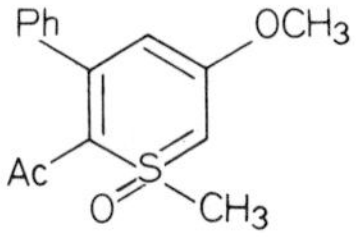

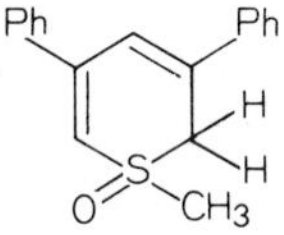

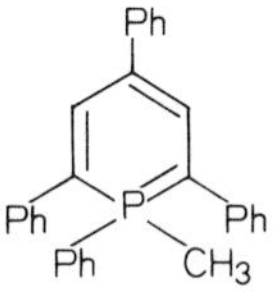

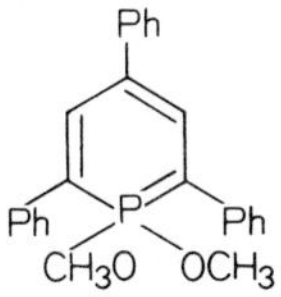

More recent work (SS&-74) has shown that the brown, amorphous solids to which *Price* (PH&-63) assigned structures I through V are actually oligomers of undetermined nature. Compounds corresponding to III could be isolated by Price's method, i.e. deprotonation of the corresponding thiopyrylium salts with aryllithium, for R = aryl and R' = methyl, but their solutions decomposed upon standing, and their NMR spectra showed them to be ordinary sulfonium ylides (SM&-74). Further support for the ylide structure is given by the stabilization of III by an electron-withdrawing pentafluorophenyl group on the ylide carbon (III, R = C$_6$F$_5$) to such an extent that this compound could be isolated in optically active, crystalline form, with a free energy barrier to pyramidal inversion at sulfur of at least 23.7 kcal/mole (MS&-74, MS&-75).

Note, however, that the ylide character of the oxothiabenzenes does not eliminate $dp-\pi$ bonding, which always requires a strong electron donor and a strong positive charge on the central atom. Furthermore, $dp-\pi$ bonding is understood to be of the type proposed by *Dewar* (DLW-60, sec. II.F) for the phosphonitriles, in which case there is a node on sulfur; thus, there can be no through-conjugation and no aromaticity. This is supported by Hückel MO calculations (*VBS-68*) on 1,1-dialkylphosphabenzene, which must have bonding very similar to that in the oxothiabenzenes. Of four models tested, the Dewar model constituted the best fit to the observed basicity properties and ultraviolet spectra.

The 1,1-dialkylphosphabenzenes probably have considerably weaker $dp-\pi$ bonding than do the oxothiabenzenes, as indicated by the high susceptibility of the former to autoxidation in air and to nucleophilic substitution at carbon-2 and carbon-6. The dialkylphosphabenzenes also undergo proton exchange at these centers in aqueous acid, and the equilibrium is much more displaced toward the onium salt than it is in the case of the oxothiabenzenes (*M-63*). However, the properties of phosphabenzenes vary strongly with the substituent at phosphorus; thus, 1,1-dialkoxyphosphabenzenes are not alkylated by oxonium salts and are only protonated by strong acids such as trifluoroacetic (*DS-68*). In the NMR, the carbon-3 proton chemical shift fell at $-7.93\,ppm$ in XI compared to $-7.42\,ppm$ in X (*DS-68, MM-68*). Based on the finding that $dp-\pi$ bonding in acyclic vinylsulfonium salts tends to transmit positive charge to the β carbon (*CPH-66*), this would seem to indicate stronger $dp-\pi$ bonding in XI. These chemical shifts are considerably different than those for the carbon-2 and carbon-4 protons in oxothiabenzenes, but CNDO (complete neglect of differential overlap) calculations on 1,1-dimethoxyphosphabenzene show negative charge on carbon-2 and carbon-4 and positive charge on carbon-3, (*OS-70*). Unfortunately, no NMR spectra are available for phosphabenzenes unsubstituted at carbon-2 and carbon-4; consequently, no comparison with the spectra of the oxothiabenzenes is possible.

More recent CNDO calculations (*SS&-76*) support these conclusions and give a coherent explanation of the bonding in phosphabenzenes (λ^5-phosphorins). Inclusion of *d*-orbitals in the basis set decreased the positive charge on phosphorus by 40–50%, depending on the exocyclic substituents on phosphorus, and decreased the negative charge on the ring carbons α and γ to phosphorus by 20–25%. Changing the exocyclic substituents from the methyl to fluorine increased the positive charge on phosphorus by 36% but had much less effect on the negative charge on the ring carbons. Use of a conjugation decoupling method showed that 72.3% of the total phosphorus-ring carbon conjugation was due to $dp-\pi$ bonding and the rest to hyperconjugation; of the $dp-\pi$ bonding, 69% involved the d_{yz} and the rest the d_{xz} orbital. Thus, *d*-orbital contributions to the "aromaticity" are primarily of the Hückel rather than the Dewar type (see p. 13). However, the ionization potentials and therefore the photoelectron spectra were little affected by *d*-orbitals, reflecting instead the basic polar structure, i.e. a tetravalent phosphorus radical cation and a pentadienyl radical anion formed by donation of one of the P(III) lone pair electrons to the ring carbons.

By comparison, the NMR spectrum of $Me_2S(O)=CH_2$ (*ST-68a*) showed a single peak at -2.1 to $-3.0\,ppm$ depending on solvent and temperature. Phosphorus and arsenic ylides are known to undergo intermolecular proton exchange at the methyl and carbanionic methylene groups. At 30 °C the methyl protons of $Me_3P=CH_2$ fell at $-1.22\,ppm$ and the methylene protons at $0.78\,ppm$, whereas at 100 °C only one broad peak is found (*ST-68a*). When traces of acid are present, only one sharp peak is found, even at 30 °C. Therefore, it is deemed highly probable that the $Me_2S(O)=CH_2$ was contaminated with acid which promoted rapid proton exchange (*ST-68*). Since there are three times as many methyl as methylene hydrogens, the single peak observed must be fairly representative of the shielding on the methyl groups. The rather high shielding on the methyl groups in $Me_3P=CH_2$ is noteworthy in view of the ^{31}P chemical shifts, which showed much greater shielding on phosphorus in the free ylide than in the corresponding phosphonium salt, i.e. tetramethylphosphonium iodide, or in the lithium complex of the ylide, for which the ^{31}P chemical shift was almost the same as in the phosphonium salt (*AS-76*). This suggests that $dp-\pi$ bonding is greatly reduced by association of the charge on the methylene carbon with the lithium, a result supported by CNDO calculations (*AS-76*), which also showed that the actual amount of charge on the methylene carbon is little affected by complexation with lithium in agreement with the ^{13}C chemical shifts. It seems apparent now that any attempt to prepare an unstabilized ylide by reaction of a phosphonium salt with an alkyllithium will give the lithium-ylide complex rather than the free ylide, the reference substance which was prepared by a method involving desilylation of $Me_3P=CHSiMe_3$ (*ST-68a*).

Both phosphorus atoms in $[(Ph_3P)_2CH]^+Br^-$ were equivalent, since only one peak was found in the ^{31}P NMR spectrum (*DG&-64*); this is to be expected by comparison with the π bonding in the phosphonitriles. The proton chemical shift for the methylidene carbon was -1.92 ppm; comparison with the $-.368$ ppm chemical shift (*T-70*) for $Ph_3P=CHCOMe$ clearly indicates the lesser stabilization available to ylides from $dp-\pi$ bonding compared to enolic resonance. This is all the more apparent in view of the positive charge on the bis-ylide. In another interesting example, ^{31}P chemical shifts for a series of trivalent phosphorus compounds, X_3P, increase in the order X = methoxy > dimethylamino > phenyl > methyl, whereas the opposite result was found for the oxyphosphorane XII (*T-70*). Thus, oxygen and nitrogen donate electrons to pentavalent phosphorus by $dp-\pi$ bonding but withdraw electrons from trivalent phosphorus by induction, a clear illustration of the comparative lack of importance of $dp-\pi$ bonding in the lower valence states.

The ^{31}P NMR chemical shift, which is a measure of the positive charge at phosphorus, is considerably shifted to higher field in $Ph_3P=CHCOOMe$ (A) compared to $Ph_3P=CHCOPh$ (B), and the difference in chemical shift between the ylide and the corresponding phosphonium salt is also much less for (A) than for (B). A linear relationship was observed between the phosphorus $2p$ binding energies determined by

X-ray photoelectron spectroscopy and the ^{31}P chemical shifts (*ST&-75*). The carbonyl stretching frequencies showed the same effect, the bathochromic shift upon going from phosphonium salt to ylide being much less for (A) than for (B). Consequently, (A) has much less enolic character than (B). This is in agreement with the lower acidity of ethyl acetoacetate compared to acetylacetone (*G-59*), which results from *pp*–π electron donation by methoxyl in the enolate anion. Apparently, the difference in proton chemical shifts between the phosphonium salt and ylide was larger for (A) than for (B) because of the much greater negative charge on the ylide carbon in (A). Measurement of the heats of combustion showed the P=C bond energy to be much larger for (A) than for (B). This agrees with X-ray diffraction work previously discussed in Section III A. suggesting that *dp*–π bonding increases with increasing negative charge on the ylide carbon and is hindered by enol-carbonyl resonance stabilization of the ylide.

NMR spectroscopy showed that the product of reaction XIII is XVI, not XV, whereas the carbon analog gives the opposite results, XIV (*SM-70*). This proves that ylide stabilization by α-silyl groups is not steric in nature. *Schmidbaur* (*SM-70*), however, has pointed out that this does not prove the stabilization is due to *dp*–π bonding; it may just as well be due to the length of the C–Si bond, which minimizes repulsions between the carbanionic lone pair and α substituents. The long bond may also be responsible for the greater stability of Cl_2 compared to F_2, in explanation of which *dp*–π bonding has also been proposed (*M-55*).

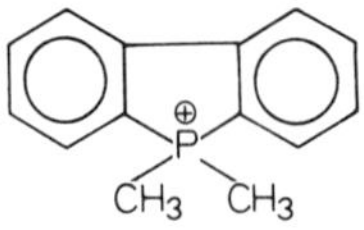

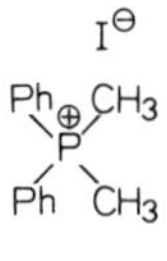

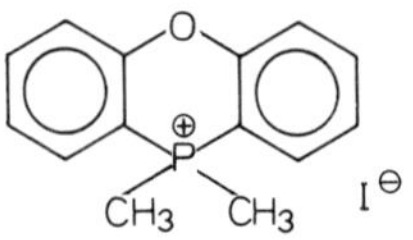

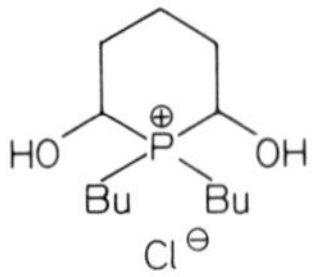

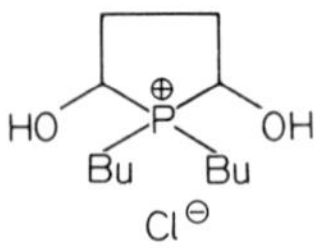

There is an apparent correlation between ^{31}P chemical shifts and hydrolysis rates for certain phosphate esters; the 18 *ppm* deshielding of the five-membered cyclic esters compared to their acyclic analogs agrees with MO calculations which indicate weaker *dp*–π bonding for the strained cyclic esters (*C-66a*; see also Sec. III A). This correlation, if indeed it exists, cannot be extended to phosphonium salts; ^{31}P chemical shifts relative to phosphoric acid are -27 *ppm* for XVII, -19 *ppm* for XVIII and 7.5 *ppm* for XIX, whereas hydrolysis rates vary in the order XVII > XIX > XVIII. *Allen* and *Tebby* (*AT-70*) suggest that these variations in chemical shift are probably related to deviations from tetrahedral symmetry rather than to changes in *dp*–π bonding, small deviations from tetrahedral symmetry causing somewhat increased shielding and large deviations producing considerable deshielding. Thus, ^{31}P chemical shifts were -33 *ppm* for tetraethylphosphonium iodide, which is tetrahedrally symmetric, -29 *ppm* for cyclohexyltrimethylolphosphonium bromide and -46 *ppm* for XXI both of which are greatly distorted from the tetrahedral. This cannot be due to changes in $dp-\pi$ bonding since all nonaromatic phosphonium salts should have negligible $dp-\pi$ bonding. On the other hand, hydrolysis rates are a function of the amount of distortion needed to form the trigonal bipyramid, being naturally less for the cyclic than for the acyclic structures.

Increasing substitution of phenyl for fluorine in hexafluorocyclotriphosphazene increases the ^{31}P chemical shifts to higher field (*ATM-68*). This could be due either to the smaller electron-withdrawing inductive effect of phenyl, or to an increase in exocyclic P–F, $dp-\pi$ bonding. Indication that the latter is the case is given by the decrease in ^{19}F chemical shifts with increasing phenyl substitution. Apparently, exocyclic π bonding decreases; since the d_{xz} orbital is known to be involved in both endocyclic and exocyclic bonding (*P-64*) this seems quite reasonable. Using a correlation similar to that developed by *Rakshys, Taft,* and *Sheppard* (*RTS-68*) between the difference of the meta and para ^{19}F chemical shifts in FC$_6$H$_4$X, and the Taft *R* value for X, *Chivers* and *Paddock* (*CP-72*) found that $\delta_p-\delta_m$ for mono(pentafluorophenyl) substituted cyclophosphazenes, C$_6$F$_5$N$_n$P$_n$F$_{2n-1}$, ranged from -14.7 *ppm* to -16.1 *ppm* depending on n, implying that the phosphonitrilic rings are about as electron-withdrawing as the cyano group, for which $\delta_p-\delta_m = -15.7$ *ppm*. The variation with n is due to the fact that there are two π bonding systems at right angles to each other (*M-69, P-64*); see Sec. II. The stronger of these is heteromorphic and degenerates in the limit of equal contributions from the d_{xz} and d_{yz} orbitals to the three-center or "island" model of *Dewar, Lucken* and *Whitehead* (*DLW-60*). Thus, for this π system the energy of the highest occupied MO and the delocalization energy per electron do not alternate with ring size. The weaker π' system is homomorphic and follows the *Hückel* $4n + 2$ rule. Since exocyclic $dp-\pi$ bonding involves mainly the d_{z^2} orbital, which cannot contribute at all to the π system and only very weakly to the π' system, it is understandable that exocyclic $dp-\pi$ bonding should vary with ring size and be strongest for the 6 and 10 electron systems (*CP-72*).

Rakshys, Taft and *Sheppard* (*RTS-68*) found that $\delta_p-\delta_m = -1.35$ for FC$_6$H$_4$–SiMe$_3$, indicating that silyl groups are weakly electron-withdrawing by $dp-\pi$ bonding, whereas phosphino and thio groups bearing electron-donating substituents such as alkyl, phenyl, or amino had $\delta_p-\delta_m > 0$ (*RTS-68, ST-72*). Phosphino groups with electron-withdrawing substituents were capable of $dp-\pi$ bonding, with $\Delta\delta$ as low as -6.2 for –PF$_2$ and -5.7 for –P(CF$_3$)$_2$; by comparison $\Delta\delta = -1.8$ for –SCF$_3$ indicating that thio groups are capable of only very weak $dp-\pi$ bonding even under favorable circumstances. Sulfinyl groups other than SOF, SOCF$_3$, and SOCl also showed only very weak $dp-\pi$ bonding. Although sulfones show strong $dp-\pi$ bonding, in general $dp-\pi$ bonding is weaker for sulfur than for phosphorus in the high oxidation states; thus, $\Delta\delta = -7.1$ for ArPMePh$_2$ and -9.4 for ArPF$_4$ vs -4.9 for ArSMe$_2$ and -5.9 for ArSF$_3$. The σ_R value for ArPF$_4$ is one of the largest known but is not necessarily characteristic. For instance, σ_R values suggested $dp-\pi$ bonding to the fluorophenyl (Ar) group to a negligible extent in Ar$_3$P(OEt)$_2$ and ArSF$_5$, (*ST-72, CDD-71*). For ArSF$_5$ this is perhaps best explained by strong S–F, $dp-\pi$-bonding, which completely "saturates" the three d-orbitals available for $dp-\pi$ bonding. This is supported by the short S–F bond length (*ES-63*) in comparison to that expected for a three-center bond, and by the geometry, which is very favorable for $dp-\pi$ overlap. In ArPF$_4$ the geometry is less favorable, and there is one more d-orbital and one less fluorine; consequently, electron donation to the central atom is diminished. In Ar$_3$P(OEt)$_2$ the π-electron density on the basal fluorophenyl groups is probably too low for strong $dp-\pi$ bonding, and the ethoxy groups are held by apical bonds in which $dp-\pi$ bonding is inherently weaker.

The use of dynamic (temperature-variable) NMR spectroscopy to probe inversion and rotation barriers is based on two assumptions which have seldom been tested (*RMA-70*). The first of these is simply that there are distinct barriers to rotation and inversion, and the second is that the difference in chemical shift between the diastereotopic groups is large enough to be detected. Measurements on N-chloroaziridine at 20.53 MHz showed all four protons to be equivalent (*BK&-65*), whereas at 60 or 100 MHz an AA'BB' pattern was found which did not coalesce even at 120 °C. Thus N-chloroaziridine, like most aziridines, shows a high inversion barrier, as would be expected from the strong electron-withdrawing inductive effect of chlorine, which should increase the p character of the N–Cl bond, and, consequently, the s character of the lone pair. In fact, the inversion barrier for 1-chloro-2-methyl-aziridine is so high that the compound can be separated into invertomers by gas chromatography at 25 °C (*B-68*).

Coalescence temperatures corresponding to inversion barriers have been determined for several aziridines with N–S bonds; they were − 30 °C for N-phenylsulfonylaziridine, 0 °C for N-phenylsulfinylaziridine, − 11 °C for N-phenylsulfenylaziridine, and − 108 °C for N-diphenylphosphinoylaziridine (*ATC-67*). The low inversion barrier for the Ph$_2$P(O)-group must be due to reduction of the electron density on nitrogen by P–N, $dp-\pi$ bonding. The barriers for the sulfonyl- and sulfinylaziridines are in agreement with nonempirical SCF–MO calculations on $^\ominus$CH$_2$SHO and $^\ominus$CH$_2$SHO$_2$, using Gaussian basis sets. The results showed the former carbanion to have a barrier 3.7 kcal/higher than that for the latter (*RWC-69, WRC-69*). Both barriers are attributed to repulsion between the carbanionic lone pair and the sulfur lone-pair (in the sulfoxide, or the polar S–O bond in the sulfone), and assuming $dp-\pi$ bonding to be negligible. However, these calculations refer to a combined inversion-rotation between conformers XXII and XXIII, which are in fact identical. Conclusions about substituent effects on inversion or rotation barriers may not apply to such combined barriers.

The ketimines PhX(O)$_n$N=(C$_6$H$_4$Me)$_2$ provide an interesting illustration of the effect of $dp-\pi$ bonding on inversion barriers. As has been pointed out by *Lehn* (*L-70*),

XXII   XXIII   XXIV

imines can be considered as showing the maximum possible angle strain in a pyramidal system (zero angle at nitrogen); thus the inversion barriers (26 kcal for $CH_2=NH$ vs 18 kcal for aziridine) and their sensitivity to electronic effects of substituents should be even larger than for aziridines. In the case of X = S, the inversion barrier decreased from the 18.4 kcal value for $n = 0$ to 14.1 kcal for $n = 1$ and 13.0 kcal for $n = 2$ (*DK-76*), but where X=Se, the inversion barrier was 20.5 kcal, about the same as in $t-BuN=C(C_6H_4Me)_2$, and varied little with n. Although negative hyperconjugation (overlap with σ^*) could explain the failure of Se to show inversion barrier reduction via electron-accepting conjugation, it does not explain the variation with n since the antibonding orbitals on sulfur change in character and become much less capable of overlap on going from sulfenyl ($n = 0$) to sulfinyl ($n = 1$) or sulfonyl ($n = 2$). Similarly, steric or inductive effects would be expected to be much larger for sulfonyl than for sulfinyl, whereas the decrease in inversion barrier is much larger on going from $n = 0$ to $n = 1$ than from $n = 1$ to $n = 2$.

There is a great interest in these studies in connection with the almost complete retention of configuration during deuterium exchange α to sulfonyl groups in optically active sulfones. As will be discussed later in Section IV.B, the ease of formation of α-sulfonyl compared to α-sulfinyl carbanions may well be a result of $dp-\pi$ bonding, but these SCF calculations clearly show that $dp-\pi$ bonding plays no part in the configurational stability of these carbanions. This agrees with, e.g., MO calculations on trisilylamine (*P-67*), which show $dp-\pi$ bonding to have almost no conformational requirements. Lone-pair/polar S—O bond repulsions probably also explain the 11.4 kcal/mole rotational (or rotational-inversional) barrier in Et_2NSO_2Cl and in ET_2NSO_2Ph (*JS-70*). The low configurational stability of optically active carbanions derived from phenyl 2-octyl sulfoxide compared to that from phenyl 2-octyl sulfone, as observed by *Cram* (*CTS-66*) in disagreement with these calculations, is probably a solvent effect; other α-sulfinyl carbanions have been observed to show lower conformational stability in nonpolar than in polar solvents (*FSW-72*).

The 18.9 kcal/mole inversion barrier in iPrPhPSiMe$_3$ (*BM-70a*) is considerably less than that in trialkylphosphines (*BM-70*). It might be tempting to attribute this to $dp-\pi$ bonding, but a plot of the inversion barriers for iPrPhPMMe$_3$ and for BzMePPBzMe correlated linearly with Allred electronegativity for M = C, Si, Ge, Sn, and P, (*BM-71*), proving thereby that the decreased inversion barrier is due entirely to the electron-donating, inductive effect. This is in agreement with SCF calculations (*LM-70*) on H_2NSiH_3, for which the inversion barrier was found to be almost entirely unaffected by the presence or absence of $dp-\pi$ bonding, even though the highest occupied orbital was of the π type. However, $dp-\pi$ bonding may well be involved in the lowering of the C—N rotational barrier between cisoid and transoid forms from 22.6 kcal/mole in $HC(O)NHCMe_3$ to 19.8 kcal/mole in $HC(O)NHSiMe_3$, (*KY-72*). In this case the electron-donating, inductive effect of silicon should favor the resonance form XXIV, thus increasing the rotational barrier.

NMR studies of several silylamines including bis(dimethylamino)methyl-phenylsilane and trimethylsilylaziridine showed equivalent N-methyl and methylene groups at temperatures down to $-120\,°C$, corresponding to a free energy barrier < 8 kcal/mole (*CD&-73*). Many phosphinoamines, on the other hand, show much larger barriers, ranging from 14.6 kcal/mole for PhPClN(*i*Bu)$_2$ ($T_c = 15\,°C$) to

< 8 kcal/mole for F_2PNMe_2 (*CD&-70*). The free energy barrier increase on going from $PhPClNMe_2$ to $PhPClN(iPr)_2$ indicates that the barrier is one of rotation, since increasing the size of groups on nitrogen should lower the energy of the planar transition state for nitrogen inversion. In fact, nitrogen inversion is probably quite fast, whereas inversion at phosphorus is too slow to be measured on the NMR time scale. Even higher free energy barriers are found for sulfenamides, ranging as high as 17.8 kcal/mole for $ArSN(Ph)CH_2Ph$ (*RKJ-69*). As with the phosphinoamines, these barriers increase with increasing size of the N-alkyl groups and are thus rotational in nature.

These results suggest lone pair repulsion as the dominant factor affecting rotational barriers. This explains the low rotational barrier in silylamines, where there is only one lone pair, and the increase in the free energy barrier in going from phosphinoamines to sulfenamides, increasing the number of interacting lone pairs from two to three. However, this cannot be the whole story. The much lower barriers found in symmetrical analogs, e.g. $MeNPClCF_3$ (*CD&-70*), have been attributed to $dp-\pi$ bonding on the assumption that the conformational dependence of σ overlap by the d_{xz} or d_{yz} orbitals hinders π overlap by these orbitals, the extent depending on conformation. There is, however, a simpler explanation, one involving the preference for *gauche* conformations, as observed in α-sulfinyl and α-sulfonyl carbanions (*RWC-69*) and in most cases of rotation around bonds to atoms with lone pairs (*W-72*). For a general aminophosphine, R_2NPXY, the *trans* conformation XXV can be eliminated by the observation of diastereotopic methyl protons in Me_2NPCl_2 at temperatures below $-113\,°C$. In this conformation the methyl protons would be equivalent, and the NMR spectrum would consist of a single line at all temperatures. For symmetrical aminophosphines (S = Y) the two gauche conformations XXVI and XXVII are identical, but for unsymmetrical analogs, the barrier must be increased by the free energy difference between XXVI and XXVII.

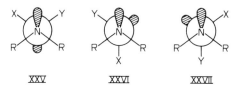

XXV XXVI XXVII

The low rotational barrier in $Me_3P=NMe$ (*BC&-73*), for which the high infrared P=N stretching frequency (1239-cm^{-1}) indicates strong $dp-\pi$ bonding, has been used as an argument against a conformational dependence in $dp-\pi$ bonding, but the lack of a lone pair on phosphorus makes a high barrier unlikely here. However, the increase in rotational barrier in R_2NSX with increasing electronegativity of X (*RKJ-69*) must indicate some extra conformationally-dependent contribution to the S–N barrier, either negative hyperconjugation ($n-\sigma^*$ overlap) (*RJ-71*) or $dp-\pi$ bonding. On the other hand, although the S–N rotational barrier in XXVIII (*KZ-75*) increased by 7.1 kcal on going from Ar = p-tolyl to Ar = 2,4-dinitrophenyl, there was no corresponding decrease in the N—C(O) rotational barrier as would be expected if $dp-\pi$ bonding were reducing the lone pair electron density on nitrogen. In fact, there is no way in which electron-accepting conjugation by sulfur could change the S—N bond without affecting the C—N rotational barriers; clearly, this too may be some kind of *gauche* effect.

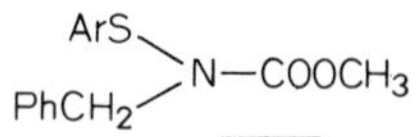

ArS\
    N—COOCH$_3$
PhCH$_2$/
XXVIII

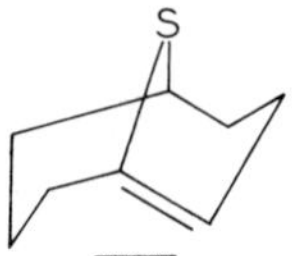

XXIX 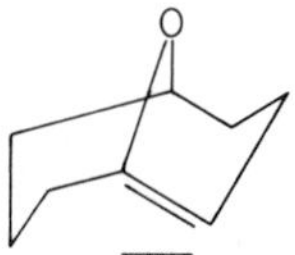   XXX

The chemical shift of the β vinyl hydrogen in XXIX was at lower field (deshielded) compared to that for the vinyl H in cyclohexene, whereas the corresponding value for the oxygen analog XXX was at higher field. Since both vinyl ethers and sulfides show shielding of the β vinyl H compared to ethylene, *Wiseman (QW-73)* has argued that the lowers shielding of the vinyl sulfides compared to vinyl ethers is due to competition between $pp-\pi$ and $dp-\pi$ bonding. In XXIX twisting due to the bridgehead bonding hinders $pp-\pi$ but not $dp-\pi$ bonding. However, extended Hückel calculations by *Batich (B-75)* for vinyl mercaptan and vinyl alcohol show that the greater positive charge on the terminal carbon for the former, as observed experimentally, can be reproduced without *d*-orbitals in the basis set. This is not surprising since the larger size of sulfur $3p$ orbitals should make —S—C much weaker than —O—C,

$pp-\pi$ bonding. On the other hand, this still does not explain Wiseman's results, which if not an artifact due to anisotropy, definitely imply the operation of acceptor orbitals on the bridging sulfur atom.

Survey of Relevant Articles on Physical Properties Related to $dp-\pi$ Bonding – NMR (Table III C)

System and Results	Conclusions	References

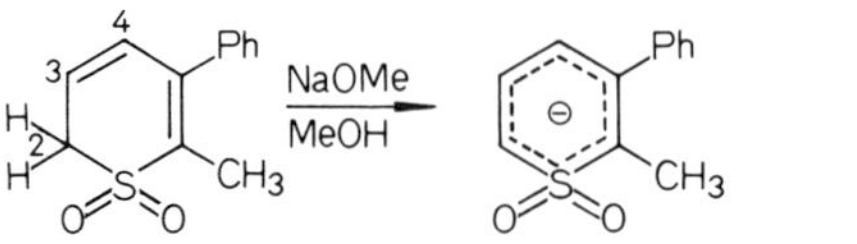

	(H-2)τ (H-3)τ (H-4)τ	
neutral species	5.82 3.80 3.80	
anion	5.20 3.67 5.28	

The downfield shift for H−2 upon anion formation proves that carbon-2 is trigonal, since acyclic sulfones show an upfield shift upon carbanion formation. The downfield shifts for H−3 and H−4 indicate that the carbanion is a completely conjugated system. Apparently the ring geometry enforces carbanion planarity, thus enhancing conjugation and increasing stability.   (BMP-70)

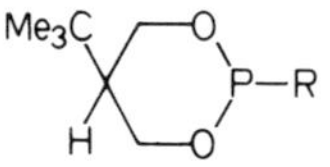

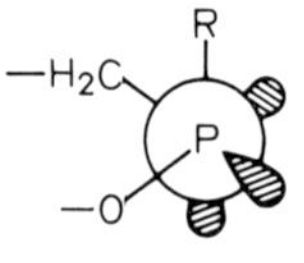

R = dimethylamino exists as an equilibrium mixture of *trans* and *cis* isomers in 83:17 ratio, respectively. R = isopropyl prefers the *cis* configuration.

The dimethylamino group R prefers to be equatorial even though an axial configuration would maximize *gauche* interactions between the oxygen and phosphorus lone pairs (I). However, if the nitrogen is planar, as is likely since the two oxygens are strongly electron withdrawing, (see CS-70) it would suffer eclipsing interactions in the axial position. Thus, in this system $dp-\pi$ bonding is conformationally dependent.   (BT-72)

The difference in proton chemical shifts between the β and α olefinic protons (with respect to the heteroatom) is − 92.3 cps (β more shielded) in PhCH=CHNMe$_2$ and − 19.6 cps in PhCH=CHSMe$_2^\oplus$BF$_4^\ominus$; the S-methyl protons were equivalent at − 60° to + 114 °C.

The sulfur-olefin conjugation in PhCH=CHSMe is $pp-\pi$ (albeit much weaker than in the corresponding enamine) rather than $dp-\pi$ as in phenyl vinyl sulfide, probably because of the greater positive charge on sulfur resulting from the greater inductive effect of phenyl than of methyl. Formation of the ammonium salt eliminates $pp-\pi$ bonding so the proton is influenced mainly by the inductive effect of the adjacent phenyl. Thus, the corresponding sulfonium salt must have fairly strong $dp-\pi$ bonding with no dependence on conformation.   (CPH-66)

System and Results	Conclusions	References
The $2s$ contribution to the N–H bond as measured by $J(^{15}NH)$, increases from 27.9% for $CF_3(Me)PNH_2$ to 30.8% for $(CF_3)_2PNH_2$.	Increasing inductive electron-withdrawal should increase the p character and decrease the s character of the N–H bond; thus, two trifluoromethyl groups apparently do increase the positive charge on phosphorus so much as to induce sufficient P–N, $dp-\pi$ bonding to make the nitrogen planar (ideally sp^2, 33% s character).	(CS–70)
The N-methyl groups became nonequivalent in the NMR when the temperature was reduced below $-46\,°C$; this corresponds to $\Delta H^* = 7.6$ kcal/mole and $\Delta S^* = -19$ eu. $Cl_3C-\overset{\overset{O}{\|}}{S}-NMe_2$	From the highly negative entropy of activation this can be deduced as a rotational barrier, probably with sulfur pyramidal and nitrogen planar in the transition state. In comparison to (M–64) note how the rotational barrier increases with increasing electronegativity of substituents on sulfur or phosphorus, suggesting some sort of $dp-\pi$ contribution to the rotational barrier. (RJ–71)	(JS–67)
The linear Hammett correlation of ^{19}F chemical shifts of m- and p-FC_6H_4SiABZ with σ_I and σ_R does not hold when A = B = Z = amino, alkoxy, or F.	Interference with the Ar–Si, $dp-\pi$ by $dp-\pi$ bonding to A, B and/or Z results from changes in the electron density on Si. Apparently $dp-\pi$ bonding is a low-energy effect; therefore, small changes in electron density can have a major influence on it.	(L–72)
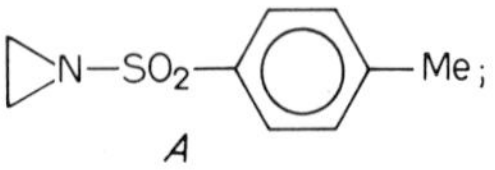 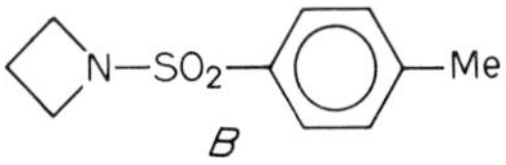 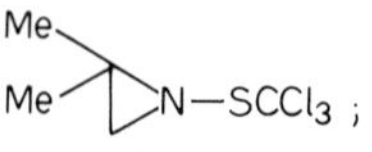 $\triangleright N-SO_2-\langle\bigcirc\rangle-Me$; A $\diamondsuit N-SO_2-\langle\bigcirc\rangle-Me$ B $\underset{Me}{\overset{Me}{\diagdown}}N-SCCl_3$; $\underset{Me}{\overset{Me}{\triangleright}}N-SCCl_3$ C 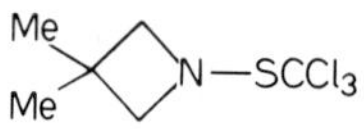 D Activation barriers (ΔG^*) for equivalence of the ring methylene protons in A and B or the gem dimethyl protons in C and D are 12.4 and 6.2 kcal for A and B, respectively but 9.1 and 12.1 kcal for C and D respectively.	Equivalence here requires both inversion around N and rotation around S–N bond. Thus, inversion must be rate-determining (lowest possible ΔG^*) for A and B but rotation is rate determining for C and D. Since $dp-\pi$ bonding reduces inversion much more than rotation barriers, this suggests $dp-\pi$ bonding in A and B but not C and D.	(LPO–71)

Physical Properties Relative to $dp-\pi$ Bonding

System and Results	Conclusions	References
For rotation around P–N bond, A: [bicyclic O,O-P–NMe₂]; $\Delta G^{\ddagger}$ = 7.8 kcal/mol B: [bicyclic Me, S,S-P–NMe₂]; $\Delta G^{\ddagger}$ = 8.9 kcal/mol C: [Me, S,S-P–NMe₂]; $\Delta G^{\ddagger}$ = 7.6 kcal/mol	The greater electronegativity of O compared to S reduces the lone pair electron density on P, thus lowering the rotational barrier. The higher barrier for B compared to C might be explained by stronger P–N, $dp-\pi$ bonding in B, which shortens the bond and thus increases repulsion between the P and N lone pairs.	(BGM–75)
Menthyl methylphosphinate [structure of menthyl methylphosphinate]	Conformation affects $dp-\pi$ bonding. Molecular models indicate that the COP system is planar for the S but not the R configuration at P.	(MBK–70)
$\underset{\text{R-S-NMe}_2}{\overset{\overset{O}{\|}}{}}$ In all cases the N-methyl groups were equivalent in the NMR down to –60 °C.	Either $dp-\pi$ bonding in these and in the analogous α-sulfinyl carbanions is negligible, or it has no conformational requirements.	(M–64)
Reaction of phenyl vinyl and allyl benzyl sulfides with diazomethane gave A and B, respectively, whereas reaction with allyl benzyl ether gave a mixture of C and D. A: PhS-[pyrazoline, N=N]; B: Ph-S-[pyrazoline, N=N] C: Ph-O-[pyrazoline, N=N]; D: Ph-O-[pyrazoline N=N]	When diazomethane acts as a 1,3-dipolarophile the carbon has partial negative charge and attacks the end of the olefinic double bond, which has less negative or more positive charge. Thus, the preference for terminal attack with the sulfides is probably due to $dp-\pi$ bonding as postulated in (KKF–73) and (FK&–73).	(OK–73)
A: [phospholene Me, P-Me]; B: [phospholene Me, P-Me] |	A	B | |---|---|---| | IR ν(c = c) | 1658 | 1613 cm^{-1} | | NMR $\delta\,^{31}$P | +32.6 | +15.2 ppm |	Electron-donating $pp-\pi$ conjugation reduces the electron density on P, thus causing a downfield shift of the ^{31}P NMR chemical shifts.	(QBM–71)

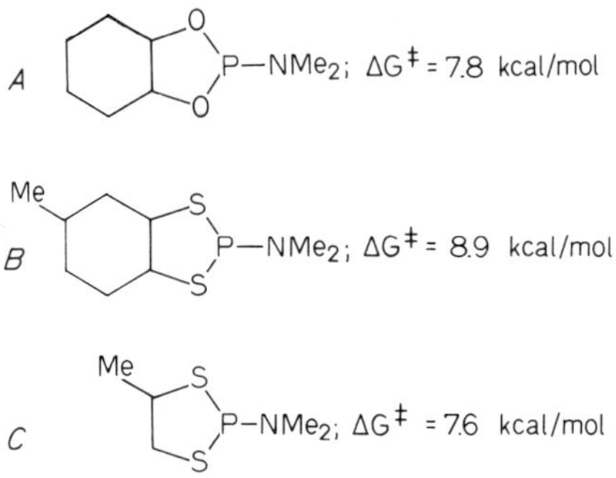

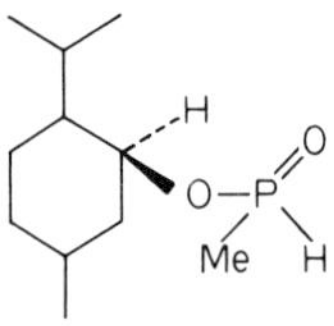

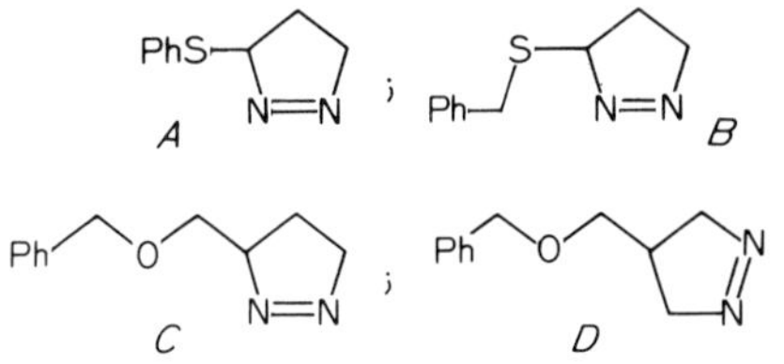

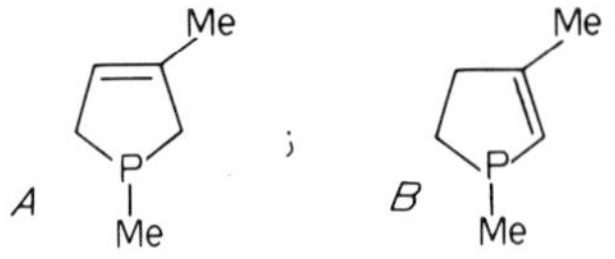

System and Results	Conclusions	References

Ar = Ph or p-tolyl
R = Ph, MeC$_6$H$_4$, (NO$_2$)$_2$C$_6$H$_3$, or CCl$_3$ 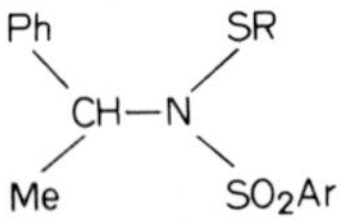
 	
Ph   SR
  \   /
   CH—N
  /   \
Me   SO$_2$Ar
 	
The α-methylbenzyl group is a chiral center so rotation around the N–S bond, as measured by dynamic NMR spectroscopy, is accompanied by epimerization. Both ΔG* for rotation around the N–S bond and the equilibrium constant for epimerization, as measured kinetically, increased with increasing electronegativity of R.
	The *dp–π* bond induces a preference for a particular diastereomer.	(RC&–72)

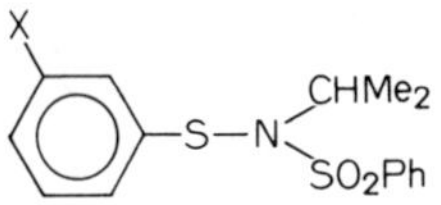 [Structures A, B, C, D showing arenesulfonyl N-isopropyl sulfenamides with varying X substituents] 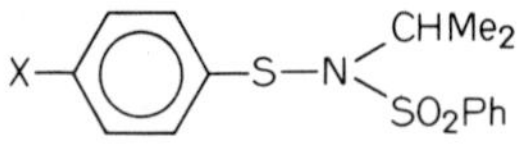 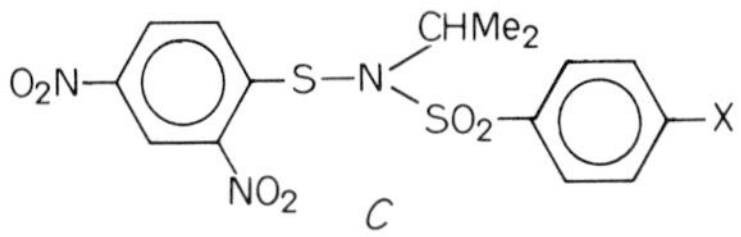 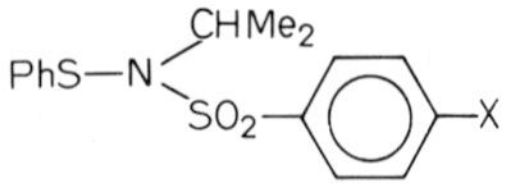 A: X–C$_6$H$_4$–S–N(CHMe$_2$)(SO$_2$Ph) (meta-X) B: X–C$_6$H$_4$–S–N(CHMe$_2$)(SO$_2$Ph) (para-X) C: 2,4-(O$_2$N)(NO$_2$)C$_6$H$_3$–S–N(CHMe$_2$)(SO$_2$–C$_6$H$_4$–X) D: PhS–N(CHMe$_2$)(SO$_2$–C$_6$H$_4$–X) In all four series, the free energies of activation for rotation around the S–N bond plotted linearly against Hammett σ constants. $p_A = -282$, $p_B = -582$, $p_C = 11$, $p_D = 246$. Thus, for A and B increasing electronegativity of X increases ΔG*, whereas for D the opposite is observed; for C the variation of ΔG* with σ is negligible.	ΔG* is affected by a conformationally-dependent *dp–π* bond, which is negligible for X = less electronegative than H. This *dp–π* bond is hindered by increasing electron withdrawal by the arenesulfonyl group. Negative hyperconjugation (n–σ*) is not an acceptable explanation because that would make $p_A < p_B$.	(RJ–71)
In Ph$_3$PO the ^{13}C NMR chem shifts of α and para carbons are very similar. On protonation the ^{31}P Chem Shift was deshielded 27.9 *ppm* and the para carbon was 12.8 *ppm* deshielded relative to the α carbon.	Protonation of oxygen reduces its electron donating ability thus increasing the positive charge on phosphorus by reducing *dp–π* bonding. Increased P–O *dp–π* bonding in	(AFS–75)

47

Physical Properties Relative to $dp-\pi$ Bonding

System and Results	Conclusions	References
	the unprotonated species hinders $dp-\pi$ bonding to phenyl which would cause increased positive charge on the para compared to the α carbon.	
[Structures: Me₂M with two N-Et groups in a ring; and a bicyclic M with four N-Et groups]	The methyl chemical shifts would seem to indicate that $dp-\pi$ bonding is greater for silicon than tin, whereas the nitrogen electronegativities would seem to indicate the opposite. Thus, both methods are influenced by anisotropy and cannot give a valid indication of $dp-\pi$ bonding.	(RYZ–67)
In $CH_2(NEt)_2$, $Me_2M(NEt)_2$, where M = C, Si, Ge or Sn, the chemical shifts of the methyl and methylene protons increased to higher field in the order C < Sn < Ge < Si, whereas the electronegativity of nitrogen, as determined from the difference in chemical shift between the α and β protons of the ethyl group, increased in the order C ≪ Si < Ge < Sn.		
^{31}P NMR chemical shifts vs H_3PO_4 $P\Phi_2-NHP\Phi_2S-$; $\delta = -55.1$ ppm $P\Phi_2-N=P\Phi_2S-$; $\delta = -35.6$ ppm $MeSP\Phi_2=N-P\Phi_2S-$; $\delta_A = -29.2$ ppm (A) (B) $\delta_B = -42.4$ ppm $MeSP\Phi_2-N=P\Phi_2S-$; $\delta = -38.4$ ppm	The positive charge produced on phosphorus by S-methylation should cause a large deshielding; that it does not must be due to increased P–N, $dp-\pi$ bonding.	(SBG–67)
$(Me_3M)_nX$, M=C, Si, Ge, or Sn; X = halogen ($n = 1$) X = O, S, or Se($n = 2$) The variations of chemical shift and coupling constant with atomic weight of X are the same for all except M=C.	The authors concluded that, since $dp-\pi$ bonding and hyperconjugative overlap do not play a role in tin bonding, these results must be due to magnetic anisotropy, dispersion forces, etc.	(SR–64)
Difference between ^{19}F chem shifts for 2-fluoronaphthalene and 2-fluoro-6-X-naphthalene: X = Me – 1.45 ppm; Me_3Si – 0.77 ppm; Me_3Ge – 0.36 ppm; Me_3Sn – 0.33 ppm; Me_3Pb – 0.11 ppm	$dp-\pi$ bonding decreases with increasing atomic weight of the Group IV elements, but it is not completely negligible, even for Pb.	(AK–70)
For R = isopropylphenylphosphano the inversion barrier around phosphorus is 2.4 kcal/mol less in $RSi(OCH_3)_3$ than in $RSi(CH_3)_3$, and 2.8 kcal/mol less in $RC(OCH_3)_3$ than in $RC(CH_3)_3$.	Although the results for the silylphosphanes could be explained by the improved $dp-\pi$ bonding going from $RSi(CH_3)_3$, this explanation could not apply to the carbon substituted phosphanes. The most plausible explanation in both cases is probably negative hyper-	(BM–72)

System and Results	Conclusions	References
	conjugation, i.e., overlap of the P lone pair with antibonding MO's of the methoxylated substituents.	
^{29}Si chemical shifts of ArSiX$_3$ fit Hammett plots with $\rho > 0$ for X=F, OEt, or Cl and $\rho < 0$ for X=Me or H. Chemical shifts correlated with the sum of electronegativities of all four substituents on Si, shielding being maximum for X=H or F and minimum for X=H or Me.	Although the change in Hammett ρ with X might be attributed to $dp-\pi$ bonding, the existence of a correlation with total electronegativity eliminates such an explanation; the shape of the correlation explains the change in sign for ρ.	(ES&–74)
^{13}C–NMR chemical shifts of vinyl- and ethynyl-triphenylphosphonium salts show strong deshielding of the β carbon compared to ethylene or acetylene, much greater in magnitude than the shielding effect of the methyl group in propene or propyne. ^{31}P chemical shifts showed greater shielding (greater electron density) at P than in ethyltriphenylphosphonium salts.	The deshielding at the β carbon is probably stronger than can be explained by the electron-withdrawing inductive effect, and hence is probably best attributed to $dp-\pi$ bonding between P and the double or triple bond; this would also explain the increase in electron density at P.	(AFS–75a)
The acidity of the methylene carbon, as inferred from proton chemical shifts in (EtO)$_2$P(O)CH$_2$COR is much greater than in the corresponding alcohol (ETO)$_2$P(O)CH$_2$CH(OH)R and increases in the order R= – OET < – CH$_3$ < – CH$_2\overset{\oplus}{N}$(CH$_3$)$_2\overset{\ominus}{I}$. The decrease in positive charge on phosphorous compared to the alcohol, as inferred from ^{31}P chemical shifts, followed the same order but the changes were far smaller than those on the methylene carbon.	The strength of $dp-\pi$ bonding to phosphorus increases as the acidity on carbon, i.e., the stability of the corresponding base (carbanion) increases but is actually quite weak in all cases.	(MMS–68)
The total positive charge on phosphorus determined from ^{31}P NMR chemical shifts decreased in the orders Ph$_3$PO $\geqslant$ Ph$_3$PSe > Ph$_3$P and (PhO)$_3$PS > Ph$_3$PS > (PhS)$_3$PS. The π electron density on phosphorus, determined by substracting (from the total positive charge) the σ charge calculated from the electronegativities of the substituents, decreased in essentially the same order, except that it was greater for (PhS)$_3$PS than for Ph$_3$PS.	As might be expected, increasing positive charge on phosphorus results in increased $dp-\pi$ bonding, but $dp-\pi$ bonding is never enough to completely counteract the σ inductive effect. However, the compensating $dp-\pi$ and σ effects probably do explain the lack of variation of phosphorus $2p$ binding energies measured by X-ray photoelectron microscopy in these compounds.	(MS&–71)
^{31}P NMR chemical shifts of phosphoric acid and alkyl phosphates show little change upon protonation. P in P(OH)$_3$ is deshielded by	All the P–O bonds in phosphates have the same $dp-\pi$ bonding; thus the change in	(OM–71)

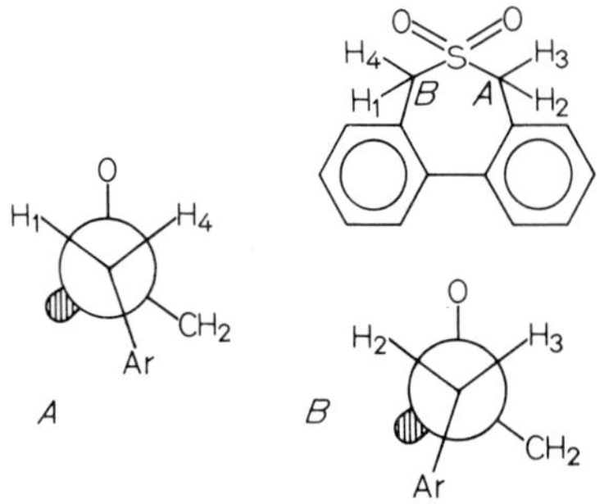

System and Results	Conclusions	References
protonation, but that in trialkyl phosphites is shielded, although the protons of the alkyl group are deshielded.	positive charge on P upon protonation is small. Loss of the lone pair in P(OH)$_3$ protonation results in a large increase in positive charge, but in phosphite esters this is compensated by increased $dp-\pi$ bonding, which in turn results in greater oxygen electronegativity and greater electron withdrawal from R in $^+$HP(OR)$_3$, as indicated by the case of R–O cleavage when R can form a stable carbonium ion.	
XC$_6$H$_4$SiMe$_3$ Separate linear correlations for the ortho, meta, and para series were found between the methyl protons chemical shift, ^{13}CH coupling, and the Hammett σ constants; in all cases correlation coefficients were > 0.82. X=NO$_2$ and X=CF$_3$ did not fit any correlation. For J vs. $\sigma\rho$, ρ = 2.89 compared to 2.68 for p–XC$_6$H$_4$OMe and 1.08 for p–XC$_6$H$_4$NMe$_2$.	Because of its $dp-\pi$ bonding, silicon is an excellent transmitter of electronic effects, but the magnitude of Ph–Si, $dp-\pi$ bonding is quite small, as indicated by the small magnitude of the changes in J. Nitro and trifluoromethyl groups reduce the electron density in the benzene ring so greatly as to entirely eliminate Ph–Si, $dp-\pi$ bonding.	(FS–71)
k_1/k_2 (D exchange rates at H–1 and H–2) $\geqslant k_3/k_4$ in all solvents. k_3/k_4 = 1/4 in t–BuOD/K$^+$(Ot–Bu)$^-$ but 250/1 in CH$_3$OD/Na$^+$ $^-$(OCD$_3$).	Carbanion stability is controlled more by solvent than by lone pair repulsions (WRC–69).	(FSW–72)
In PhXHMMe$_3$ (X=N or P), J(^{15}NH) and J(^{31}PHO) both increase in the order M = Sn < Ge < Si < C.	There is no M–X, $dp-\pi$ bonding. See text and (RZ–68).	(HUZ–71)
Linear correlations were found between the chemical shifts of the acetylenic protons in Me$_3$MC≡CH and Ph$_3$MC≡CH and [(A/ν^2) C≡C + (A/ν^2) C–H] where A is the optical density and ν is the IR stretching fre-	Since all M (other than C) have the same electronegativity, this must be due to the decrease in $dp-\pi$ bonding in going from Si to Pb, thereby increasing the	(GLS–73)

NMR and Related Indices

System and Results	Conclusions	References
quency in cm^{-1}. The correlations are valid for M = Si, Ge, Sn, or Pb but not for M = C. The ^{13}C chemical shifts did not correlate linearly with the IR results but did decrease with increasing electron density in the acetylenic proton bond.	electron density of the acetylenic bond.	
CNDO calculations of the charge distribution in $(CH_2{=}CH)_2Y$ and $PhYCH{=}CH_2$ show less negative charge on the β than the α carbon for Y=S, whereas the opposite result is found for Y=O, or for Y=S with no d-orbitals in the basis set.	This supports the existence of vinyl-sulfur, dp–π bonding. See also (FK&–73) and text.	(KKF–73)
^{13}C–NMR chemical shifts of the phosphonate carbon in $(EtO)_2P(O)CH_2X$ fit an excellent linear correlation with those of CH_3X. Less perfect correlations were obtained between $J(^{13}C-^{31}P)$ and J(CH), or the calculated s character of the C–P bond.	There is no specific interaction between X and the phosphorus substituent, which implies that P–O, dp–π bonding is constant. This is also indicated by the invariance of the ethoxy carbon chemical shifts.	(G–71)
^{31}P chemical shifts of $(EtO)_2P(O)CH_2CH{=}CHR$ show very little variation with R = – OEt, – SEt, or halogen.	There is no dp–π bonding to an allylic double bond.	(MM&–68)
^{29}Si chemical shifts of $Me_{4-n}SiX_n$ (X = halogen, alkoxy, vinyl, phenyl or dimethylamino) agree with calculations on an MO model excluding d-orbitals.	The difference between the dependence on n of the ^{29}Si and ^{13}C chemical shifts of $Me_{4-n}SiX_n$ and $Me_{4-n}CX_n$, respectively, are due to the low electronegativity of Si and not dp–π bonding.	(ER&–73)
SCF–MO calculations of the inversion barrier in phosphine show that the use of d-polarization functions increases the barrier height by 6.6 kcal/mol but has little effect on the electron distribution.	d-Functions polarize the basis set but are not involved in bonding.	(LM–72)
At room temperature, ^{19}F NMR of A shows equivalent fluorines, and proton NMR shows all methyl protons equivalent. At – 100° ^{31}P NMR showed two peaks corresponding to two conformers. The major conformer showed equivalent fluorines but two types of methyl groups; the minor conformer showed equivalent methyl groups. *A* (structure with Me groups, P, Ph, F substituents)	The major conformer is A, with rapid switching of the apical and basal fluorines and CMe_2 groups by pseudorotation or turnstile rotation. The minor conformer has the four membered ring basal-basal, which illustrates the magnitude of the electronegativity preference of fluorine for the apical position. If interconversion of the two conformers occurred there would have to be an intermediate with both fluorines basal; this would explain the relatively high barrier to the interconversion.	(DD&–76)

System and Results	Conclusions	Reference
Correlation of ^{13}C chemical shifts for various 2- and 3-substituted thiophenes with charge density calculated by the ω-Hückel MO method, which takes the σ skeleton into account, was somewhat closer to linear with than without d-orbitals in the basis set.	Since ^{13}C chemical shifts are known to correlate linearly with charge density, this may prove that d-orbital conjugation does affect the charge density in substituted thiophenes, as predicted earlier (*C–68*, p. 52).	(*OSN–76*)

D. UV, ESR, and Dipole Moments

Ultraviolet spectroscopy is an important source of information about $dp-\pi$ bonding since the d-orbital interaction should be strongest with the highest occupied and lowest unoccupied π orbitals (HOMO and LUMO) of the donating ligand. Since the ultraviolet spectra of silicon and phosphorus have been thoroughly covered in a recent book by *Ramsey* (*R-69*), this review will concentrate on sulfur compounds and on very recent work in order to supplement Ramsey's work.

The question of $dp-\pi$ bonding in divalent sulfur has been much discussed, (*PO-62*). An early attack on the problem was based on substituent interference, (*GKS-65*) in which a para substituent in toluene may increase or decrease the intensity of the ^1L_B band, depending on whether it is electron donating or withdrawing, respectively, compared to the methyl group. Based on measurement of the specific absorbance ϵ (*GT-65*), thiol was initially found to be electron withdrawing compared to methyl, but a later study utilizing the oscillator strength (*JO-62, LZ-69*), which depends on the integrated absorbance over the entire band, gave the opposite result, proving that methyl is not electron donating enough to introduce $dp-\pi$ bonding between thiol and p-tolyl groups. This is supported by photoelectron spectroscopy (*FH&-72*), which showed the first three ionization potentials of benzenethiol to lie at 8.28, 9.38, and 10.65 eV. The second of these can be assigned to the noninteracting $a_2\pi$, one of the two degenerate benzene HOMOs, which are split by substituents. The first and third result from a linear combination of the interacting $b_1\pi$ and the sulfur $3p_z$ containing one lone pair, i.e. $pp-\pi$ bonding. Since the average of these two bands is 9.46 eV compared to 9.38 for the $a_2\pi$, there is apparently little stabilization of the $b_1\pi$ by electron donation to sulfur.

Present knowledge would suggest it is quite possible that thiols and sulfides do not engage in $dp-\pi$ bonding. The ultraviolet spectra of thiophene and dibenzothiophene (*ZP-65, Z-65, GL-66*), were in much better agreement with Hückel MO calculations using only the p orbitals on sulfur than with the *Longuet-Higgins* model (*L-49*), in which sulfur can bond to two carbons simultaneously by use of two pd^2 hybrids. CNDO calculations on thiophene with and without d-orbitals (*C-68*) showed that $d-\sigma$ bonding is at least as important as $d-\pi$ bonding and that d-orbital inclusion had a considerable effect on the calculated dipole moments, electron distribution, and NMR spin coupling constant but almost no effect on the total energy.

Thus, in spite of some dipole moment evidence to the contrary (*BU-63, BG-63*), it is probable that dp–π bonding is of little importance in the ground state of organic sulfides. However, except for thiophenes, aromatic sulfides, or other species where there are low-lying unoccupied olefinic or aromatic π MOs, d-orbitals apparently contribute considerably to the excited state. Thus, CNDO calculations (*WK-71*) on I, II, and III could be made to fit the UV spectra only when d-orbitals were included. Without d-orbitals the longest wavelength transition was predicted to lie at 3–5 times higher energy than observed. Inclusion of d-orbitals gave first transitions at up to 1.5 times lower energy than observed, but the variation of wavelength with ring size was correctly predicted. As might be expected, the longest wavelength transition involved sulfur lone pairs in a $3p$ orbital excited to a LUMO (lowest unoccupied MO) with up to 36% $3d$ character and 44–66% C–S contributions.

Oae (*OTO-64*) has used dp–π overlap across space to explain the red shift of alkyl mercaptals (gem disulfides, 235 nm) with respect to the corresponding monosulfides (210 nm), but blue shifted with respect to the corresponding disulfides, (*BCS-62, PO-62*). Since such an overlap would mean that the excited state has resonance contribution from a three-membered ring, IV, we may expect conformational requirements. This would explain the 250 nm absorption observed for V and VI; apparently the fixed S–C–S angle in these rings is at an optimum value for overlap (*OTO-64*). Bathochromic shifts of up to 20 *nm* were observed experimentally (*WK-71*) for VII, VIII, and IX compared to I, II, and III, respectively, but the CNDO calculations showed no such shifts and indicated 1.3 interactions to be minimal.

Extended Hückel calculations (*CC&-71a*), on the tetrathioethylenes, X, XI and XII, gave similar results; calculations without d-orbitals predicted the first transition of X to be a factor of three too high in energy. All three compounds showed three or four low-energy vacant π or σ orbitals with extensive d contributions. However d-orbitals are probably not essential to explain the chemistry of tetrathioethylenes, which are noteworthy for their lack of resemblance to the analogous tetraamino- and tetraalkoxyethylenes; thus, for example, X is inert to such reagents as sulfuric acid, dimethyl acetylenedicaboxylate, phenyl azide, and other electrophiles which attack the double bond of electron-rich olefins. This indicates that the electron density on the bridging carbons is small, which agrees with the extended Hückel calculations of the charge distributions. This charge density distribution may be viewed as a consequence of contributions from twelve dipolar resonance structures (*PO-62*).

Similarly extended Hückel calculations (*MS-72*) on catenated sulfur species, cyclo-S_n, linear S_n diradicals, and sulfanes H_2S_n were in excellent agreement with the observed UV spectra when d-orbitals were included. As expected, d-orbital inclusion had a major effect on the stability of the vacant orbitals, but provided a negligible stability increase to occupied orbitals. However, 3 d-orbitals per se are probably not necessary; calculations on dialkyl polysulfides using 4 s orbitals also gave good agreement with experiments (*R-59*). In addition, the qualitative features of the spectra of elemental sulfur species and of sulfanes were reproduced by calculations using only the 3 s and 3 p-orbitals (*MG&-72*), even though the predicted transition energies were too high. For the sulfanes this is readily understood; the energies of the d-stabilized LUMO varied little with n whereas the HOMO energies increased continuously with n, just as in conjugated polyenes.

The high energy HOMO also explains much of the surprising chemistry of elemental sulfur, in particular the tendency of cyclooctasulfur to behave as a Lewis acid (*S-65a*) and its reaction with mercaptide ions (*MP-72*) to give RS_9^-, which undergoes further reaction to give lower polysulfides. The failure of these species to behave as Lewis bases is understandable when one realizes that the lone pair containing p orbitals on each sulfur will inevitably overlap, resulting in a series of π orbitals similar to those in conjugated polyenes but with the difference that even the antibonding π^* orbitals will be occupied; thus, each lone pair is "spread out" over all the sulfur atoms in the ring or chain. Simple Hückel (*S-62a*) calculations on butadiene indicate that the highest antibonding MO places less electron density on the terminal carbons than on the interior carbons; this agrees with the observation (*MP-72*) that attack of nucleophiles on a polysulfide occurs at the terminal sulfurs.

In the presence of oxygen, the $\sim 200\,nm$ band of sulfonium salts, attributed to excitation of the lone pair into a vacant 3 d orbital, suffers a bathochromic shift of about 10 nm; this has been attributed by Ohkubo (*OY-71*) to interaction of the unoccupied, antibonding π_g orbital of the oxygen molecule with these "partially occupied" d-orbitals; this electron withdrawal causes a slight decrease in the LUMO energy. This interaction has been used to explain the catalysis by sulfonium and other onium salts of hydrocarbon oxidation (*OK-72*) and hydroperoxide decomposition (*OK-72a*); the explanation is supported by extended Hückel and SCF calculations.

The correlation observed between the rates of oxidation of a particular hydrocarbon (e.g. cumene) under the influence of a given onium salt catalyst (*OK-71*) and the bathochromic shift in the UV spectrum of that catalyst upon addition of oxygen strongly suggests that the catalysis involves electron donation from the LUMO of the onium salt to the unoccupied half of the doubly degenerate π_g orbital of oxygen. This stabilizes the LUMO and destabilizes the oxygen π_g enough to permit direct hydrogen abstraction from the hydrocarbon by oxygen, as indicated by the loss of the induction period in the autoxidation of cumene at 85 °C upon addition of triphenylsulfonium chloride (*OYF-69*). That d-orbitals are involved in this complexation with oxygen is further supported by the relatively weak catalytic effects of ammonium and oxonium salts. Thus, the maximum rate of cumene oxidation increased by a factor of 2.0 upon addition of tetraethylammonium tetrafluoroborate and 3.94 upon addition of triphenyloxonium tetrafluoroborate; the corresponding catalysis factors were 6.58 for triphenylsulfonium tetrafluoroborate and 15.8 for triphenylsulfonium chloride and 9.9 for tetraphenylphosphonium chloride (in tetralin) (*OY-70*). The large effect due to changing the anion from tetrafluoroborate to chloride indicates that the same factors which favor covalent bonding to the anion also favor catalysis by complexation with oxygen, and *d*-orbitals are strongly involved in both; on the other hand changing the anion had very little effect on the catalysis factors of ammonium and oxonium salts since without *d*-orbitals the bonding to the anion must be mostly ionic.

Although SCF calculations indicate 29% and 5.7% greater electron density (*OK-72*) in the d_{xz} and d_{yz} orbitals, respectively, of tetramethylphosphonium than of trimethylsulfonium chloride, (these two *d*-orbitals being geometrically most suited for overlap with the oxygen π_g), sulfonium salts were consistently better autoxidation catalysts than phosphonium salts. Thus, the sulfonium lone pair must also be involved in the complexation with oxygen, but apparently not to too large an extent, since otherwise the resulting stabilization of the lone pair HOMO would cause a blue shift in the UV spectrum; whether sulfonium salts are planar as indicated by the SCF calculations (*OK-72*) or pyramidal as indicated by the isolation of optically active sulfonium salts is irrelevant, since in either case the lone pair is in an orbital of high *p* character capable of overlap with the oxygen π_g, or the σ LUMO of hydroperoxides.

Jaffe and *Orchin* (*JO-62*) have discussed the UV spectra of aryl sulfones, in which $dp-\pi$ bonding should be fairly strong. Although the bathochromic effect of a methanesulfonyl group on the ^1L_a and ^1L_b bands of benzene was small, apparently illustrating the relative weakness of $dp-\pi$ bonding to phenyl, substitution of electron donors on the benzene ring produced large bathochromic shifts. Thus, the ^1L_a band at 217 *nm* in phenyl methyl sulfone occurred at 269 *nm* in *p*-aminophenyl methyl sulfone (*JO-62*). Similarly, the primary band of diphenyl sulfoximine $Ph_2S(O)NH$ at 231 *nm* suffered a 57 *nm* bathochromic shift upon substitution of a *p*-amino group on one of the benzene rings. A further bathochromic shift of 8 *nm* occurred on going from neutral to acidic solution; protonation of the sulfoximine nitrogen increases the formal positive charge near sulfur, thus enhancing $dp-\pi$ bonding. At the same time, no change occurred in the S—O and S—N infrared bands, proving that the sulfoximine obeys Case II conjugation as described by *Koch* and *Moffitt* (*KM-51*). As expected for $dp-\pi$ bonding, there was no steric inhibition of resonance; the ^1L_b band at 264 *nm* in phenyl methyl sulfone occurred at 267 *nm* in *p*-tolyl methyl sulfone and 280 *nm* in mesityl methyl sulfone (*JO-62*). This has been confirmed by later work which showed that the intensity of the ^1L_b band varied much less with *n* in XIII than in XIV (*KA-69*). Similar bathochromic shifts, indicating similar $dp-\pi$ conjugation in the excited state, are also found in aromatic sulfonium salts and sulfoxides, although in the latter case $pp-\pi$ bonding also occurs, leading to extremely complex behavior (*JO-62*).

In order to study $dp-\pi$ bonding between phenyl and silyl groups, it would be best to determine separately the effects of silyl substitution on the HOMO and LUMO of benzene. *Bock* and *Alt* (*BA-70a*) studied the HOMO by means of the ultraviolet spectra of the charge-transfer complexes of substituted benzenes with tetracyanoethylene; here the excited state is a property of the tetracyanoethylene alone and so does not vary with the substituent. Electron-donating substituents destabilize the HOMO and thereby cause a bathochromic shift in the charge-transfer band. As a result, the tetracyanoethylene charge-transfer bands fall at 384 *nm* for benzene, 405 *nm* for toluene and 416 *nm* for tert-butylbenzene, whereas that for trimethylsilylbenzene falls at 405 *nm*, apparently indicating that $dp-\pi$ bonding is just slightly stronger than the electron-donating inductive effect, which should otherwise have given a charge-transfer band at longer wavelengths than that for *t*-butylbenzene. In methanol, nitrobenzene shows a 258 *nm* band resulting from intramolecular charge transfer to the nitro group. The occurrence of this same band at 273 *nm* in *p*-nitrotoluene, 270 *nm* in *p*-nitrophenyltrimethylsilane, and 298 *nm* in *p*-nitrobenzyltrimethylsilane has been attributed by *Bock* and *Alt* (*AB-69*) to ground-state Ar–Si dp-bonding, being slightly stronger than the inductive effect.

On the other hand *Schiemenz* (*S-71*) has shown that no combination of + I and − M effects can explain the UV spectra of substituted nitrobenzenes. For 11 *p*-substituted nitrobenzenes having groups with strong $pp-\pi$ (e.g. –COOH) or $dp-\pi$ (e.g. –POPh$_2$) –M effects, the hypsochromic shift relative to nitrobenzene itself was $<700\,\text{cm}^{-1}$, and the –I effect of the dimethylammonium chloride group resulted in a $2150\,\text{cm}^{-1}$ hypsochromic shift, compared to a $5050\,\text{cm}^{-1}$ bathochromic shift for the supposedly weaker + I effect of the –CH$_2$SiMe$_3$ group. This must therefore have partial carbanionic character resulting from Si–C hyperconjugation, as originally proposed by *Pitt* (*P-70*) and elaborated with CNDO calculations (*P-71*) on various silyl-substituted acetylenes, and indicating that the groups –SiF$_3$ and –SiH$_3$, but not –SiMe$_3$, can act as π electron acceptors even without $dp-\pi$ bonding. This has been confirmed by recent photoelectron spectra (*M-73*), which showed the ionization potentials of benzene and silylbenzene to be 9.24 eV and 9.43 eV, respectively; however, quantitative agreement of MO calculations with these results could only be obtained by including *d*-orbitals in the basis set.

The LUMO can be studied by half-wave reduction potentials, which provide a measure of its energy, and by ESR spectroscopy, from which its electron distribution may be analyzed. Substitution of a trimethylsilyl group in the para position of nitrobenzene decreased the half-wave reduction potentials to less negative values. Alkyl substitution had the opposite effect, suggesting that in the excited state $dp-\pi$ bonding is considerably stronger than the inductive effect (*AB-69*). Thus, the bathochromic shift in the longest wavelength ultraviolet transition of PhC≡CY, which shift is greater for Y = Me$_3$Si than for Y = Me$_3$C (*BA-70a*) even though the absorption maxima of the tetracyanoethylene charge-transfer complexes are in the order Y = H < Me$_3$Si < Me$_3$C, must be due primarily to $dp-\pi$ stabilization of the excited state.

The doubly degenerate LUMO of benzene is split by substitution in the 1 position into a lower symmetric orbital, which concentrates the electron density on C$_1$ and C$_4$ and a higher antisymmeric orbital, which concentrates the electron density on C_2 and C_5. In the radical anions derived from trimethylsilylbenzene and 1,4-bis-(trimethylsilyl)benzene, the ESR coupling constants for the 2 and 5 protons are much less than in the radical anions derived from toluene and *p*-xylene (*GH&-68*). The methyl group is a net electron-donor and causes the unpaired electron in the radical anion to occupy the higher antisymmetric orbital, whereas the trimethylsilyl group is a net electron-acceptor and causes the unpaired electron to occupy the lower symmetric orbital. The coupling constants at any position are proportional to the spin density, i.e. the unpaired electron density, at that position. Consequently, Birch reduction of *p*-bis(trimethylsilyl)benzene gives the 1,4-dihydro product, whereas *p*-xylene gives the 2,5-dihydro product. Similarly, Birch reduction of 1,4-bis(trimethylsilyl)-naphthalene gives the 1,4-dihydro product, whereas reduction of 1,4-dimethylnaphthalene gives the 5,8-dihydro product (*AFB-69*).

Similar UV and photoelectron spectral studies on silylacetylenes (*BA-70*) and silylethylenes (*BS-68, MB&-72, WS-72*) have indicated a small degree of $dp-\pi$ stabilization in the former, but a large extent of hyperconjugative destabilization of the HOMO and a similarly large extent of $dp-\pi$ stabilization of the LUMO in the latter. However, CNDO calculations on silyl- and disilylacetylene (*EBB-74*) could be made to agree with the photoelectron spectra, which showed the π bond energy of $HC{\equiv}CSiH_3$ to lie between those for acetylene and methylacetylene, only when *d*-orbitals were *not* included. Although the photoelectron spectra could be accounted solely by hyperconjugation, the involvement of *d*-orbitals is not totally disproven since CNDO calculations on molecules with Si—H bonds are known to exaggerate the role of *d*-orbitals. This was demonstrated by means of a comparison of the spin densities at silicon and the ring carbons in the anion radicals derived from $PhSiMe_nH_{3-n}$ calculated by this method with hyperfine coupling constants (*R-75*) measured experimentally in the ESR spectra. Here the calculations showed an increase in the spin density at the silyl group and a corresponding weakening of the Si—Ph bond as the number of Si—H bonds increases. For the silylethylenes, the results have been discussed in detail by *Ramsey* (*R-69*) and are in good agreement with MO calculations (*NV-71*) using the Del Re method for the σ skeleton and an SCF method for the π system.

The *m*- and *p*-trimethylsilyl-N,N-dimethylanilines contain both electron donating and electron withdrawing substituents which can affect the HOMO and LUMO (*DWJ-72*). The energy of the HOMO increased in the order

$$p - Me_3Si < H < m - Me_3Si < Me_3C,$$

which would seem to indicate, in agreement with *Bock*, that the trimethylsilyl group is capable in the ground state of weak $dp-\pi$ bonding, or at least of electron acceptance by some other mechanism, (see e.g. *P-71*). ESR spectra and CNDO calculations on the corresponding radical cations (*JD&-72*) suggest that delocalization of the unpaired electron to the silyl group occurs both by hyperconjugation and by $dp-\pi$ bonding, with the latter more important. In fact, the two mechanisms are probably not independent, for only one *d*-orbital has the proper symmetry for overlap with the aromatic system, but all five can overlap with Me—Si σ bonds; thus, charge transferred to silicon via $dp-\pi$ bonding is probably stored in Me—Si σ^* orbitals with *d*-

orbital contributions. However, the radical cation HOMO has its node close to silicon, which implies that neither effect is particularly large. On the other hand, ESR spectra of the corresponding radical anions suggest mixing to the extent that the unpaired electron is confined to a LUMO having both symmetric and unsymmetric characteristics, and this is supported by CNDO calculations (*JJ-75*). Thus, in the radical anions derived from *o*- and *m*-trimethylsilyl-N,N-dimethylaniline, the highest spin density was at the position para to silicon; in all cases the lowest spin density was at the position para to nitrogen. The radical anion derived from the para isomer was 5 kcal/mole less stable than those derived from the ortho and meta isomers. The results indicate that electron acceptance by d and σ^* orbitals on silicon is slightly greater than electron donation by nitrogen.

Studies in this system were extended to $H_n Me_{3-n} SiC_6 H_4 NMe_2$ (*JJ-75a*) on the assumption that hyperconjugation should be more sensitive to substituents at silicon than is $dp-\pi$ bonding. The silyl group SiH_3 stabilized the HOMO (measured as the oxidation potential) to a much greater extent than the trimethylsilyl group, but stabilization was very similar for SiH_3, $SiH_2 Me$, and $SiHMe_2$. The same was true for LUMO stabilization which was even more affected by increasing n, with spin delocalization in the radical anion derived from $H_3 SiC_6 H_4 NMe_2$ being almost as great as in that from $O_2 NC_6 H_4 NMe_2$. This suggests that the primary interaction is hyperconjugation involving aniline π^* (symmetric) and Si–H σ^* orbitals rather than $dp-\pi$ bonding, a conclusion supported by CNDO calculations (*JJ-75*).

The same methods have been used to study the HOMO and LUMO of phenylphosphanes and anilines. The lone pair ionization potential occurred at 8.5 eV in trimethylamine and 7.6 eV in dimethylaniline, proving a considerable interaction of the nitrogen lone pair with the benzene system. On the other hand, the lone pair ionization potentials were 8.6 eV in trimethylphosphane and 8.1 eV in phenyldimethylphosphane, proving the essential absence of $pp-\pi$ bonding (*SS-72*). However, the ESR spectra of the radical anions of phenyldimethylphosphane and phenyltrimethylsilane were quite similar, with coupling constants for the phenyl protons in the order $a_p > a_o > a_m$, thus proving occupation of the lower symmetric LUMO in both cases (*GPB-70*). No comparison is possible with the radical anion of dimethylaniline since in simple anilines the LUMO is of too high energy to permit formation of a radical anion. However, in N,N-bis(trimethylsilyl)aniline the LUMO was sufficiently stabilized by $dp-\pi$ bonding and/or steric hindrance to $pp-\pi$ bonding to permit formation of a radical anion by reduction with potassium in dimethoxyethane; here the unpaired electron occupied the higher antisymmetric LUMO (*GKB-69*).

Comparison with the ultraviolet spectra is hindered by difficulties in assigning the various bands. Assuming *Ramsey* (*R-69*) is correct in assigning the 266 *nm* band of phenyldimethylphosphane and the 251 *nm* band of dimethylaniline to configuration interaction between the ^1L_a and $n \rightarrow \pi^*$ transitions, it would seem that $dp-\pi$ stabilization of the LUMO outweighs the stabilization of the HOMO in the absence of phosphorus $pp-\pi$ bonding. However, this band fell at 234 nm in both phenylphosphane and aniline (*R-69*). This can probably be explained entirely by changes in the HOMO, which is destabilized by alkyl substitution, more so for phosphorus than for nitrogen, since in aniline the lone pair is already of high energy even without alkyl substution. For the phosphanes this band probably has more ^1L_a than $n \rightarrow \pi^*$ character.

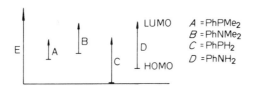

This same band fell at 297 nm in triphenylamine and 262 nm in triphenylphosphane; in the former, $pp-\pi$ bonding permits delocalization across nitrogen into all three benzene rings, but this is impossible for $dp-\pi$ bonding.

ESR spectra have also been determined for a number of other silyl-substituted radical anions. Substitution of trimethylsilyl groups in the 2 and 5 positions of benzosemiquinone increased the spin density at the 3 and 6 positions and decreased the spin density at the carbonyl groups, whereas tert-butyl substitution had the opposite effect (*GKB-69&GKB-69a*). The trimethylsilyl group is a net electron acceptor and stabilizes the LUMO of benzoquinone, as predicted by Hückel MO calculations and confirmed by half-wave reduction potentials (*GKB-69&GKB-69a*). On the other hand electron acceptance by silicon reduced the spin density in the benzene ring and increased that in the carbonyl group of the $PhCOSiMe_3$ radical anion by 10–20% compared to benzaldehyde.

The relative importance of hyperconjugative and $dp-\pi$ bonding is well illustrated by the ESR spectra of β-substituted ethyl radicals $\cdot CH_2CH_2X$. For X = RS, R_3Si, R_3Ge, or R_3Sn (R alkyl) the low values of the β proton hyperfine coupling constant and its large temperature dependence indicate restricted rotation and a preference for the eclipsed conformation XV, which can be considered as an intermediate state between free rotation and the bridged structure XVI (*KK-72*). This can be attributed either to hyperconjugative or to $p-d$ homoconjugation (overlap across space), as discussed earlier for gem disulfides. Calculations (*KK-72*), of the magnitude of hyperconjugation from the metal hyperfine coupling constant of $R_3MCH_2\dot{C}H_2$, (M = Group IVA) and of $p-d$ overlap from the decrease in g value with increasing atomic weight of M led to the conclusion that both effects contribute equally. This has been disputed by Symons (*LS-72, S-72*), who denies that the changes in g value have any significance as regards d-orbital participation. Later work (*GI-73*), however, on the ^{13}C hyperfine coupling constants in $(Me_3C)_2\dot{C}CH_2X$ showed equal magnitudes of hyperconjugation for X = CF_3 and X = Me_3Si and about twice that magnitude for X = $SiCl_3$. Since the conformation of these radicals is fixed by the bulky tert-butyl groups and since the $\cdot CH_2CH_2CF_3$ radical shows free rotation, it would seem that hyperconjugation alone cannot explain the preferred conformation. Similar results were found for $XCH_2CH=CHCH_2$ (*KK-71*); both the substituted allyl and substituted ethyl radicals gave a linear correlation of the rotational energy barrier with the ionization potential of the Group IVA metal substituent, as expected for a hyperconjugative interaction. The rotational energy barriers decreased linearly with increasing free atom d-orbital energy level, which is the reverse of what one would expect; but this may be due (*KK-71*) to variations in the magnitude of the Hamiltonian for $p-d$ or $\pi-d$ overlap with the different metal substituents. The differences in the d-orbital energy levels are small.

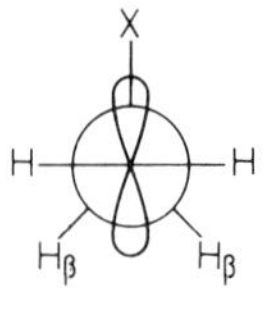

UV spectroscopy can contribute significantly to the question of $dp-\pi$ bonding in ylides. The two cyclopentadienylides, XVII and XVIII, showed three UV bands each in hexane; these fell at 4.19 eV, 4.5–5.5 eV, and 6.36 eV for XVII and 4.63 eV, 5–6 eV, and 6.33 eV for XVIII (*IYY-71*). Comparison with semiempirical MO calculations (*IYY-71*), based on the cyclopentadienyl anion or radical for the hydrocarbon portion, and a free sulfur or phosphorus atom having an effective nuclear charge of 1.4, and Slater-type 3 d-orbitals showed that the two high energy bands could be attributed entirely to cyclopentadienyl anion transitions and the low energy band to intramolecular charge transfer from a cyclopentadienyl anion ground state to the 3 d-orbital, i.e. an ylene excited state. This assignment is supported by the large red shift observed for this band but not for the two high energy bands upon changing from hexane to methanol solvent. Fitting the calculated eigenvalues to the observed band energies gives 1.4 eV and 1.33 eV resonance energies, respectively, for XVII and XVIII, corresponding to 87% and 84% ylide contribution, respectively, to the ground state for XVII and XVIII. This is in good agreement with the 86% ylide character calculated for triphenylphosphonium cyclopentadienylide on the basis of the differences in C–C bond lengths within the cyclopentadiene ring, as measured by X-ray diffraction (*AWW-73*). The P–C bond length for this ylide was 1.718 Å, considerably longer than the 1.66 Å (*B-69*) in triphenylphosphonium methylide. Measurements of the phosphorus 2p binding energies by X-ray photoelectron spectroscopy indicated the positive charge on phosphorus to be 0.69 for triphenylphosphonium cyclopentadienylide, XXII, and 0.55 for the corresponding fluorenylide, XX. This is in agreement with the greater basicity of fluorenyl than of cyclopentadienyl carbanion, which results in a greater excess of π electron density in the ring available for $dp-\pi$ bonding to phosphorus in the fluorenylide.

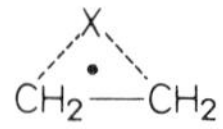

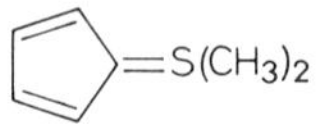

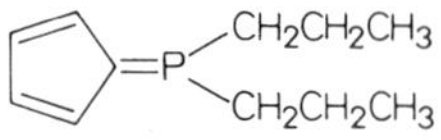

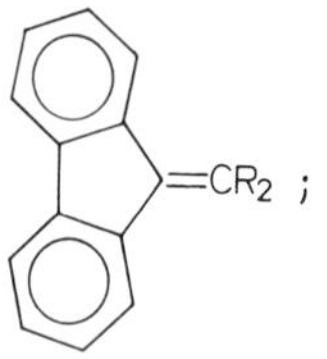

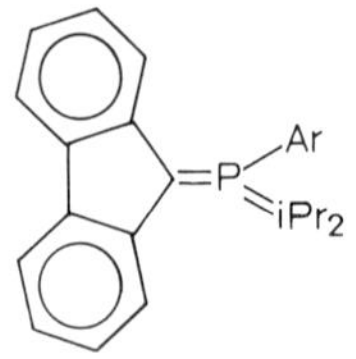

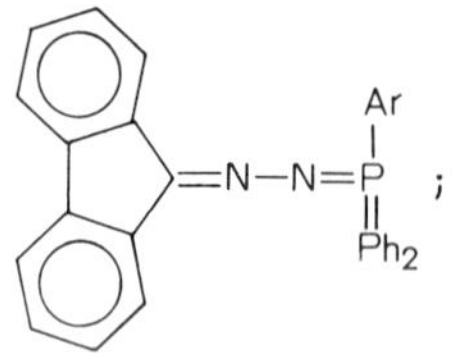

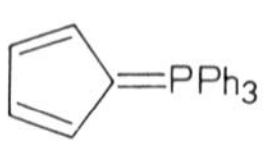

For XX the lowest energy UV band at 384 *nm* for Ar = phenyl was almost completely unaffected by going to Ar = *p*-anisyl or Ar = $p-Me_2NC_6H_4$, suggesting that different *d*-orbitals are involved in bonding to the aryl and fluorenylidene groups. *Goetz* and *Klabuhn* (*GK-69*) assigned this band to a $\pi \rightarrow \pi^*$ transition of the $dp-\pi$ bond, but *Ramsey* (*R-69*) notes that the ultraviolet spectra of ylides show a much closer resemblance to intimate ion-pairs (e.g., 9-fluorenyllithium) than to ordinary *pp*-double bonds, e.g., XIX. Since all available evidence indicates that $dp-\pi$ bonds do not in any way resemble $pp-\pi$ bonds, this is not particularly surprising.

Interesting enough, the lowest-energy transition in XXI (*GKJ-70*) also fell in the same region, 380–384 *nm*, which would seem to indicate very weak conjugation, if any, over the extended $P^+-N=N-\bar{C} \leftrightarrow P=N-N=C$ system.

Similarly, the main ultraviolet band of $p-MeC_6H_4SO_2N=SRMe$ fell at 227–230 *nm*, regardless of whether R was alkyl or aryl (*TFO-70*). For R = *p*-anisyl the spectrum changed from a main band at 232 *nm* and a shoulder at 242 *nm* to two bands at 234 and 267 *nm*, on going from water to 75% sulfuric acid solvent (*TFO-70*). The 242 and 267 *nm* bands can probably be assigned to the ^1L_a transition of the *p*-anisyl group, with Ar–S, $dp-\pi$ bonding; this would seem to indicate stronger $dp-\pi$ bonding to the aryl group when the nitrogen is protonated, eliminating any N–S, $dp-\pi$ bonding, but of course strong sulfuric acid might also cause protonation of the lone pair on the tetravalent sulfur. Good evidence for weak N–S, $dp-\pi$ bonding is found in the large bathochromic shift (154 *nm*) on going from $PhN=SMe_2$ to $p-NO_2C_6H_4N=SMe_2$ (*CV-70*), clearly indicating electron donation by the amino nitrogen. At the same time, the bathochromic shift in the infrared stretching frequency (905 cm^{-1} in $PhN=SMe_2$) was less than 15 cm^{-1}. By comparison, most sulfonamides and sulfonylsulfimines show somewhat higher N–S stretching frequency. This evidence for a weak N=S bond in the sulfilimines, $ArN=SMe_2$, which are analogous to ylides and show 40–60% ionic N–S character depending on Ar (*EK&-76*), is in line with the ESR spectrum of the radical cation $Ph_3\overset{+}{P}-\overset{\cdot}{C}H_2$ (*LM-67*) prepared by X-ray irradiation, with loss of HCl, of tetramethylphosphonium chloride. This showed very little delocalization of the unpaired electron from the methylene group to phosphorus.

Ramsey (*R-69*) has discussed extensively the problem of the large red shifts in the $n \rightarrow \pi^*$ transitions of silyl ketones. Based on first ionization potentials determined by extrapolation to zero of the mass spectral ion current/appearance potential curve (*BAS-69*), *Ramsey* concluded that the red shift of the $n \rightarrow \pi^*$ transition from 292 *nm* for $PhCOCMe_3$ to 424 *nm* for $PhCOSiMe_3$ is due about 25% to stabilization of the LUMO, not necessarily due to $dp-\pi$ bonding, and 75% to destabilization of the HOMO by the electron-donating inductive effect. However, the photoelectron spectra of $PhCOSiMe_3$ (*B-75a*) and $MeCOSiMe_3$ (*RB&-74*) show that the peak corresponding to the first ionization potential is up to 0.5 ev in width, much wider than expected for ionization of a lone pair electron. In conjunction with CNDO calculations (*RB&-74*), this is strong support for Pitt's proposal (*P-70*) of hyperconjugative destabilization of the HOMO ($n-\sigma$ overlap). For $(R_3Si)_2CO$, in fact, the first UV band has more $\sigma \rightarrow \pi^*$ than $n \rightarrow \pi^*$ character. This σ character of the HOMO also explains the ease of photolysis of silyl ketones (*DB-73, RB&-74*). The photoelectron spectral ionization potentials differ from the mass spectral appearance potentials and indicate that less than 65% of the red shift of $PhCOSiMe_3$ and

MeCOSiMe$_3$ compared to the corresponding carbon compounds is due to hyperconjugative HOMO destabilization.

On the other hand, the LUMO stabilization (> 35%) is more likely due to negative hyperconjugation (σ^*–π^* overlap) than to dp–π bonding. Similar results were found for RN=NXMe$_3$ (R = Ph, CMe$_3$, SiMe$_3$), but here the photoelectron spectral ionization potentials (*BW&-76*) show that the red shift on going from X = C to X = Si, which is of the same order of magnitude as in silyl ketones, has only a 55–60% contribution from HOMO destabilization. These conclusions gain qualitative support from the ESR spectra of the corresponding radical anions (*GKB-69a*), which permits direct observation of the LUMO; thus the total spin density on the ortho and para positions of the benzene ring was about 10–20% less in PhCOSiMe$_3$ than in PhCOCMe$_3$ and acetophenone, for which the coupling constants were fairly similar. This is clear proof of electron acceptance by the trimethylsilyl group in the excited state.

Similarly the $n \to \pi^*$ transition of PhCOPO(OEt)$_2$ shows a 50 nm red shift relative to benzaldehyde (*TS-69*). The small size of this shift compared to that in the silyl ketones is due to the electron-withdrawing inductive effect of the phosphoryl group, which prevents destabilization of the lone pair HOMO. A red shift of about the same magnitude was found for the $n \to \pi^*$ transition of PhCOPEt$_2$; a slightly smaller red shift was found for [PhCOPMeET$_2$]I and a slightly larger one for PhCOPSEt$_2$ (*OOK-73*). On the other hand the frequency of the $\pi \to \pi^*$ transition decreased in the order PhCOCMe$_3$ > PhCOSiMe$_3$ > PhCOPEt$_2$ > PhCOPO(OEt)$_2$ > [PhCOPMeEt$_2$]I (*OOK-73*). Since these changes in the $\pi \to \pi^*$ transition are probably due almost entirely to dp–π stabilization of the LUMO, these results may indicate some destabilization of the lone pair HOMO by the –PEt$_2$ and –PSEt$_2$ groups.

In α silyl ketones the destabilization of the lone-pair containing p orbital by overlap with the electron-rich Si–C, σ bond has been suggested (*R-69*). For β-silyl ketones hyperconjugation of the type represented by the resonance form XXIII is unlikely to affect the lone pair, but could conceivably affect the carbonyl (bonding) π energy level. However, the observed $\pi \to \pi^*$ transitions are 240 nm for PhCOCMe$_3$, 243 nm for PhCOCH$_2$SiMe$_3$, and 245 nm for PhCOCH$_2$PO(Et)$_2$. These small bathochromic shifts are probably due to stabilization of the LUMO by weak 1,3 d-overlap across space (*TS-69*). Similarly, CNDO calculations (*BD-73*) on methyl- or silyl-substituted α, β unsaturated ketones showed little change in the lone pair HOMO energy level, but major dp–π stabilization of the π and π^* for silyl compared to methyl substitution. Experimentally, silyl substitution resulted in a bathochromic shift for the $n \to \pi^*$ transition, but the changes in the $\pi \to \pi^*$ were small and depended on the position of silyl substitution, as exemplified by XXV, which in methanol showed a 500 cm^{-1} red shift for the $n \to \pi^*$ transition compared to XXIV (*BD-73*).

It should be pointed out that acylsilanes are photochemically unstable, which is a further indication of the high HOMO energy level. In alcohols under a mercury vapor lamp (300–420 *nm*) benzoyltriphenylsilane is converted to a triphenylsiloxyacetal, apparently *via* an intermediate siloxycarbene, which is highly nucleophilic (*DB-73*). The carbene formation is fast and reversible and probably proceeds by a mechanism similar to that of the rearrangement of silylcarbinols (*BLM-67*), as discussed later in Section V.E. The conversion of optically active silylketones in alcohols to the corresponding siloxyacetals with retention of configuration supports this hypothesis (*DB-73*).

As pointed out in *Ramsey's* (*R-69*) extensive discussion of the subject, explanations of the large red shift in the UV spectra of polysilanes with increasing chain length in terms of d-orbital involvement really boil down to the possible existence of a partially delocalized LUMO with extensive stabilization by d-orbitals. This is suggested by the ESR spectrum of the radical anion of decamethylcyclopentasilane (*CWG-69*), which was split by ten methyl groups, showing the unpaired electron to be delocalized over all five silicon atoms. Hyperfine splitting by carbon-13 was larger than that by silicon-29, which agrees with the idea of a LUMO made up of diffuse $3d$, $4s$ and/or $4p$ orbitals, so that the spin density at the silicon nucleus is low. The HOMO is completely delocalized, and the various Si–Si bands in the photoelectron spectra of linear, branched, and cyclic permethylpolysilanes correlated excellently with Hückel MO calculations using a simple linear combination of the σ Si–Si orbitals. There was no need to assume any contribution to the HOMO from unoccupied atomic orbitals on silicon (*BE-71*). The photoelectron spectra of the polysilanes themselves could not be so simply interpreted because of interaction between Si–Si and Si–H bonds (*BE&-76*), but again there was no need to assume d-orbital involvement.

On the other hand, the bathochromic shift with increasing n in the phenyl polysilanes, $Ph(SiMe_2)_nPh$, is due entirely to the increase in HOMO energy level. The LUMO is an aromatic π^* reduced in energy by $dp-\pi$ bonding to the first silicon, but very little affected by the other atoms in the silicon chain (*PCT-72*). Chain length is even less important in the naphthylpolysilanes, for here the two lowest energy UV bands are naphthalene ^1L_a and ^1L_b transitions, subject to inductive and mesomeric effects, which, of course, attenuate rapidly with increasing distance from the naphthalene ring. This is due to the higher energy of the naphthalene π HOMO than of the delocalized polysilane σ HOMO, as measured by the ionization potentials (*PCT-72*).

It is well known that the high energy Walsh-type σ orbitals of the cyclopropane ring behave as pseudo-π-orbitals. Extended Hückel calculations (*MH-75*) show that these pseudo-π orbitals are as capable of overlap with acceptor (i.e., d) orbitals on silicon as ordinary olefinic bonds. To illustrate, the Si–C overlap population was calculated as 0.60 in silacyclopropane but 0.695 in XXVI, one of the few silacyclopropane derivative then known; the analogous carbon compounds showed an overlap decrease in going from the cyclopropane to XXVII. For silicon this same bond-strengthening effect occurred even without d-orbitals in the basis set but was less than half as strong. Apparently, Si–H antibonding orbitals are also capable of electron acceptance.

The recent successful, general synthesis of silacyclopropanes (*S-75*) has resulted in an elegant confirmation of this theory. Thus, XXVIII was prepared in good yield from XXIX by reaction with magnesium in tetrahydrofuran at room temperature, and

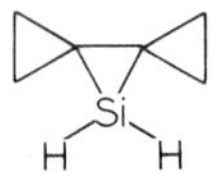

Physical Properties Relative to dp–π Bonding

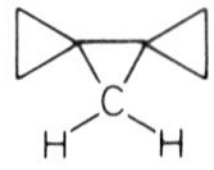

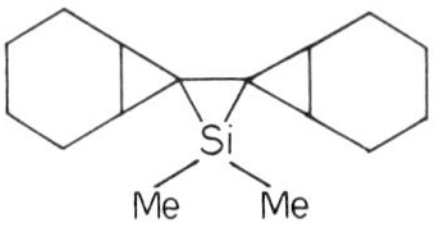

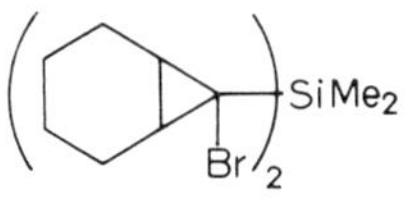

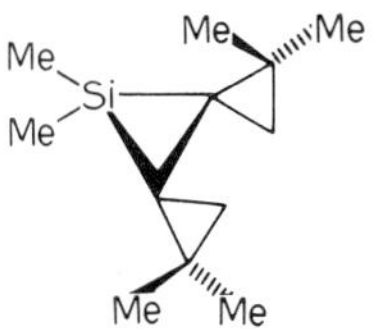

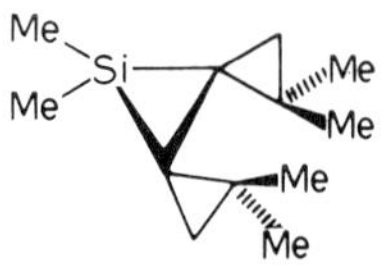

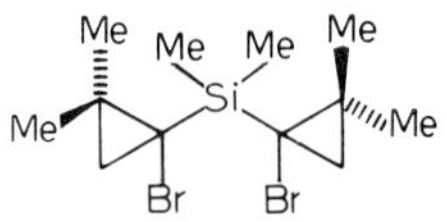

the enantomeric pair XXX and XXXI were similarly prepared from the corresponding dibromides; the meso dibromide XXXII did not undergo ring closure. Good results were also obtained with nonspirocyclic silacyclopropanes; for example, hexamethylsilacyclopropane was prepared from $Me_2Si(CBrMe_2)_2$. X-ray diffraction of XXVIII, which gives an endocyclic C–Si-C angle of 49.2° (DW&-76), and ^{29}Si–NMR spectroscopy of XXVIII, XXX, and XXXI (S-75), which give a > 50 ppm upfield shift with respect to tetramethylsilane, analogous to a similar upfield shift in phosphiranes, confirm the high degree of strain in silacyclopropane and explain the extraordinary reactivity. All silacyclopropanes react instantly and exothermally with water, alcohols, and even dry oxygen in tetrahydrofuran at room temperature, and reaction with amines, ketones, thiols, sulfur, SO_2, $LiAlH_4$, and other reagents occurs readily under conditions considerably milder than for reaction with 1,1-dimethyl-1-silacyclobutane. On the other hand, hexamethylsilacyclopropane underwent ready thermal decomposition; it could not be isolated by normal means and its half-life at 63 °C was only 5 hours, whereas XXX had a half-life of > 7 days at 63 °C and could be distilled unchanged at 100 °C (.01 mm Hg). This confirms the stabilization imparted to this highly strained system by overlap of cyclopropane orbitals with acceptor orbitals on silicon (which could be Si–C, σ as well as d-orbitals).

The existence of such a stabilizing overlap between silicon and electron-rich carbon implies that silacyclopropenes (silirenes) should be even more stable than silacyclopropanes (siliranes). This suggestion was first advanced by Vol'pin (*VK&-62*), but his attempted preparation of XXXIII by reaction of diphenylacetylene with dimethylsilylene, generated from dimethyldichlorosilane and sodium in refluxing xylene, actually resulted in the dimer XXXIV, as was proven by more accurate molecular weight determinations (*WB-63*) and mass spectroscopy (*JGN-63*). It is possible that XXXIII was an intermediate in the formation of XXXIV (*GCA-64*). Seyferth has recently (*SAV-76*) succeeded in preparing the silylated silirene XXXV by reaction of bis(trimethylsilyl)acetylene with dimethylsilylene generated by thermal decomposition of hexamethylsilirane at 66 °C. The structure of XXXV was proven by mass spectroscopy and by ^{29}Si NMR, in which the signal from the ring silicon was 106.2 *ppm* upfield from tetramethylsilane, compared to 50–55 *ppm* upfield for siliranes. This evidence is clearly indicative of strong shielding on the silicon caused by $dp-\pi$ bonding, as suggested by Vol'pin. The half-life of XXXV was 60 hr at 70°–75 °C, making it more than ten times as stable as hexamethylsilirane (*S-75*). The reasons for the isolation of XXXV but not XXXIII are probably to be correlated with the much lower temperatures used by Seyferth than by Vol'pin to generate dimethylsilylene, and to the strong electron donating effect of the trimethylsilyl groups in XXXV, which increases $dp-\pi$ bonding in XXXV compared to XXXIII and thus increases the stabilization of this highly strained ring system.

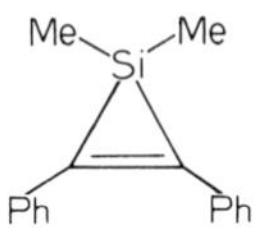

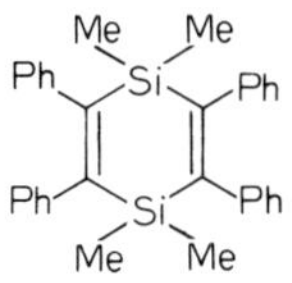

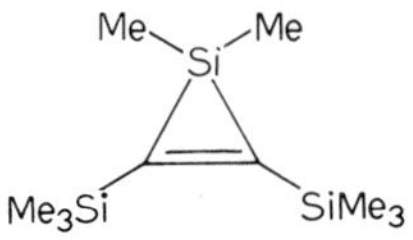

Physical Properties Relative to $dp-\pi$ Bonding

Survey of Relevant Articles on Physical Properties Related to $dp-\pi$ Bonding – UV, ESR and Dipole Moments (Table III D)

System and Results	Conclusions	References
The dipole moments of RXCOCl and RXCOCF$_3$ are larger for X = S than for X = O, the difference decreasing in the order R = Ph < Me < Et, Pr, Bu. The rates of hydrolysis were larger for X = S when R = Ph or Me, but were very similar for S and O when R was a higher alkyl group.	Since the dipole moment depends on $pp-\pi$ electron donation by X to the carbonyl group, this means that S has more electron density to donate than does O, implying Ph–S, $dp-\pi$ bonding or Me–S, hyperconjugation. Since hydrolysis involves electrophilic attack on the electron-deficient carbonyl carbon, strong electron donation by the PhS-group also explains the hydrolysis rate results.	(BB&-74)
Me–⟨○⟩–SO$_2$–N–S⟨R/Me⟩ R = alkyl or aryl *A* The UV spectrum of A was independent of R, and, for R = – C$_6$H$_4$$\overset{+}{\text{N}}H_3$, was the same in 2% and 75% sulfuric acid.	The SNS chromophore is a single $dp-\pi$ system. Due to the size and shape of the d-orbitals, the resulting $d-\pi$ MO is unsymmetrical, with most of the electron density on nitrogen.	(TFO-70)
SCF calculations on ethylene sulfide, dimethyl sulfide, and hydrogen sulfide and comparison with the UV spectra showed three unoccupied σ^* orbitals with the same order of increasing energy in all three cases. The dependence of these orbitals on sulfur $3d$ and $4s$ contributions was fairly small. The excited states for 7 of the 9 assigned UV transitions had $4s$ rather than $3d$ contributions.	The $4s$ orbitals are more important than the $3d$ in the excited states of organic sulfides, which is reasonable since the $4s$ is of lower energy in the free sulfur atom. However, even the $4s$ contributions are not major.	(BGP-74)
Dipole moments of SiH$_3$X increase in the order X = Cl < X = Br < X = I, whereas the opposite was found for CH$_3$X and EtX. Dipole moments of CH$_3$SiH$_2$X varied little with X. C–X bond moments in SiH$_3$CH$_2$X varied little with X; for X = I it was close to that for EtI, but for X = Cl it was substantially less than for EtCl.	The results for SiH$_3$X can be explained by $dp-\pi$ electron donation from X to Si, counteracting the electron-withdrawing inductive effect of X and decreasing in order Cl > Br > I. For CH$_3$SiH$_2$S, $dp-\pi$ bonding must be substantially weaker, (not strong enough to outweigh the inductive effect), and about as strong as the β Si–X, $dp-\pi$ or $dp-\sigma$ interaction in SiH$_3$CH$_2$X.	(BM-70b)
Deviations between calculated and observed dipole moments for 10 ArSAc and ArOAc, including Ar = 4-methoxy-3,5-dimethylphenyl were small; the only exception was *p*-dimethylaminophenylthiolacetate.	See (BU-63) (below)	(BG-63)

Systems and Results	Conclusions	References
The observed dipole moment for 40 aryl methyl sulfides agreed with those calculated from group moments except for Ar = *p*-nitrophenyl and Ar = *p*-aminophenyl, for which the deviations were 0.53 D and 0.39 D, respectively; the deviation was 0.58 D for Ar = *p*-dimethylaminophenyl. Yet for *p*-aminoanisole the calculated and observed dipole moments were in good agreement.	The deviations are due to resonance across the ring, with sulfur donating electrons to the nitro group and accepting electrons from sufficiently powerful electron donors. Thus, electron-accepting dp–π conjugation can be significant in the ground state of aromatic sulfides.	(*BU-63*)
The UV and photoelectron spectra of anisole, thioanisole, benzyl methyl sulfide, and benzyl methyl ether correlated with each other and with SCF calculations excluding *d*-orbitals.	Divalent sulfur does not stabilize the aromatic π^* LUMO by use of *d*-orbitals.	(*BWK-72*)
Going from CH_3X to SiH_3X increases the first ionization potential by 0.33 eV and broadens the peak in the photoelectron spectrum, with appearance of vibrational fine structure for X = Cl or Br. For SiH_3Cl the separation between the first two vibrational states is 520 cm^{-1} compared to a 545 cm^{-1} IR band for the Si–Cl bond.	For X = Cl or Br the HOMO consists of lone pair electrons on the halogen; the appearance of the vibrational fine structure is indicative of partial delocalization of the lone pair via dp–π bonding, which also explains the increase in ionization potential. This does not occur for X = F because the lone pair electrons on fluorine are of very high energy.	(*CE-71*)
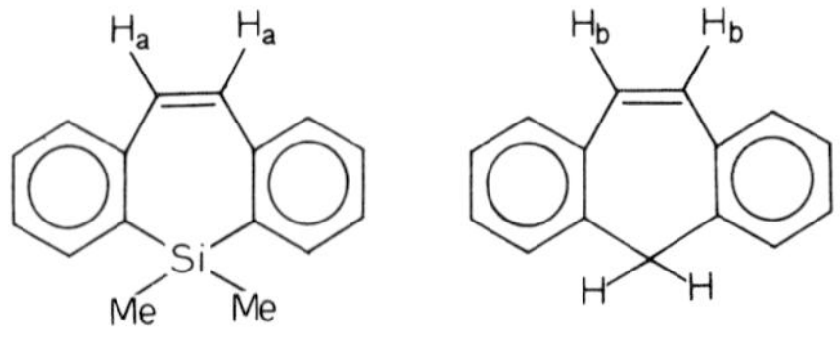 [Structures A (dibenzosilepine with Si(Me)₂ bridge, H_a labels) and B (dibenzocycloheptene with CH₂ bridge, H_b labels)] The UV spectra of A und B were quite similar; the highest energy band fell at 291 *nm* in A and 284 *nm* in B; and both spectra were quite different from that of anthracene. In the NMR, H_a and H_b were olefinic, and no ring current was found. [Structure: Ni complex with two benzotriazolyl ligands bearing Et substituents on N and SiPh₃ groups] 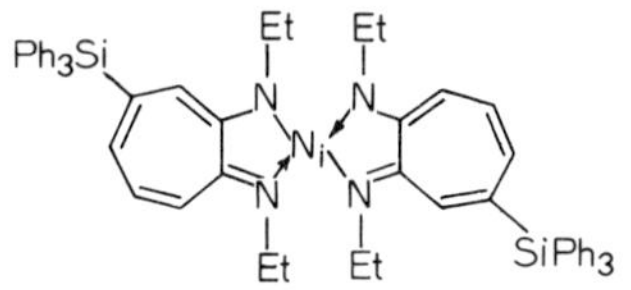	The lack of aromaticity of A proves that silicon is not capable of significant amounts of through-conjugation. This is further support for Dewar's "island" model of dp–π bonding, in which silicon is nodal.	(*CM-71*)

Systems and Results	Conclusions	References
The α and β protons in the 7-membered ring show very large NMR chemical shifts compared to the corresponding unchelated structure due to spin delocalization from the paramagnetic nickel; no such effect was noted for the phenyl protons.	Silicon is not capable of through-conjugation; this is further support for the Dewar "island" model of dp–π bonding.	(EM-67)
ESR spectra of the anion radicals of A and B show the free electron g value, 2.0023, proving zero spin density on sulfur. Molecular models of A show that direct overlap across space of the π systems of the two dicyanoethylene moieties is extensive.	In these symmetric systems the π^* LUMO is nodal at sulfur, and the coefficients at adjacent atoms are of opposite sign, so that the direct overlap is bonding. This is further proof that d-orbitals are important neither to the HOMO nor the LUMO in thiophene and related cyclic, unsaturated sulfides.	(ER-73)
A study of the silicon-29 coupling constants in the ESR spectra of the anion radicals from a series of trimethylsilyl-substituted ethylenes, butadienes, benzenes, and naphthalenes showed that the contribution of spin density at silicon to the coupling constant was negligible.	This is due to the diffuse nature of the dp–π bond, which places most of the electron density over the donor atom and ensures little interaction with the nuclear spin.	(GHB-70)
In A, the lone pair and π electron ionization potentials determined by photoelectron spectroscopy overlap at 9.0 eV. In B the lone pair ionizes at 8.5 eV, the electrons at 9.0 eV.	In B the lone pair on the nitrogen is not affected by dp-bonding to silicon and feels the full inductive effect of the trimethylsilyl group, thus explaining the low ionization potential.	(HH&-69)
CNDO calculations on H_3CCOOH and $H_3SiCOOH$ with and without d-orbitals show that dp–π bonding has a major effect on the LUMO, changing it from σ^* to π^*, but has no effect on the HOMO nor on the total energies of the acids and the corresponding anions.	The greater acidity of silyl carboxylic acids than of their carbon analogs is best explained in terms of the larger size and polarizability of Si; dp–π bonding is not involved because acidity is a ground state property.	(KPB-72)
$PhOSiMe_3$ showed a UV absorption maximum at 273 nm, close to that of anisole, and did not show the correlation (linear) of UV absorption frequency with electronegativity of M observed in C_6H_5OM for M = H, C, Na^+ but not M = P.	The effective electronegativities of Si and P are here controlled by dp–π bonding with oxygen.	(NH-70)

Systems and Results	Conclusions	References
SCF–MO calculations on H$_3$SiCl show that inclusion of Si–3 d-orbitals in the basis set has a major effect on the π HOMO (the Cl, p–π lone pairs), stabilizing the orbital and greatly increasing the electron density on Si. Electron density plots for each MO show that d-orbitals on Si and Cl have similar effects, increasing the electron density in and around the Si–Cl bond, but Cl–3d has its major effect on the σ MO lying just below the π HOMO.	The effect on the π HOMO explains the photoelectron spectrum of (CE-71). Overall, the main effect of d-orbitals is to reduce charge polarization.	(HV-74)
First ionization potentials determined from UV spectra of the tetracyanoethylene charge-transfer complexes. A: (indane with M(Me)$_2$ at ring junction) M = Si, 8.54 eV B: (indane with Si(Me)$_2$) 8.41 eV C: (indane with H, SiMe$_3$) 8.13 eV	In C the HOMO is destabilized by hyperconjugation; in B the Si is in a nodal plane, thus eliminating hyperconjugation. In A the higher energy HOMO for M = C than for M = Si can be explained without reference to d-orbitals by the longer length of the Si–C bond. This reduces σ–π overlap, which increases the electron density in and thus destabilizes the π HOMO.	(P-70)
The bathochromic shift of the $n \rightarrow \pi^*$ transition of Me$_3$SiCOOH compared to Me$_3$CCOOH was about half that for the analogous methyl ketones.	The dp–π stabilization of the LUMO is the same for the ketones and carboxylic acids, but the more delocalized carboxyl group HOMO is less influenced by the Si inductive effect.	(SDJ-71)
Comparison of the photoelectron spectra of vinyltrimethyl and divinyldimethylsilane showed that the π HOMO of the monovinyl silane is split 0.2 eV in the divinyl silane.	Silicon permits conjugation of two vinyl groups through itself by hyperconjugation and/or dp–π bonding.	(WS-72a)
The difference in the first ionization potentials of trimethylvinyl- and trimethylallylsilane was better predicted by CNDO calculations without than with d-orbitals in the basis set. These calculations showed that the conformation of trimethylallylsilane which permits σ–π overlap (hyperconjugation) is more stable than the one which does not by 5.1 kcal/mole (sp basis set) or 3.3 kcal/mole (spd basis set).	Since the large UV red shift and the high nucleophilic reactivity that characterize the "β effect" are all connected with the high HOMO energy, the "β effect" must be due to hyperconjugation rather than to 1,3 dp–π overlap.	(WS-72b)

Physical Properties Relative to $dp-\pi$ Bonding

Systems and Results	Conclusions	References
Quantum chemical calculations were carried out on $Me_3SiNHPh$, $(Me_3Si)_2NPh$ and $(Me_2HSi)_2NPh$. The reliability of the results achieved by Del Re and Pariser-Parr-Pople methods has been evaluated by comparing calculated ultraviolet spectra and dipole moments with experimental values.	The $dp-\pi$ bond between silicon and nitrogen reduces the Ph–N π bond order from 0.346 in $PhNMe_2$ to 0.275 in $(Me_3Si)_2NPh$ and $(Me_2HSi)_2NPh$.	(NH-72)
Photoelectron spectra of $PhCl$ and $p-ClC_6H_4SiMe_3$ show very similar splittings between the two Cl lone pair bands, whereas this splitting is zero for $p-ClC_6H_4CMe_3$.	The splitting is due to the ability of only one Cl lone pair to conjugate with benzene π orbitals. This conjugation is favored by electron-withdrawing substituents para to Cl and impaired by electron donating substituents, thus explaining the zero split for $p-ClC_6H_4CMe_3$. These results support Bock's conclusion of weak $dp-\pi$ bonding between Si and the π HOMO, just cancelling out the electron-donating inductive effect (AB-69).	(LM-75)
Photoelectron spectra of SiH_nX_{4-n} (X = Halogen) show strong bonding character for some of the halogen lone pairs in all cases except SiH_3F; much of this is of the $dp-\pi$ type. This effect stengthens the Si–X bond by an amount which decreases in the order $SiH_3Br > SiH_3Cl > SiH_3F$.	The lone pairs on fluorine are too tightly bound to interact with Si, d-orbitals as a result of the high electronegativity of fluorine. However, chlorine and bromine have electronegativities about the same as for nitrogen, i.e. strong enough to produce enough of a positive charge on silicon, and lowering the d-orbital energies enough to permit $dp-\pi$ bonding, yet not strong enough to hold the lone pairs too tightly; thus, $dp-\pi$ backbonding occurs.	(FH&-71)
Half-wave reduction potentials were 0.759 and 0.864 v respectively for 4- and 3-nitrophenyl-triphenylphosphonium iodides compared to 0.890 and 0.880 v. respectively for the analogous trimethylammonium chlorides. ESR spectra of the reduction products indicate phosphorus-phenyl conjugation.	The $dp-\pi$ bonding between phosphonium phosphorus and the singly occupied benzene LUMO in the reduction product is favored by the lengthened conjugated system possible when the nitro group is para, thus stabilizing the reduction product from the 4-nitrophenyl isomer. These conclusions are strongly supported by CNDO calculations.	(RWM-76)

IV. Effect of dp-π Bonding on Chemical Properties and Reactivity

In this section we try to evaluate the effect of dp–π interactions on chemical properties and reactions, generally on the molecule as a whole, or at sites adjacent to sulfur, phosphorus or silicon.

A. Acidity-Basicity

The extent of delocalization of nonbonding electron pairs can be determined by infrared measurements of acidity and basicity. This has been proposed as a criterion of dp–π bonding. Probably one of the earliest applications of this approach to the study of dp–π bonding in silicon is that of *West* and *Baney* (WB-59), who measured the acidity of carbinols and silanols by the bathochromic shift in the O—H stretching frequency upon addition of Lewis bases such as ethyl ether or mesitylene, and the basicity of silanols and carbinols by the shift in the O—H stretching frequency of phenol upon addition of the silanol. Some of the results were as follows:

	$\Delta\nu$ PhOH	$\Delta\nu$ ether
trimethylcarbinol	271 cm^{-1}	122 cm^{-1}
trimethylsilanol	216	238
triphenylcarbinol	172	174
triphenylsilanol	175	311

Delocalization of the nonbonding pairs on oxygen by dp–π bonding weakens the O—H bond so that the acidity of silanols is intermediate between that of carbinols and phenols. The basicities, however, as measured with phenol, appear not to be reduced by dp–π bonding, probably because of the competing inductive effect of silicon, which on the basis of the electronegativities of carbon and silicon should be quite large. An alternative is that only one of the two unshared pairs on oxygen is involved in dp–π bonding since one of the unshared pairs should be in an s orbital and would not overlap the silicon d-orbitals.

The measurement of basicity by changes in the O—H stretching frequency of phenol was extended to siloxanes by using the series $Me_{4-n}M(OEt)_n$, where M is C, Si, or Ge. For carbon and germanium the basicities (*UJC-68*) decreased steeply and linearly with increasing n, as expected from the inductive effects of these substituents, changing from electron donation when the number of methyl groups is large, to electron withdrawal as the number of ethoxy groups increases. For silicon the same results were obtained but with a much smaller slope. Thus $MeSi(OEt)_3$ and $Si(OEt)_4$ are more basic than $MeC(OEt)_3$ and $C(OEt)_4$, but Me_3SiOEt is less basic than Me_3COEt. The low basicity of Me_3SiOEt relative to $Si(OEt)_4$ supports Cruickshank's supposition (*C-61*) that the strength of the $dp-\pi$ bond to each oxygen decreases as the number of Si—O bonds in the molecule increases. However, the higher basicity of $Si(OEt)_4$ than of $C(OEt)_4$ (*WWL-61*) can only be accounted for by assuming negligible $dp-\pi$ bonding, a distinct contradiction of *Cruickshank's* theory (*C-61*) which predicts a π bond order of 1/2 for orthosilicate esters.

Proton *nmr* spectroscopy may also be used to measure basicity, using the change in chemical shift or ^{13}C—H coupling constant of a reference acid (e.g. chloroform) upon addition of base. A study (*RYZ-67*) of the changes in the chemical shift of chloroform upon addition of siloxanes gave results in qualitative agreement with those of West et. al. (*WWL-61*). However, an attempt to obtain a linear correlation between the changes in chloroform chemical shift and C—H stretching frequency upon addition of I, II, or of $R_2M(NEt_2)_2$ was a complete failure. This is strong indication that all of these methods for measurement of basicity are of qualitative significance only, particularly in the absence of attempts to relate the observed parameters to the equilibrium constant for formation of the hydrogen bonded acid-base complex.

I

II

However, the contradictory results for the series $Me_nSi(OEt)_{4-n}$ suggest that these results may be more than of qualitative significance. To assume that $dp-\pi$ bonding in $Si(OEt)_4$ is outweighed by the electron-donating inductive effect brings up the question of how much decrease in electron-donation is to be expected on going from a trimethylsilyl to a triethoxysilyl group, and whether this decrease is greater than the decrease in $dp-\pi$ bonding. Unfortunately, basicity measurements alone provide no means of separating inductive and π-bonding effects.

Direct formation of complexes with strong Lewis acids is obviously the most desirable method of studying basicity. Unfortunately, silicon compounds generally have such low basicities that the method can rarely be used. One case where it could be used involved measurement of the dissociative vapor pressures of BH_3 adducts with silanes of the formula $SiH_{4-n}(NMe_2)_n$ (*AP-65*). For $n = 2$, the dissociative vapor pressure was 1.7 mm Hg; for $n = 3$ it was 1.8 mm. Since the steric hindrance to complex formation of three dimethylamino groups must obviously be considerably larger than that of two dimethylamino groups, the dissociative vapor pressure for $n = 3$ should be much less than that for $n = 2$. Apparently the decrease in $dp-\pi$ bonding per Si–N bond upon going from an $Si(N)_2$ to an $Si(N)_3$ system almost exactly cancels the steric effect.

Similarly the strength of S–O bonds in sulfites and similar solvents has been studied on the basis of their solvation of halide ions, as determined from the solubility product of silver chloride and the stability constant of $AgCl_2^-$ with respect to ligand exchange with solvent. Addition of tetrahydrothiophene to propylene carbonate or of sulfur dioxide to propionitrile increased the solubility of AgCl, but in the latter case there was also a significant decrease in the stability of $AgCl_2^-$ (*S-75a*). This suggests that both tetrahydrothiophene and sulfur dioxide complex with free silver ion and that sulfur dioxide also forms a complex with chloride. In general, solvation of chloride was much greater for sulfite esters than for sulfoxides. Since $dp-\pi$ bonding is known to be much stronger in the former, charge transfer from chloride to empty d-orbitals is implied. This is supported by CNDO calculations (*S-75a*) which showed charge transfer for both tetrahydrothiophene and sulfur dioxide, but with the S–Cl bond 26 kcal more stable in the latter case. The d-orbital populations of sulfur dioxide and tetrahydrothiophene increased 29.1% and 55.5% respectively upon complexation and it was 41.7% greater for the sulfur dioxide than for the tetrahydrothiophene complex. The chloride complex with dimethyl sulfite was 200 times more stable than that with ethylene sulfite. This is attributed (*S-75a*) to the increase in electron density on the ring oxygens of ethylene sulfite upon complexation, creating significant destabilization of the ring through 1,3 nonbonded interactions.

Ylides are strong bases; evidently, therefore, direct pKa measurements on carbonyl-stabilized ylides can provide evidence of $dp-\pi$ bonding. For example (*JA-68*), the pKa values for the conjugate acids of Me_3P=CHCOPh and Ph_3P=CHCOPh, 8.6 and 5.6 respectively, provide clear evidence of the reduction in $dp-\pi$ bonding upon substitution of electron-donors for electron = acceptors on phosphorus. Also, Me_2S=CHCOPh, pKa 7.68, is less basic than the corresponding phosphonium ylide, as is to be expected from the higher electronegativity of sulfur, but Ph_2S=CHCOPh and Ph_3P=CHCOPh have almost identical basicities, indicating that both elements are equally capable of $dp-\pi$ bonding under equally favorable conditions.

B. Electrophilic Aromatic Substitution

When aryl groups are bonded to an element capable of $dp-\pi$ bonding, we should expect a decrease in aromatic substitution at the ortho and para positions due to the decrease in electron density at these positions. In fact, chlorination of benzyltrimethylsilane at 60 °C with iodine as catalyst occurred 105 times faster than that of phenyltrimethylsilane, whereas chlorination of benzyltrichlorosilane was 2360 times faster than that of phenyltrichlorosilane under the same conditions. This is most readily attributed to the much greater $dp-\pi$ bonding between phenyl and trichlorosilyl than between phenyl and trimethylsilyl. The rate for trimethylsilylbenzene was 36.6 times that for benzene (*LC-69*), indicating that $dp-\pi$ bonding is not strong enough here to overcome the electron-donating inductive effect; this should be compared to the conclusions of *Bock* and *Alt* (*BA-70&BA-70a*) based on ultraviolet spectroscopy.

In another example, nitration of phenyltrimethylammonium ion gives 11% para product, whereas nitration of phenyltrimethylphosphonium is faster but gives less para product (*GM&-68a*). The higher activity of the phosphonium ion is to be explained by the shielding of the phenyl group from the positive charge by the larger electron kernel around phosphorus, but the specific deactivation of the para position must be due to $dp-\pi$ bonding. This is in contrast to the inductive effect, which activates or deactivates in the order ortho > meta > para.

C. Alpha-sulfonyl and -sulfinyl Carbanions

Carbanions are stabilized by electron-withdrawing groups adjacent to the negative charge center; the sulfonyl and sulfinyl groups are certainly electron-withdrawing enough to qualify in this respect. Unexpectedly, however, carbanionic reactions involving optically active sulfones proceed with high retention of configuration, the same occurring to a lesser extent with sulfoxides. *Cram* (*CTS-66*) has concluded that α-sulfonyl carbanions are intrinsically asymmetric whereas α-sulfinyl carbanions are intrinsically symmetric with solvent-induced asymmetry. The intrinsic asymmetry might result from hindrance to rotation around the C—S bond of a planar carbanion, or from electrostatic inhibition of inversion around a pyramidal carbanion; this latter possibility has been discussed previously in connection with *nmr* studies of inversion around pyramidal nitrogen in sulfonamides. The experimental work leading to these conclusions has been recently reviewed (*T-69, G-69*) and will not be discussed here.

Recent work (*PFW-71*) favors the pyramidal α-sulfonyl carbanion. At 100 °C in MeOD/NaOMe, deuterium exchange of III was 331 times faster than that of phenyl 2-octyl sulfone, due to the greater resonance stabilization of the carbanion from III, but was three times slower than that of phenyl α-methylbenzyl sulfone. A planar configuration is enforced by the geometry of III. This must have very slightly higher energy than the pyramidal configuration of the phenyl dimethylbenzyl sulfone carbanion. Exchange of III proceeded with complete racemization; a similar result (*CL-65*) was found for the decarboxylation of IV and expected since planar carbanions derived from both III and IV are symmetrical.

The pyramidal carbanion is given strong support by SCF (self-consistent field)-MO calculations (*RWC-69, WRC-69*), on $^\ominus CH_2S(H)O$ and $^\ominus CH_2S(H)O_2$, which indicate two energy minima and one saddle point between them. On the energy surface the minima occur at rotational angles 180° apart and with H–C–H angles at the carbanionic carbon of 115°. They, in fact, have an identical conformation in which the carbanionic lone pair lies on the plane bisecting the O–S–O, or, in sulfoxides, the O–S-lone pair angle. The saddle point represents a combined inversion-rotation barrier. The most important conclusion for our purposes, however, is that this conformational preference is due entirely to electrostatic repulsion between the lone pairs and the oxygen atoms. This conclusion is eminently reasonable since the large size of d-orbitals should mean few steric requirements for overlap, as indicated for example by MO calculations on trisilylamine (*P-67*).

Unfortunately, these calculations are not in complete agreement with experiment. In MeOD/NaOMe the axial α proton of *cis*-4-phenyltetrahydrothiopyran-1-oxide (*HAK-69*) exchanged 7.8 times faster than the equatorial α proton. The ratio of the two exchange rates was even higher in D_2O/NaOD, but was approximately 1.0 in the less polar media, tert-butanol and dimethyl sulfoxide. Since the conformations are frozen it is clear that the favored carbanionic species, at least in polar solvents, is that of V rather than VI as expected. For the *trans* isomer, the axial α proton which is frozen *trans* to the sulfoxide lone pair, exchanged 62.5 times slower than the equatorial proton, in agreement with theory; again, however, this stereospecificity was found only in polar solvents. Similar results were found (*DF&-70*) for benzyl methyl sulfoxide, to which conformation VII may be assigned. Exchange of H–2 was favored in polar media, H–1 in nonpolar media, the two benzyl protons being distinguishable by *nmr*. For sulfones the situation is simpler, and the two *nmr*-distinguishable benzyl protons in VIII exchanged at rates which corresponded to a free energy difference of 0.76 kcal between conformations IX and X in tBuOD/NaOPh (*FS-69*), IX being favored, compared to 3.9 kcal calculated (*WRC-69*).

Similarly in XI, the relative rates of exchange of the axial and equatorial protons were 1:1.6, respectively, with NaOD/D_2O in dimethyl sulfoxide. At 36.5 °C this corresponds to a preference for XII over XIII of only 0.287 kcal in terms of free energy. Such a small difference may just as well be due to asymmetric solvation and need not imply any intrinsic difference between the two carbanions (*BC&-71*). Furthermore, no results are available for α-sulfonyl carbanions in polar solvents, and it may be that their stereochemistry is the same as for the corresponding sulfoxides, i.e. a preference for XIII over XII in polar solvents and no particular preference in nonpolar solvents. Apparently, the energy difference between stereoisomeric carbanions is controlled much more by solvation than by electron repulsions, as suggested by theoretical calculations (*RWC-69, WRC-69*).

In addition to solvation, the conformation of α-sulfinyl carbanions is also affected by the counterion, usually lithium. Thus, the ratio of diastereomers produced in exchange of benzyl methyl sulfoxide in tetrahydrofuran depended on the presence of other soluble lithium salts in addition to the organolithium base; certain of these salts (LiBr, LiI) may occur as impurities in commercial organolithium preparations (*DM-75*). There is also a recent report (*LCM-76*) challenging the common assumption of pyramidal α-sulfinyl carbanions. $J(^{13}CH)$ coupling constants in the NMR should decrease on going from a carbon acid to the corresponding carbanion; this is observed in going from methane to methyllithium. Instead J increases by as much as 23 cps in going from benzyl methyl sulfoxide or sulfone to the corresponding organolithiums, which certainly implies high s character in the hybridization at the carbanion carbon. This is in agreement with deuterium exchange results which indicate little stereochemical preference for the carbanion in nonpolar solvents. On the other hand, this does not prove full sp^2 hybridization. Thus, the increase in J upon lithiation of tert.-butyl methyl sulfoxide or sulfone is less than half that upon lithiation of benzyl methyl sulfoxide or sulfone; this implies that the conformation in α-lithiobenzyl methyl sulfoxide or sulfone is affected as much by benzylic resonance as by conjugation with the sulfinyl or sulfonyl group.

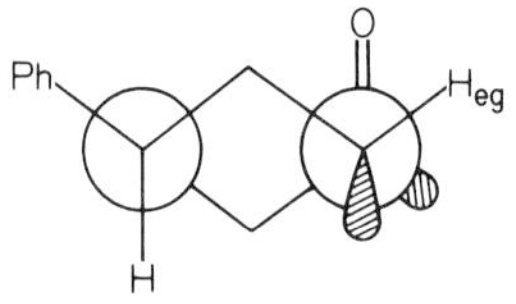

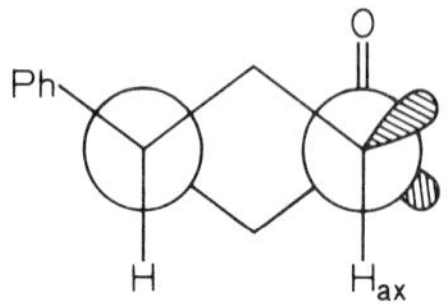

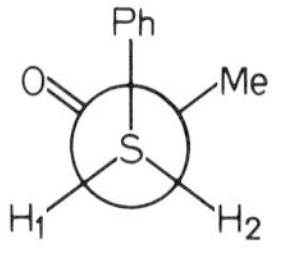

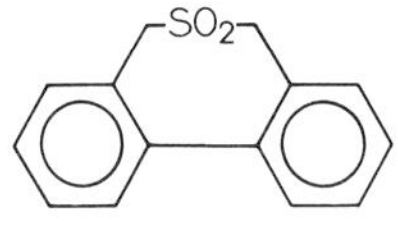

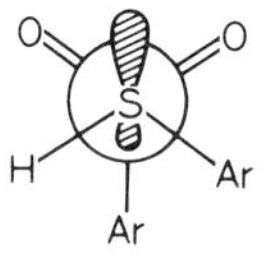

In addition, it must be remembered that MO calculations (*RWC-69*) refer to the thermodynamic acidity of the sulfoxides, which is not the same as the kinetic acidity measured in these studies. While kinetic acidity may provide an indication of thermodynamic acidity, the correlation is not always valid. In particular it is well known that internal return may render meaningless any conclusions about carbanion stabilities gleaned from measurements of isotope exchange rates. A quantitative measure of internal return is possible from the relative rates of exchange of protiated and tritiated substrate in a deuterated solvent using *Streitwieser's* technique (*SH&-71a*). A recent study (*FN-76*) has used this method to show that internal return is negligible (3–12%) in the exchange of the diastereotopic benzyl protons of benzyl methyl sulfoxide in water but is a major factor (70–80%) in tert.-butyl alcohol.

Whether or not $dp-\pi$ bonding controls the stereochemistry, it must contribute significantly to the stability of α-sulfonyl, if not of α-sulfinyl, carbanions. For instance, deuterium exchange of dimethyl sulfone (*B-68a*) was considerably faster than that of tetramethylammonium ion, in spite of the greater electron-withdrawing effect in the latter. Deuterium exchange at the α position of 2-hexenyl methyl sulfone was 700

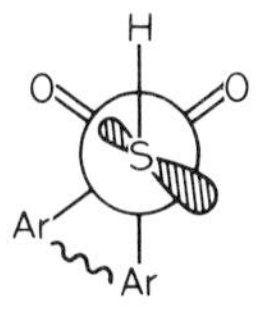

X 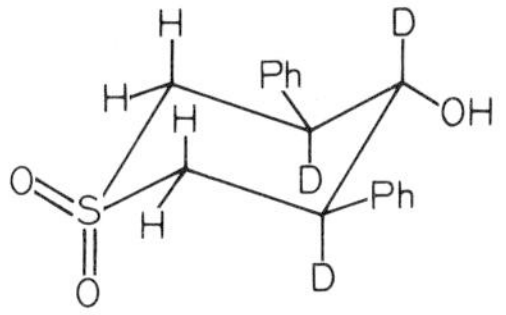   XI

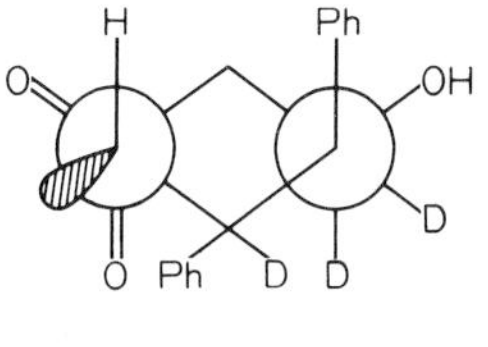

XII 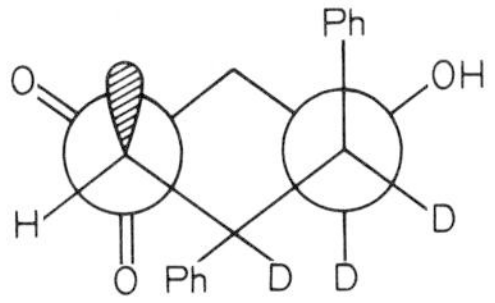   XIII

times faster than at the γ position of 1-hexenyl methyl sulfone but only 10 times faster than at the methyl group in either isomer. Two explanations have been advanced by Broaddus (B-68b) for the surprisingly high reactivity at the methyl group. The first is that the increased p character of the carbanion formed at the methyl group compared to that formed at the allylic methylene group enhances $dp-\pi$ bonding, thereby compensating to a great extent for the lack of allylic resonance at the methyl group. The inductive effect of the sulfonyl group should be experienced to the same extent by the methyl and methylene groups. The second explanation is that the strong inductive effect of the sulfonyl interferes with allylic resonance. Thus, deuterium exchange (SG-66) at the N-methylene group was only 86 times slower in $Me_3\overset{+}{N}CH_2CMe=CH_2$ than in $Me_3\overset{+}{N}CH_2CH=CHCH=CH_2$ whereas isomerization of 1,4-pentadiene in strong base was 140,000 times faster than that of 1-butene. Since the inductive effect of the trimethylammonium group should be stronger than that of a sulfone, these results imply that both effects contribute to the observed exchange rates in the hexenyl methyl sulfones, with the latter effect (interference with allylic resonance) probably more important.

The rates of deuterium exchange at the α position in XIV and XV were 15,700 and 3300 times faster, respectively, than in the corresponding acyclic system XVI and XVII (BM&-71). This would seem to indicate some sort of "aromatic-type" conjugation in these cyclic systems. Although $dp-\pi$ bonds involving d_{yz} orbitals can form a homomorphic (M-69) delocalized system which shows Hückel type alternation of properties, exchange repulsions generally destabilize the d_{yz} orbital which contrib-

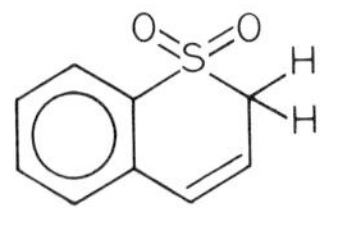

XIV 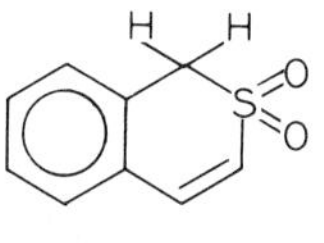   XV

Ph—CH=CH—CH$_2$SO$_2$Ph
XVI

Ph—CH=CH—SO$_2$—CH$_2$—Ph
XVII

utes very little to the bonding in cyclic phosphazenes. Another possible explanation is a 1,3 interaction between the two carbon atoms adjacent to sulfur, i.e. homoaromaticity. However, if we assume normal C—S bond distances and C—S—C angles, then the distance between these two adjacent carbons would be 2.8 Å, thus making the π overlap necessary for aromaticity almost negligible, although some $pp-\sigma$ overlap might be possible even at this distance.

The Ramberg-Bäcklund reaction (*P-68*) is a useful application of α-sulfonyl carbanions to the synthesis of olefins from α-halo sulfones. The generally accepted mechanism involves pre-equilibrium formation of carbanion, then rate-determining halide expulsion to form episulfone, followed by stereospecific sulfur dioxide extrusion. Experimentally, the decomposition of ethyl α-chloroethyl sulfone in base (*N-66*) gave *cis*- and *trans*-2-butene in 3:1 ratio independent of solvent. This result may be explained by faster formation and decomposition of the *cis* episulfone to give *cis* olefin than of the *trans* episulfone to give *trans* olefin. Woodward and Hoffmann (*WH-69*) have proposed that episulfones can undergo concerted disrotatory SO_2 extrusion. Bordwell (*BWH-68*), however, on the basis of an observed solvent dependence of the relative rates of (stereospecific) decomposition of *cis*- and *trans*-stilbene episulfones and the formation of traces of polysulfone, has argued for a two-step decomposition involving intermediate zwitterions or polar biradicals.

Episulfone formation must involve backside attack by the halogen-bearing carbon on the carbanion, that is, inversion of configuration at both carbons as is most aptly proven by the synthesis of XVIII from XIX (*PH-69*). Inversion has also been demonstrated in the attack of methyl iodide on optically active, lithiated, benzyl methyl sulfoxide in tetrahydrofuran (*DVM-71*). On the other hand, attack of deuterium oxide on the same carbanion under the same conditions proceeded with retention of configuration.

More recent work tends to prove the validity of the two-stage mechanism and also illustrates the importance of solvent effects. Thus, olefin formation from $Me_2CHSO_2CHBrPh$ in NaOMe/MeOD was accompanied by 37% deuterium exchange, whereas no exchange occurred in 40% aqueous dioxane (*BO-74*). As expected, decreasing solvent polarity may hinder carbanion formation at the isopropyl group, but it also increases the rate of attack of the carbanion on the —C—Br bond to form the episulfone. As a result, this becomes faster in dioxane than the rate of solvent attack on carbanion. Consequently, the increase in the rate of stilbene formation from $PhCH_2SO_2CHBrPh$ with decreasing water content in aqueous dioxane, ethanol, or dimethoxyethane is to be expected in view of episulfone formation being rate-determining (*BW-74*). Methyl substitution at either α-carbon had very little effect on the overall rate (*BW-71, BW-74*). At the carbanionic carbon this can be attributed to the opposite effects of the electron donating methyl group on carbanion formation and conversion to episulfone. When substituted at the carbon of the $-\overset{|}{\underset{|}{C}}-Br$ bond, the lack of rate effect indicates that the transition state for episulfone formation has very little zwitterionic character. Interestingly enough, the activation enthalpy and entropy for olefin formation from $PhCH_2SO_4CHBrPh$ and XX were identical in methanol in spite of the fact that the α hydrogen and bromine in the latter were in ideal position for concerted episulfone formation. Evidently, the high solvation energy favors carbanion formation over the concerted path (*BD-74a*).

Treatment of erythro $PhCHMeSO_2CBrMePh$ with NaOMe in MeOH gave *cis*-α,α'-dimethylstilbene with 93% selectivity, whereas the threo isomer gave the *trans* dimethylstilbene with 95% selectivity, (*BD-74b*). This stereochemistry is to be expected from inversion at both carbon atoms, which corresponds to retention of the relative configuration at the two reaction centers. The lack of complete stereospecificity can be attributed to reaction of the carbanion with solvent (internal return), every such reaction corresponding to inversion of configuration at one carbon. In view of the lesser steric repulsions in the *trans* episulfone, it is not surprising to find that for the threo but not the erythro isomer episulfone decomposition rather than its formation is rate-determining, as indicated by spectroscopic observation of the episulfone and by the overall rate being zero-order in base. For nonstereoisomeric sulfones, the preference for *cis* over *trans* olefins must be explained in terms of steric effects on the equilibrium constant for carbanion formation and the rate of its conversion to episulfone; the latter factor would favor XXI, leading to *cis* olefin (*P-68, BD-74a*), over XXII, but the former would favor XXII to an extent which depends on the size of R and R'. Thus, the selectivity for *cis* ranged from 78.8% for $EtSO_2CHClMe$ to 53% for $EtSO_2CBrPhMe$. Although these structures have been drawn on the assumption of the carbanion conformation preferred on the basis of SCF–MO calculations (*WRC-69*), the results would be the same if the carbanion were *trans* to one of the sulfonyl oxygens, as suggested by isotope exchange studies in polar solvents on conformationally rigid α-sulfinyl carbanions (*HAK-69, DF&-70*).

An interesting example of the Ramberg-Bäcklund reaction is the synthesis of diphenylthiirene dioxide XXIII from α,α-dibromodibenzyl sulfone. This species was approximately 1000 times more stable than the corresponding episulfone but still underwent decomposition to diphenylacetylene in refluxing benzene (*CA&-71*). This stability has been explained by recent CNDO–MO calculations (*MSV-75*) on 2,3-dimethylthiirene-1,1-dioxide, which show, in addition to the usual inductive interaction between the sulfonyl and ethylene moieties, hyperconjugation between the ethylene π HOMO and the sulfonyl σ HOMO, and a weak conjugation with the

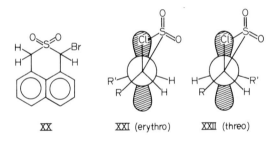

XX   XXI (erythro)   XXII (threo)

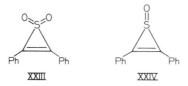

XXIII   XXIV

unoccupied sulfonyl π^* or σ^* orbitals, which are formed primarily from sulfur d-orbitals. The results of these calculations explained the observed photoelectron spectrum. Moreover, the interactions were similar to but weaker than those for cyclopropenone; the hyperconjugative interaction was about 0.5 eV in the sulfone and 1.0 eV in the cyclopropenone. In fact, the UV spectrum of XXIV (*CC-71*), prepared in the same way as XXIII, was quite similar to that of diphenylcyclopropenone. Compound XXIV was also considerably more stable than XXIII; this may, however, be due to the fact that SO is formed much less readily than SO_2. In resonance terms, the degree of stabilization of the thiirene monoxide and dioxide, and to a much greater extent cyclopropenone, can be correlated with a small contribution from an aromatic cyclopropenium structure.

The most important property of α-sulfonyl carbanions is the extremely high stereospecificity of reactions involving them. However, the MO calculations of *Wolfe* et.al. (*RWC-69*, *WRC-69*) show that $dp-\pi$ bonding need not be invoked to explain this stereospecificity. Although these calculations are not in complete agreement with experiment (*HAK-69*), the basic conclusion that the stereospecificity is due to lone pair repulsions may not have been (thereby) invalidated. Nonetheless, $dp-\pi$ bonding may contribute extensively to the stability of α-sulfonyl carbanions (*B-68a*, *B-68b*) making possible reactions like the Ramberg-Bäcklund synthesis.

D. Alpha-thio Carbanions

Since sulfide sulfur has negligible charge deficiency, α-thio carbanions would not be expected to show any significant $dp-\pi$ bonding. Oae (*YO-69*) has extensive experimental data that he interprets as evidence for $dp-\pi$ interactions in α-thio-carbanions. Nevertheless, his results are susceptible to a different interpretation.

For example, $PhSCD_3$ underwent dedeuteration in NH_3/KNH_2 4.5×10^6 times faster than $PhOCD_3$ and 2×10^8 times faster than $PhN(CD_3)_2$, but also 3.3×10^5 times faster than $(CD_3)_2S$ and 2.5×10^4 times faster than Me_3CSCD_3 (*SG-69*). The rate difference between anisole and dimethylaniline clearly reflects the electronegativity difference between oxygen and nitrogen, oxygen having a much greater electron withdrawing influence.

The greater electron withdrawal of the phenyl vs. the methyl group of itself does not account for the very greatly increased rate of exchange except as a very large difference in the extent of $dp-\pi$ bond stabilization. This could be the result of radial contraction of the *d*-orbitals on sulfur through the effect of electronegative substitution (*CM-56, CZ-62*). On the other hand, some sort of conjugation through the sulfur between the benzene ring and the carbanion center might be responsible. This would necessarily be of the $pp-\pi$ type since $dp-\pi$ orbitals are known (*M-69a, P-64, DLW-60*) to be nodal at the central atom.

Similarly, *p*-nitrophenyl-β-chloroethyl sulfide undergoes E–2 dehydrochlorination in t–BuOK/tBuOH, at 60°, 1100 times faster than the corresponding ether and 290 times faster than ethyl β-chloroethyl sulfide (*YO-69, MU&-69*). The relative rates are clearly smaller compared to those observed above in carbanion intermediated reactions. However, the same considerations appear to apply where stabilization of the incipient double bond in the E–2 transition state is involved, namely, $dp-\pi$ bonding assisted by radial contraction of the *d*-orbitals on sulfide through electronegative substitution, or conjugation through sulfur via some form of $pp-\pi$ bonding, or some variant of the 3 center-4 electron bonding of the type discussed by *Musher*, (*M-69a*).

The nature of this conjugation may possibly be seen from the variation of Q values of several substituted ethylenes for free-radical copolymerization with styrene. These values (*TT&-68*) which reflect the reactivity with the growing polystyryl radical were 0.02 for ethyl vinyl ether, zero for $CH_2=CH(OEt)_2$ (that is no copolymerization), 2.70 for $CH_2=CH(SEt)_2$, and 0.11 for methyl vinyl sulfone. The low value for the sulfone implies that both $pp-\pi$ bonding and low-lying empty orbitals, either $3d$ or $4s$, are required for this type of stabilization of carbanions or free radicals.

On the other hand, this leaves unexplained the 1200 times greater exchange rate for $DC(SCH_2)_3CMe$ compared to $DC(SEt)_3$; $DC(OCH_2)_3CMe$ does not exchange at all under these conditions (*OTO-64a*). Since the bridged ring is nonplanar, $pp-\pi$ overlap with more than one sulfur is impossible, and Oae's explanation of direct $dp-\pi$ overlap across space (*OTO-64*) may well apply.

The stabilizing effects of α-thio groups on carbanions could be due to polar effects (charge neutralization) or to polarizability, but Bordwell has amassed considerable evidence to show that while these effects do contribute, the major effect is conjugation. This evidence is based on equilibrium acidity measurements in dimethyl sulfoxide of XCH_2Y, where X is Me, MeO, MeS, Ph, PhS, or PhO, and Y is CN, SO_2Ph, or COPh (*MB&-75*); this technique avoids the problems of internal return, ion pairing, and variable Broensted coefficients, which hinder the prediction of thermodynamic acidity from kinetic measurements. Substitution of X = PhS for X = Me in these systems resulted in large increases in acidity, comparable to but slightly larger than those for X = Ph (*BB&-77*). Both X = PhS and X = Ph showed the phenomenon of "resonance saturation", i.e. a reduced acidity increase as the acidity of the parent acid $MeCH_2Y$ increased; this is to be expected since high acidity implies a high degree of charge dispersion in the corresponding carbanion, resulting in less opportunity for further charge dispersion or neutralization by X (*BB&-77a*). However, X = PhS, unlike Ph = X, showed no reduction in acidity increase in sterically hindered, e.g. 9-X-fluorene, systems; this is to be expected from the low steric requirements of $dp-\pi$ bonding.

A measure of the polar and resonance (conjugative) contributions to ΔpK the acidity increase, was obtained from the ΔpK for X = Me_3N^+, which has the largest polar effect of any X but no resonance effect. Assuming the validity of the Taft equa-

81

tion, $\Delta pK = \rho\sigma$, ρ could be determined from the ΔpK value for X = Me_3N^+ and used to predict ΔpK values for other X's. The difference $\Delta(\Delta pK)$ between predicted and measured ΔpK values is then a measure of the resonance and polarizability contributions to the acidity increase. The fact that resonance is more important than polarizability is shown by the larger $\Delta(\Delta pK)$ values obtained for X = PhS than for X = PhSe in XCH_2COPh despite the larger polarizability of selenium than sulfur (*BVV-76, BB &-77*).

It should be noted that while these results prove that the carbanion-stabilizing effects of α-thio groups are due primarily to conjugation, they do not prove (although they certainly do not disprove) that this conjugation involves $dp-\pi$ bonding. Recent MO studies (*EY&- 76*) to be discussed later suggest that the conjugation involves hyperconjugation and reduced destabilizing overlap between adjacent lone pair filled p orbitals rather than $dp-\pi$ bonding. In agreement with this is the observation of negative $\Delta(\Delta pK)$ values for X = MeO or PhO, i.e. carbanion-destabilizing conjugation. For X = PhS or MeS, $\Delta(\Delta pK)$ values were large and positive but still smaller than those for X = Ph (except in sterically hindered systems); whether or not conjugation with PhS involves $dp-\pi$ bonding, it would be expected to have less charge dispersing effect than that with the phenyl group, which provides six extra atoms over which to disperse the carbanionic charge (*BVV-76*).

CNDO (complete neglect of differential overlap) MO calculations (*CMW-70*) on the model system

$$CH_2{=}CH{-}\underset{\underset{H}{|}}{C}(SH)_2$$

indicate that $dp-\pi$ bonding may not be essential to the stability of this carbanion. Although inclusion of d-orbitals in the basis set increased the charge density on the sulfur and decreased that on the γ-carbon, 56% of the total π-electron density was on sulfur even without d-orbitals, and even with d-orbitals the charge on the γ-carbon was 11.7% greater than that on the α-carbon.

A similar conclusion was reached in extensive nonempirical SCF—MO calculations (*BC&-75*) on the thiomethyl carbanion, which showed

1) that the C—S bond length in the carbanion in the geometry which minimizes the total energy is longer than in methanethiol,
2) that the preferred conformation of the carbanion is not affected by the presence or absence of d-orbitals in the basis set, and
3) that the proton affinity of the thiomethyl carbanion is 15 kcal/mole less than that of the hydroxymethyl carbanion.

In other words, it showed that the sulfur carbanion is more stable, and is almost unaffected by inclusion of d-orbitals in the basis set. Ab-initio calculations (*EY&-76*) confirm the view that the much greater stability of the thiomethyl compared to the hydroxymethyl carbanion is due to the much greater stabilizing hyperconjugative interaction between the carbanionic lone pair and the vacant XH–σ^* orbital of sulfur, as well as the lesser destabilizing conjugation with the X lone pair and the filled XH–σ. However similar calculations on the chloromethyl carbanion, in which the chlorine has vacant *d*-orbitals but no XH–σ*, show that this species is unstable with respect to dissociation into methylene and chloride ion.

Similar calculations on the MeSCH$_2^-$ carbanion (*LW-76a*) showed that conformation XXV is 9.0 kcal/mole more stable than XXVI as a result of overlap of the carbanion lone pair with the antiperiplanar lone pair on sulfur and of hyperconjugation between the carbanion lone pair and adjacent antibonding σ* orbitals. When *d*-orbitals were excluded from the basis set, the difference in stability between XXV and XXVI decreased to 8.3 kcal. For the methoxymethyl carbanion the corresponding preference for the gauche conformation was 5.7 kcal/mole, but in the *n*-propyl carbanion the preference for gauche was only 0.7 kcal, which is to be expected since there were no lone pairs on the α atom. This provides a quantitative example of the well-known *gauche* effect (*W-72*) and also explains Oae's observation (*OTO-64a*) of rapid exchange in DC(SCH$_2$)$_3$CMe. The geometry of this bicyclic trisulfide is fixed in a way that requires formation of the carbanion in conformation XXV.

Whatever the means by which the thio group stabilizes adjacent carbanions, this stabilization can lead to interesting results. For example, treatment of the diethyl mercaptal of γ-chloropropionaldehyde with phenyl- or butyllithium gave the diethyl mercaptal of cyclopropanone (*FG-70*). Clearly, the organolithium abstracted a proton from the α-carbon, which then displaced chlorine intramolecularly. By contrast the corresponding acetal underwent only direct S$_N$2 reaction to give γ-phenyl- or γ-butyl-propionaldehyde diethyl acetal.

As an example of the synthetic utility of this gauche effect we may cite the contrathermodynamic lithiation of *m*-dithianes (*EHA-74*). Thus, lithiation at C–2 of 4,6-dimethyl-1,3-dithiane, which like *t*-butylcyclohexane is conformationally rigid, occurs entirely equatorial, so that reaction of the lithio derivative (carbanion) with D$_2$O, or with ketone, etc, results in the incoming group also being equatorial. Formation of the equatorial (gauche) carbanion XXVIIa from both XXVIIb and XXVIIc proves that the preference for equatorial over axial lithiation is at least 6 kcal/mole. These results are contrary to those obtained with 1-lithio-4-t-butylcyclohexane; the difference can be explained (*LW-76a*) by the lack of a strong gauche effect when there are no lone pairs α to the carbanion center.

E. Thiazolium Ylides

Stabilization of adjacent negative charge by sulfur apparently plays a major role in metabolism. Thiamine pyrophosphate is the active part (coenzyme) of the enzyme which decarboxylates pyruvic acid, an essential step in the energy-releasing oxidation of glucose. *Breslow (B-58)* has shown that this decarboxylation involves attack by the carboxyl carbon on the ylide derived by deprotonation of the 2-carbon of the thiazole ring of thiamine. Thiamine and thiazolium salts also catalyze the benzoin condensation by an analogous mechanism shown in XXVIII.

The decarboxylation must be controlled by the rate of ylide formation. This is true even though yeast pyruvate decarboxylase causes decarboxylation (*KO-70*) at least 10^4 times faster than the rate of ylide formation in neutral water. Manifestly, the protein portion of the enzyme serves to provide an environment in which the ylide is more stable and forms faster.

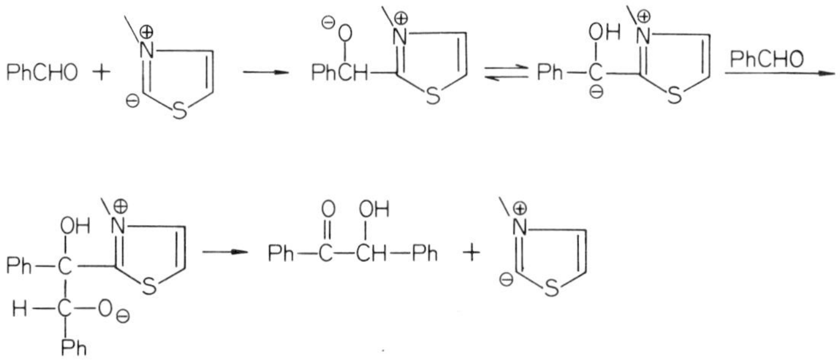

XXVIII

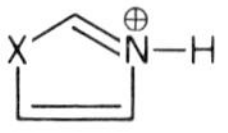

XXIXa, X = N
XXIX, X = O
XXX, X = S

The rates of ylide formation from the azolium ions XXIXa, XXIX and XXX could be correlated with the acidities at C_2 as determined from the *nmr* $J(^{13}CH)$ coupling constants (*HBM-69*). The oxazolium XXIX underwent ylide formation $10^{5.5}$ times faster than the imidazolium XXIXa and was almost as acidic at C_2 as acetylene. This suggests that the high rate of ylide formation here is due to the instability of the oxazolium ion, as indicated by its low pKa, 1.10. This is supported by simple Hückel MO calculations (*CP-64*), which show 5.7% less π-delocalization of positive charge from nitrogen in oxazolium than in thiazolium, assuming no *d*-orbital participation. However, 46.8% of this delocalized positive charge was on C_2 for oxazolium vs 31.7% for thiazolium, again assuming no *d*-orbital participation, which tended to decrease the charge deficiency on sulfur and on C_2 while increasing that on nitrogen and C_5. The thiazolium cation XXX formed ylide $10^{3.5}$ times faster than imidazolium XXIXa and had a much lower pKa, but both XXIXa and XXX had the same *nmr* coupling constant at C_2, indicating that the faster ylide formation from thiazolium does not reflect a ground state property.

Haake (HBM-69) explains both the pKa difference and the ylide formation rate difference as being due to the greater ability of the thiazole-thiazolium π-system to adjust to structural changes. The high stability of the imidazolium ion is clearly due to the equivalence of the two nitrogens, but it is quite possible that in XXX the $3p-2p$(S–C) π-overlap is less affected by small changes in hybridization upon ylide formation than is the $2p-2p$(N–C) π-overlap in XXIXa. Whether this is sufficient to explain the observed rate difference is questionable, however, and some $3d$ or $4s$ orbital participation may well be involved.

Support for such participation is found in the fact that in N-methylisothiazolium only the C_5 protons underwent deuterium exchange, even though the electron-withdrawing inductive effect of the positivelycharged nitrogen should favor exchange at C_3 *(OL-66)*. Actually, exchange probably does occur at C_3 but is slower than the rate of ring cleavage by nucleophilic attack of hydroxide (or deuteroxide). Such ring cleavage is extremely rapid with oxazolium cations and probably explains their inability *(HLI-63)* to catalyze the benzoin condensation. Like decarboxylation, which involves attack on the carbonyl (or carboxyl) carbon, the competing ring cleavage makes the equilibrium ylide concentration too low for effective catalysis. Similarly, in thiazole itself, the C_2 and C_5 protons underwent exchange *(OL&-66)* at approximately equal rates which were about $10^{-10.3}$ times that of exchange at C_2 in-thiazolium.

More recent work lends strong support to the idea of d-orbital contributions to the thiazolium system *(NH-76)*. The contributions of seven possible resonance forms to the structure of the thiazolium ion were determined by considering the experimentally observed bond lengths as the linear sum of contributions from each canonical form; the bond lengths in each canonical form were calculated from the Pauling bond order/bond length equation (see p. 14). Only three forms were found to be of importance, with contributions of 56.7% for XLIIa, 25.5% for XLIIb, and 17.7% for XLII. Structure XLII corresponds to an out-of-plane π bond formed from a $p_z/d_{xy}/d_{yz}$ hybrid at sulfur, similar to that proposed by *Longuet-Higgins (L-49)* for thiophene. CNDO calculations *(NH-76)* also suggest an in-plane π bond formed from an $s/d_{x^2-y^2}$ hybrid σ orbitals at C_2 and C_5 in the thiazolium ylide. Formation of the ylide implies strong positive charge influence at C_2 in the thiazolium ion, and this is confirmed by ^{13}C–NMR spectroscopy.

In 5-methylisothiazole the methyl protons exchanged 6,500 times faster than the H–3 ring protons and only 3.8 times slower than the H–5 ring protons in 4-methylisothiazole *(WA-69)*. This relatively fast exchange may be compared to the much greater stability of benzyl over phenyl carbanions; both cases involve the same factor, that is greater $pp-\pi$ resonance. The negative charge in the carbanion derived from 5-methylisothiazole may be delocalized into the ring *via* resonance structures such as XXXI whereas the negative charge in carbanions derived from ring deprotonation is orthogonal to the π system. This also makes plausible the synthesis of thiacyanine dyes and photographic sensitizers by base-catalyzed condensation of 2-methyl-3-alkylbenzothiazolium salts with themselves, with quinolines, or with other nitrogen heterocycles, *(H-64)*. Such syntheses almost certainly involve the dimeric "anhydrobase" XXXII, *(LD-61)*, which has been shown to exist in acetone solution in equilibrium with the unstable nucleophile XXXIII *(V-68)*.

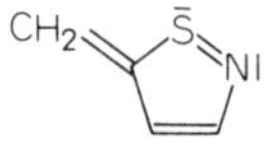

Effect of $dp-\pi$ Bonding on Chemical Properties and Reactivity

XXXI

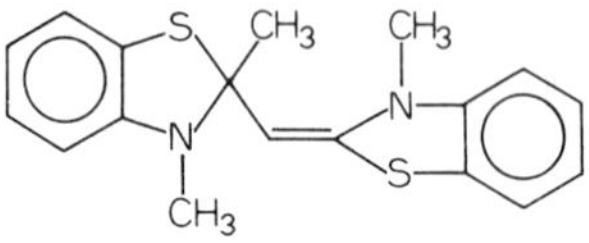

XXXII

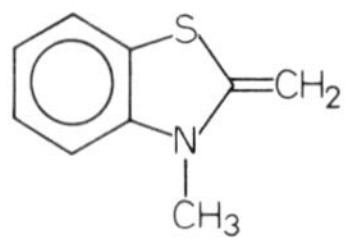

XXXIII

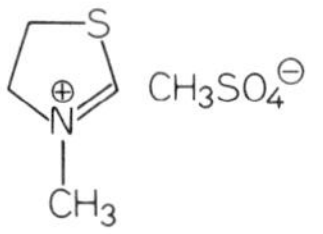

XXXIV

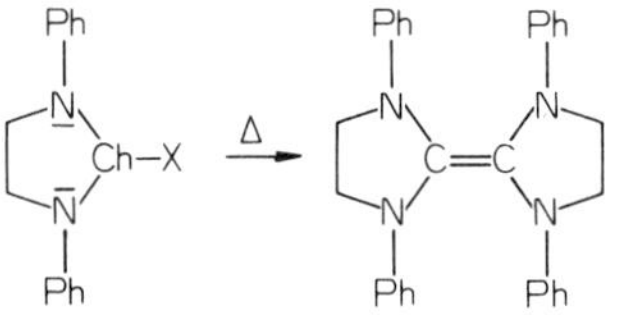

XXXV X=$-CCl_3, -OCH_3$

XXXVI

In tritiated water (T_2O) containing methosulfate anion, N-methylated dihydrothiazole, XXXIV, underwent tritium exchange at C–2 prior to methylation and ring cleavage; no exchange was noted in the corresponding acyclic cation MeS–CH=$\overset{+}{N}Me_2$ (*HLI-64*). If part of the energy necessary for ylide formation in XXXIV comes from ring strain due to the C=N bond, then the ylide must have some carbene character. Such a "nucleophilic carbene" was first proposed by *Wanzlick* (*W-62*) to explain the formation of XXXVI upon heat treatment of 1,3-diphenylimidazolidine derivatives XXXV; dimer XXXVI reacted with electrophiles such as water, oxygen or carbonyl compounds to give products containing only one imidazole ring.

Although Takamizawa (*TH&-67*) has observed what is claimed to be a carbene-type reaction, i.e. XXXVIII, in thiamine itself, which proceeds independently of the condensation reaction XXXIX, *Haake's* recent work (*NH-76*) casts doubt on the existence of a carbene contribution to the ylide. Resonance form XL was found to contribute negligibly to the thiazolium cation, which is understandable since strong electron donation by nitrogen or sulfur would convert this carbonium ion into an ammonium or sulfonium species. The carbene structure of the ylide represented by XLI seems to be an unlikely contributor on the basis of a comparison of its properties with those of the "nucleophilic carbene" derived from pyrolysis of XXXV by *Wanzlick* (*W-62*). In fact, there appears to be no reason to assume that the condensation reaction XXXVIII involves a carbene. An attractive alternative mechanism allows for direct attack of the aldehyde carbonyl on the electron-rich C_2 position of the thiazolium cation expressed by structure XLII rather than on the ylide product of prior reaction with base. Such a mechanism, XLIII, would be favored by the low basicity of triethylamine compared to sodium hydroxide.

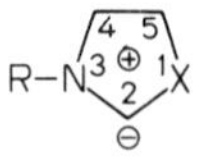

XXXVII

Whether or not the azolium ylides as represented in **XXXVII** have some carbene character, such character is not essential to their stabilization. This is proven by the occurrence of deprotonation or decarboxylation (*HBM-71*) at C_5 as well as C_2, although the carbene structure at C_5 is adjacent to only one heteroatom.

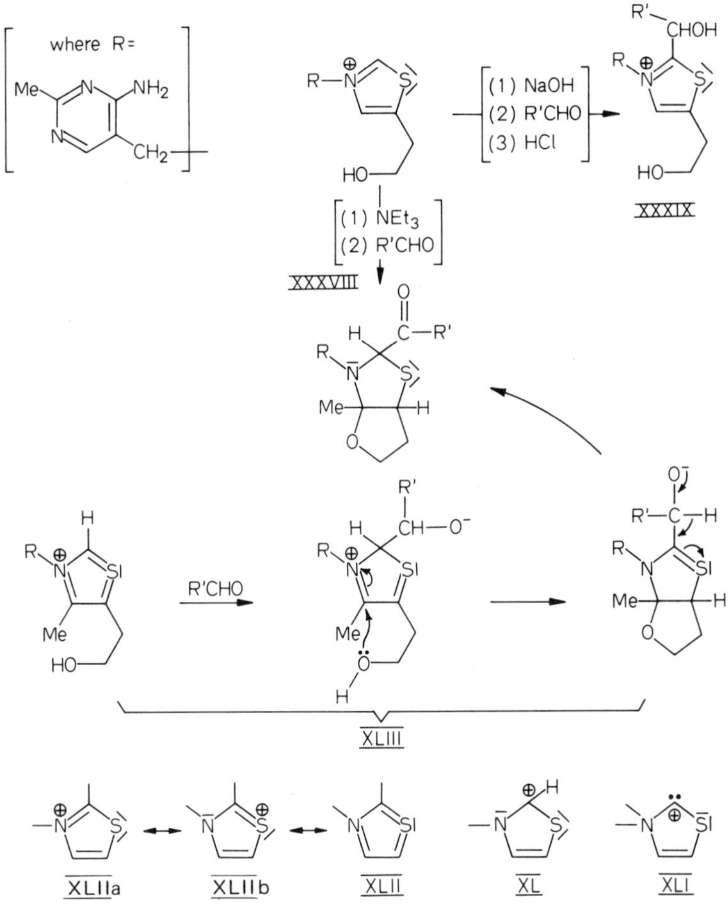

The much greater rate of formation of the C_2 ylide, reflected in the observation that C_2 loss of a carboxyl group is $> 10^6$ that of a C_5 carboxyl group (*HBM-71*), is best attributed to the greater inductive effect of nitrogen at C_2, the presence of an S–C–N, $pp-\pi$ conjugated system, and the shorter S–C_2 bond length (*KB-66*). On the other hand, the formation of ylides at C_5 in thiazolium salts (*HBM-71*) and the much greater stability of the C_2 thiazolium ylides than of the imidazolium ylides argues for some sort of d-orbital interaction between sulfur and the ylide carbon. This interaction is not necessarily $dp-\pi$ bonding and may, as *Breslow* (*B-58*) suggests, be a $d-\sigma$ overlap, since the ylide carbon is sp^2 hybridized, greatly reducing π overlap with the lone pair.

87

F. Addition to Conjugated Carbon-Carbon Double and Triple Bonds

Phosphorus-, sulfur-, or silicon-containing substituents on olefinic or acetylenic bonds affect the susceptibility of such bonds to nucleophilic and electrophilic addition as well as the direction of such addition. For instance, nucleophilic addition of e.g. amines, to acetylenic sulfides occur at the β carbon, whereas addition to acetylenic ethers occurs at the α carbon (*C-60*, *BB&-59*). Similarly, alkyllithiums add readily to vinylphosphines (*P-66*) with the alkyl group attaching to the β carbon. Additions to triphenylvinylsilane (*CB-54*) also occur readily. While phenyllithium also adds to trimethylvinylsilane (*CB-54*), the yield of adduct is only 6%, compared to 84% with triphenylvinylsilane. Aminolithiums even add to allylphosphines (*MM-72*) by a mechanism involving base-catalyzed isomerization to the corresponding vinylphosphane, but allylsilanes (*CB-54*) are inert to organolithiums.

These results could be interpreted in terms of $dp-\pi$ conjugation with the double or triple bond as suggested by NMR results (see e.g. *CPH-66*), but it is probably more profitable to consider instead the stabilization of the carbanion resulting from nucleophilic addition. Thus, in acetylenic ethers, not only is the α carbon the position of lowest electron density due to the electron-withdrawing inductive effect of oxygen, but addition at the β carbon would result in an α carbanion which is destabilized by $pp-\pi$ electron-donating conjugation with oxygen. On the other hand, stabilization of carbanions by α thio groups is well known. As discussed earlier, this stabilization has been related to the polarizability of sulfur and the long C–S bond length rather than to $dp-\pi$ bonding (*BC&-75*, *LW-76a*; see sec. IV.D.). Therefore, $dp-\pi$ bonding need not be invoked to explain nucleophilic addition β to sulfur.

Comparison between phosphanes and sulfides or amines is probably misleading since the lone pair in a phosphane seems to be incapable of $pp-\pi$ bonding (*SS-72*, sec. III.D.). For example, determination of σ_p for the diphenylphosphano group from the ionization potentials of the appropriately substituted benzoic acids (*TL&-69*) showed only weak electron-withdrawing conjugation. On the other hand both, phosphanes and α-silyl groups are apparently capable of strong $dp-\pi$ stabilization of high-energy π^* orbitals in olefinic and aromatic systems (*R-69*, sec. III.D.), as has been used to explain the $\sim 20\ nm$ bathochromic shift of the $n-\pi^*$ band in 1,2-vinylenebis-(di-n-butylphosphane) relative to monovinylphosphanes (*WP-67*, *WP-69*). Although the lone pair of carbanions is not in a π^* orbital, it is in a high-energy orbital of high p character and should be capable of significant overlap with adjacent d-orbitals.

Diisobutylaluminium hydride adds to PhC≡CCMe₃ to give a *cis* adduct with aluminium α to phenyl (*ER-75*). Similar results are obtained with other alkynes, and this is also the normal mode of addition in hydroboration (*B-62*). On the other hand, addition to PhC≡CSiMe₃ also proceeds with complete stereospecificity but gives the *trans* adduct with aluminium β to phenyl (*ER-75*). Similar but less extreme results were found in hydroboration, with the amount of addition α to the silyl group depending on its electron-donating ability, ranging from zero for $Me_3SiCH_2CH=CH_2$ (i.e. addition entirely at the terminal carbon) to 47% for vinyltrimethylsilane and 88% for vinyltrichlorosilane (*JM-72*).

Since the trimethylsilyl group is strongly electron-donating, we should expect a significant acceleration in the rates of these electrophilic additions, and this is indeed found. PhC≡CSiMe$_3$ reacts with diisobutylaluminium hydride 16 times faster than PhC≡CCMe$_3$ at 35 °C and 431 times faster than diphenylacetylene (*ER-75*). The direction of addition can be explained by the assumption that the addition involves a high-energy intermediate π complex, which decomposes to product *via* a 4-center transition state with considerable carbonium ion character. For example, in PhC≡CSiMe$_3$, carbonium ion character is preferentially formed β to the silyl groups as a result of strong electron-donating hyperconjugation by the Me$_3$SiCH$_2$-group, as observed in the red shift in the UV spectra of allyl-relative to vinylsilanes and proven by *Schiemenz* and *Pitt* (*R-69, S-71, P-70*); see sec. III.D.). Apparently, the carbon β to silicon is the point of addition of the hydridic hydrogen. Of course, a four-center transition state necessarily requires *cis* addition, but careful kinetic studies (*ER-75*) of the reaction of PhC≡CSiMe$_3$ with diisobutylaluminium hydride have shown that the initial addition is indeed *cis*, followed by rapid *cis-trans* isomerization. The isomerization seems to involve complexation of the *cis* adduct with excess diisobutylaluminium hydride; steric factors make such complexation difficult for the *trans* adduct. Such complexation implies the development of negative charge on the carbon α to silicon, which is obviously favored by $dp-\pi$ bonding but may not demand it, as indicated by *Pitt's* CNDO calculations on silylacetylenes (*P-71*). Subsequent *ab-initio* SCF calculations (*D-75*) showed the proton affinity of the silylmethyl carbanion with optimized geometry to be 50 kcal/mole less than that of the ethyl carbanion, but 75 kcal/mole greater than that of nitromethyl carbanion. In the optimum geometry the silylmethyl carbanion was pyramidal, but the SiCH angle was 112° compared to 102° for the CCH angle in ethyl carbanion. In addition, the Si–C distance bond length was 1.772 Å, considerably less than in neutral methylsilane. A similar but less extreme shortening was found in CNDO calculations on the trimethylsilylbenzene anion radical (*R-75*). While this stabilization could just as well be due to $p-\sigma^*$ as to $dp-\pi$ overlap, it is entirely possible that *d*-orbitals contribute to the low energy of the σ^* antibonding orbitals. Actually, the stabilization of carbanions by adjacent silicon may be even greater than indicated by these calculations. For instance, in Me$_2$SO/H$_2$O)/KOH at 50 °C (*EE&-75*), (Me$_3$Si)$_3$CT (T = tritium) is detritiated 5–7 times faster than Ph$_3$CT. Of course, kinetic acidity as measured here is not the same as thermodynamic equilibrium acidity, but it is still clear that α-silyl and α-trimethylsilyl groups are capable of very considerable carbanion stabilization.

Survey of Relevant Articles on the Effect of $dp-\pi$ Bonding on Chemical Properties and Reactivity (Table IV)

System and Results	Conclusions	References
A. Acidity-basicity		
$Cl_{4-n}Si(OEt)_n \cdot SnCl_4$, Dissociation vapor pressure 9.54 mm, Hg, for $n = 4$, 1.96 for $n = 3$, zero for $n = 1$.	Si–O, $dp-\pi$ bonding decreases very rapidly with increasing n. Si–Cl, $dp-\pi$ can be neglected.	(KOC-66)
pKa 5.3 for $Et_2S{=}CHCOC_6H_4NO_2$ and 9.1 for $Et_3N{=}CHCOC_6H_4NO_2$ in spite of higher electronegativity of N and shorter bond length of ammonium ylide.	$dp-\pi$ not negligible in sulfonium ylides even with only electron-donating groups on sulfur.	(PR-70)
Complexes of $(RO)_4Si$, (R = Me, Et, nPr), formed by electron donation to iodine were much weaker, as determined by the hypsochromic shift in the visible absorption spectrum of iodine, than similar complexes with EtOEt.	$dp-\pi$ bonding weakens the electron donor properties of oxygen.	(KT-75)
B. Electrophilic Aromatic Substitution		
Nitration: $(Ph\overset{\oplus}{S}Me_2)$ gives 6% para compared to 11% for $(Ph\overset{\oplus}{N}Me_3)$ and 100% for $(Ph_3\overset{\oplus}{O})$.	Sulfonium salts have $dp-\pi$ rather than $pp-\pi$ bonding.	(GDV-71)
$(Ph\overset{\oplus}{P}Me_3)$ is more reactive in nitration than $(Ph\overset{\oplus}{N}Me_3)$ but gives less para product.	Specific deactivation of para by $dp-\pi$ order (see text) bonding.	(GM&-68)
Nitration rates in aqueous sulfuric acid at 25 °C were 2.86 times slower for dimethylphenylphosphane oxide than for trimethylphenylphosphonium cation, but were 68.8 times faster for phenylphosphonic acid. Both the phosphine oxide and the phosphonic acid exist as protonated P(IV) cations under the reaction conditions.	P–O, $dp-\pi$ bonding in the protonated phosphonic acid species is so extensive as to decrease the positive charge on phosphorus sufficiently to render it less deactivating for nitration of the benzene ring. However, P–O, $dp-\pi$ bonding to one hydroxyl group (protonated phosphane oxide) is insufficient to overcome the electron-withdrawing inductive effect of oxygen.	(MPM-75)
C. Alpha-sulfonyl and -sulfinyl Carbanions $$PhSO_nCH\begin{matrix}CH_3\\n{-}C_6H_{13}\end{matrix}$$	α-sulfonyl but not α-sulfinyl carbanions are intrinsically asymmetric (see Text).	(CTS-66)
k_e/k_a (ratio of exch/rac rates) varies greatly with solvent, always > 10 for sulfone, but $1 < k_e/k_a < 3$ for sulfoxide.		
$PhCH_2SOCH_3$	See text, p. 75	(DF&-70)

System and Results	Conclusions	References
Relative rates of D exchange (24°) of two NMR distinguishable benzyl protons; range, 0.22–14.2, depending on solvent and base.		
ESR shows nitrogens equivalent, but non-equivalent in corresponding α-sulfonyl carbanion. 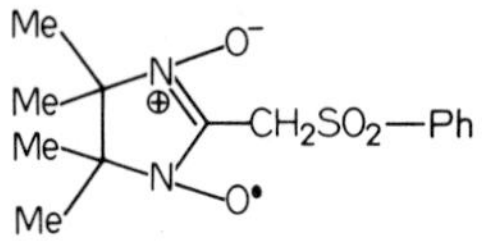 [structure: Me₂-N⊕-O⁻ / Me₂-N-O• nitroxide with –CH₂SO₂–Ph]	Pyramidal α-sulfonyl carbanion	(DU-69)
Relative exchange rates correspond to 0.76 kcal free energy difference between the two NMR-distinguishable protons, compared to 3.9 kcal by SCF–MO calculations (WRC-69). 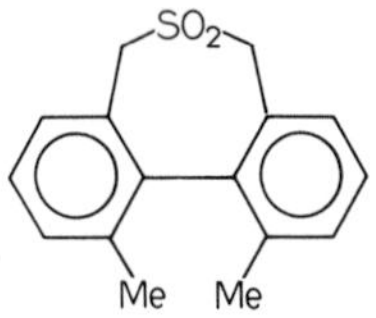 [structure: dibenzo-fused cyclic sulfone with two Me groups, –SO₂–]	MO calculations overestimate conformational effects of electrostatic repulsions.	(FS-69)
$k_1 : k_2 : k_3 : k_4 = 1 : 1100 : 310 : 1250$ 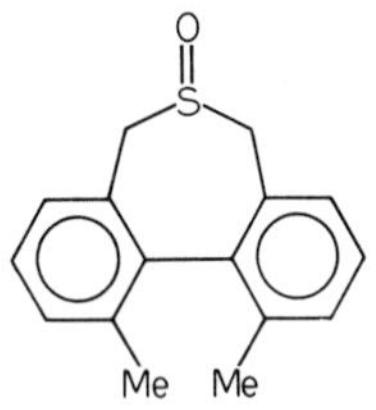 [structure: dibenzo-fused cyclic sulfoxide with two Me groups, S=O]	Disagrees with (DF&-70), (DVM-71), (HAK-69), using benzyl proton assignments of (FS-70).	(FS-69a)
$(Me_2SO_2)_2CH_2$; Rate of attack by DO^- is 3×10^8, at least 100 times less than rate of diffusion.	Slow attack argues for planar carbanion, with structure quite different from that of starting material. (Seemingly unwarranted conclusion since all carbanions, being soft bases, will form relatively slowly in a hard solvent like water.)	(HPM-70)
Heats of atomization of several vinyl sulfones differed negligibly from those calculated from bond energies.	No conjugation of sulfur d-orbitals with C=C double bonds.	(MMS-69)
$PhCH_2SO_2Me$ Benzyl protons exchanged at rates differing by a factor of 13.7 in D_2O/NaOD at 15°.	First report of asymmetric α-sulfonyl carbanion	(RB&-65)
$PhCHMeSO_2CBrMePh \rightarrow MePhC{=}CPhMe$; erythro → 93% cis; threo → 95% trans, but at a rate 27 times less than rate of bromide expulsion.	Steric hindrance between phenyl groups in cis episulfone causes its decomposition to be faster than its formation.	(BDC-70)

Effect of $dp-\pi$ Bonding on Chemical Properties and Reactivity

Systems and Results	Conclusions	References
OH group equatorial Deuterium exchange rate in $Me_2SO/D_2O/NaOD$ $k = k_{eq}/k_{ax}$. 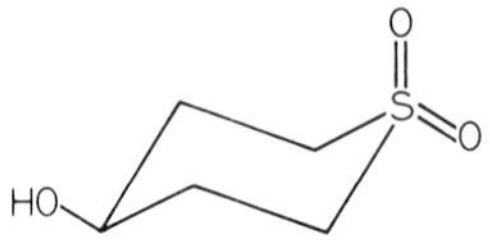 [structure: cyclohexane with HO and SO₂ groups]	Ratio too small to indicate any intrinsic conformational preference as suggested by MO calculations of (*RWC-69*). May be very different in polar solvent; See (*DF&-70*).	(*BC&-71*)
(S)–PhCH(Li)SOMe + D₂O → PhCHDSOMe $\xrightarrow{[O]}$ (S)–PhCHDSO₂Me (S)–PhCH(Li)SOMe + MeI → PhCHMeSOMe $\xrightarrow{[O]}$ (R)–PhCHMeSO₂Me (S) and (R) refer to absolute configuration of optical enantioners.	α-Sulfinyl carbanions react with CH_3I with inversion of carbanion configuration.	(*DVM-71*)
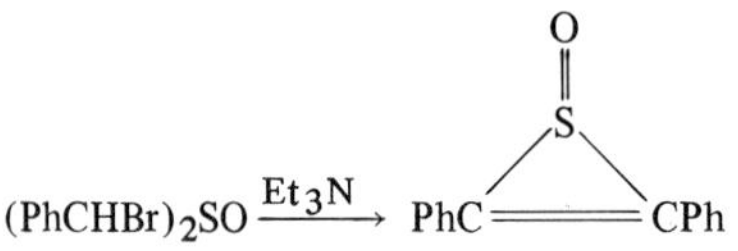 $(PhCHBr)_2SO \xrightarrow{Et_3N}$ PhC⏃CPh (with SO bridge) Product considerably more stable than corresponding sulfone; UV resembles that of diphenylcyclopropenone.	Aromatic $pp-\pi$ conjugation with lone pair in $3d$ or $4s$.	(*CC-71*)
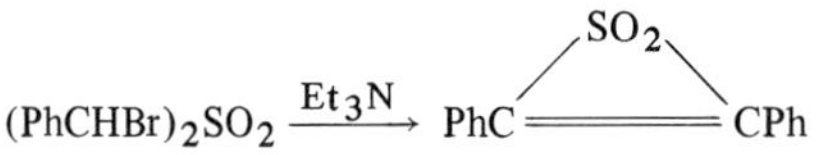 $(PhCHBr)_2SO_2 \xrightarrow{Et_3N}$ PhC⏃CPh (with SO₂ bridge) Product 1000 times more stable than stilbene episulfone; UV and NMR show little conjugation with double bond.	No $dp-\pi$ aromaticity; stability may simply be due to reluctance to form vinylcarbonium-type zwitterion on rupture of C–S bond.	(*CM&-71*)
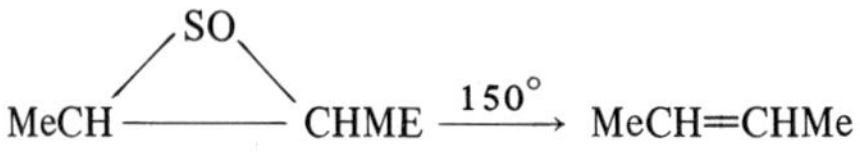 MeCH–CHME (with SO bridge) $\xrightarrow{150°}$ MeCH=CHMe cis $\xrightarrow[\Delta]{89\%}$ cis-2-butene, trans $\xrightarrow[\Delta]{58\%}$ trans	Episulfoxides decompose nonconcertedly.	(*HP-67*)
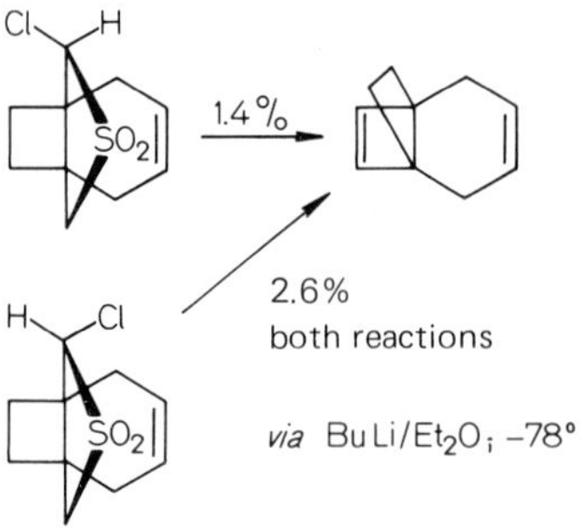 [bicyclic structure with Cl, H, SO₂] $\xrightarrow{1.4\%}$ [bicyclic alkene] [bicyclic structure with H, Cl, SO₂] 2.6% both reactions via BuLi/Et₂O; –78°	Ramberg-Bäcklund reaction is useful even in this highly strained system.	(*PH-71*)

Systems and Results	Conclusions	References

[structure: chlorocyclic sulfone] KO^{t}Bu / dry THF → [bicyclic diene] 41%	Aqueous solvation hinders carbanion inversion in this system.	(PPW-71)
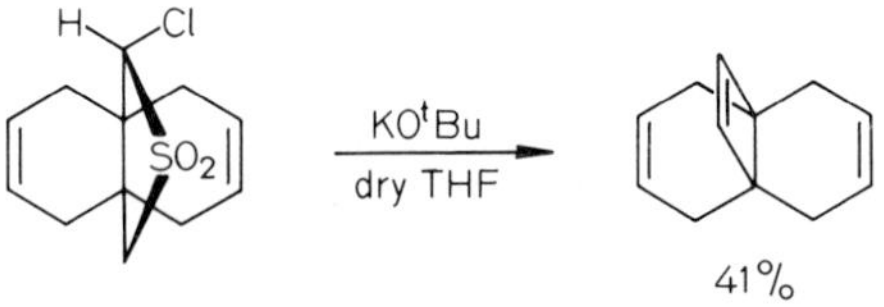 But comparable reaction does not occur with NaOH in aqueous dioxane.		
Darzens condensation of α-halo sulfones with aldehydes or ketones to give α-sulfonyl epoxides is general, but sulfones capable of Ramberg-Bäcklund reaction require nonprotonic media, e.g. methyllithium in tetrahydrofuran.	Aqueous solvation of carbonyl C hinders attack on carbanion.	(TEB-70)
MeCH—S—CHMe + BuLi → BuSLi + MeCH=CHMe *cis* → 99% *cis* *trans* → 100% *trans*.	Episulfide decomposition is concerted.	(TZ-69)
Broad, extensive review of Woodward-Hoffman rules.	Concerted disrotatory episulfone decomposition is possible if OSO plane rotates as sulfur dioxide is extruded.	(WH-69)
Heats of combustion of divinyl sulfone, several vinyl aryl sulfones, and dipropenyl sulfone, as measured calorimetrically, differed negligibly from those calculated from the sum of bond energies.	There is no stabilization of carbon-carbon double bonds α to the sulfone group and, in fact, some slight destabilization relative to isomers in which the double bond is γ to the sulfone, This suggests that dp–π bonding between vinyl and sulfone groups is negligible.	(MMS-69)
D. Alpha-thio Carbanions; A could be dilithiated 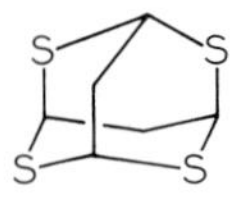 [structure A: tetrathia cage]	The existence of this non-planar carbanion implies dp–π rather than pp–π stabilization.	(BC-69)
[dithiane]=CHCH$_2$Ph ⇌ [dithiane with CH=CHPh] KOtBu	Equilibrium favors right by 2:1. Qualitative agreement with CNDO calculations on CH$_2$=CHC(SH)$_2$H (text).	(CMW-70)

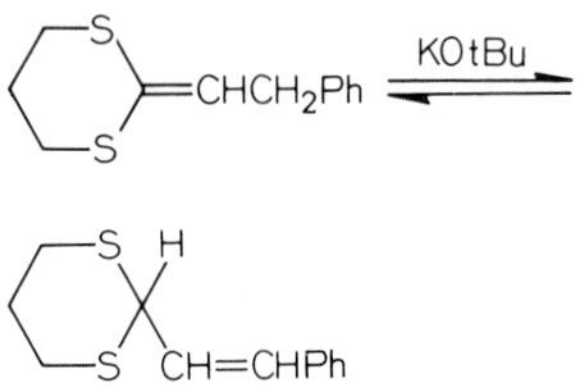

Effect of $dp-\pi$ Bonding on Chemical Properties and Reactivity

Systems and Results	Conclusions	References
$ArZ(CH_2)_3Br \rightarrow ArZCH_2-CH=CH_2$; k_0/k_S = 2.7–4.2 depending on Ar, but Hammett $\rho >$ for S than for 0.	Greater carbanionic character for S may indicate some $dp-\pi$ overlap across space, but inductive effect (for 0) is main rate-determining factor.	(YO-70)
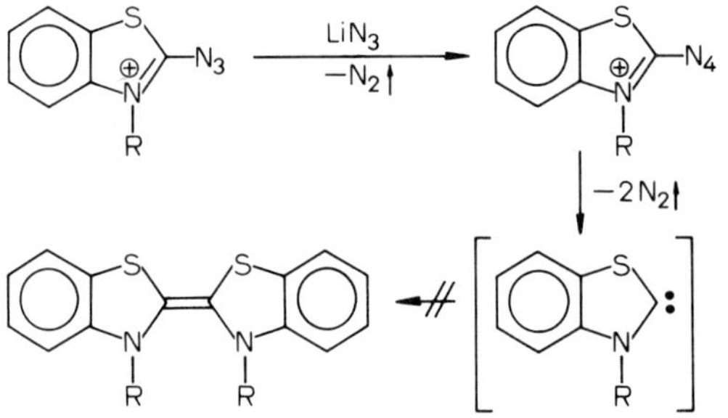 [benzothiazolium-N₃ → benzothiazolium-N₄ via LiN₃, –N₂; then –2N₂ giving dimer ⇌ carbene]	Ylide has no carbene character.	(B-64)
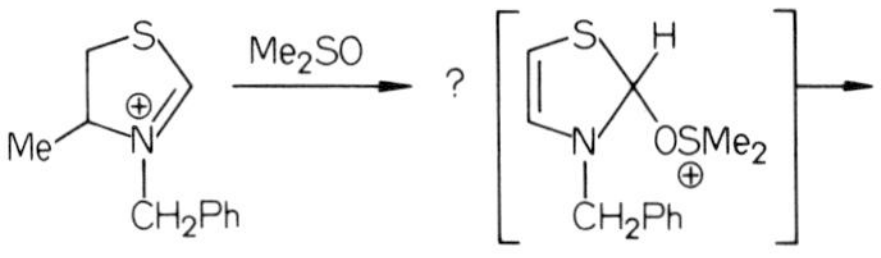 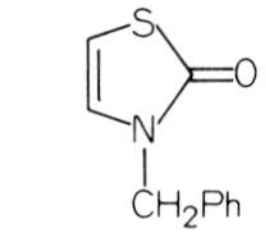 [thiazolium ylide + Me₂SO → ? → intermediate A with OSMe₂, CH₂Ph; A could not be detected]	Ylide has carbene character.	(SI-69)
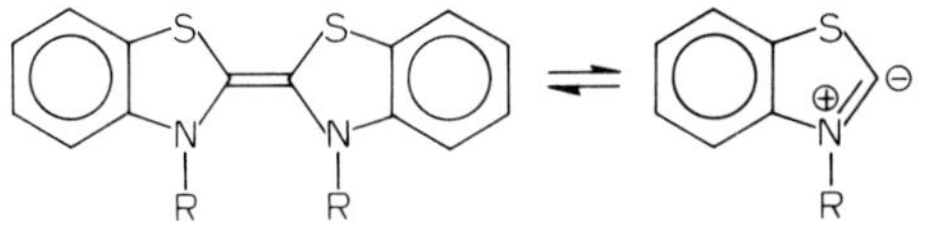 [benzothiazolium dimer ⇌ monomer ylide; Existence of monomer proven spectroscopically in acetone.]	Ylide is a carbene	(V-68a)
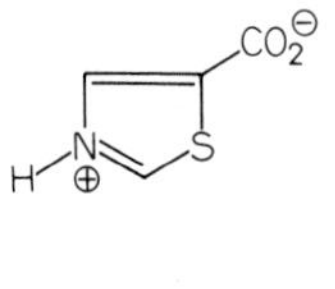 Relative decarboxylation rates at 60° $k_1 : k_2 : k_3 : k_4$ 1.0 : 1.79 × 10^6 : 0.25 : 1.58 × 10^5 [structures 1, 2, 3, 4 shown] 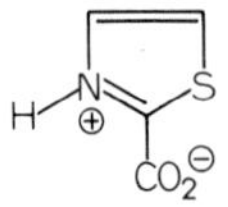 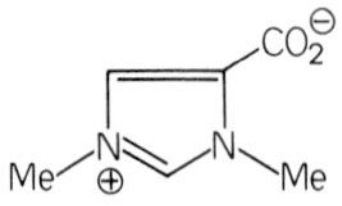 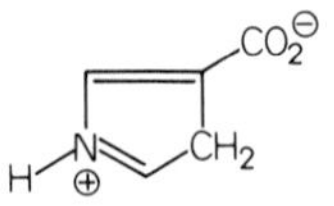	Special stabilization of thiazolium ylide by S–C, $d-\sigma$ overlap or by mixing of $3d$ to aid $pp-\pi$.	(HBM-71)

System and Results	Conclusions	References
X-ray diffraction of thiamine; S–C$_2$ = 1.68 Å, S–C$_5$ = 1.75 Å.	The shorter S–C$_2$ bond length in thiazolium ions may be partly responsible for properties; viz (*HBM-71*).	(*KB-66*)
In D$_2$O/NaOD B undergoes exchange 8.6 times faster than A. Deuterium exchange of A is 1.3 times faster than epimerization to B. 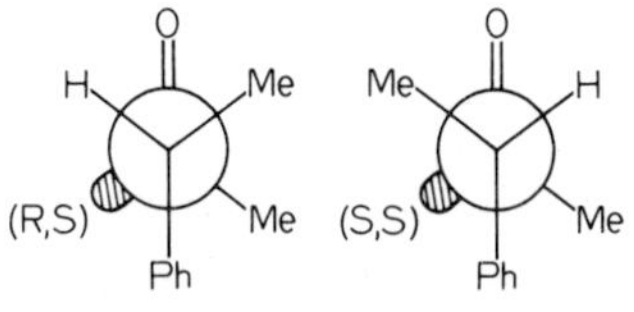 A (R,S) / B (S,S) — Newman projections with H, Me, Ph, and C(=O)Me groups	The relative exchange rates are in agreement with (*DF&-70*) and indicate that at 10 °C the carbanion derived from B is 1.2 kcal/mole more stable than that from A. However, the free energy barrier for interconversion of the two carbanions is much smaller than their energy difference, only about 0.15 kcal/mole.	(*AB-73*)
When A is reacted with D$_2$O/NaOD, the α hydrogen *trans* to the sulfinyl oxygen is preferentially exchanged; when A is reacted with MeLi and then quenched with DCl the *cis* hydrogen is preferentially exchanged. In both cases A reacts faster than benzyl methyl sulfoxide B. 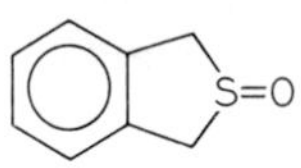 A: benzo-fused cyclic S=O; B: PhCH$_2$SOCH$_3$	If the two rings in A are planar then the conformation along C–S bonds must be fully eclipsed. Since B can assume fully *gauche* conformations it should form carbanions much more readily than A; thus, these results cannot be reconciled with the theoretical calculations of (*RWC-69*). The large solvent effect is in agreement with previous work, e.g. (*DF&-70*), but the actual stereochemistries cannot be compared due to the eclipsed conformation of A.	(*KD-73*)
Equilibrium acidities measured in dimethyl sulfoxide were greater for isopropyl sulfones than for cyclopropyl sulfones.	This indicates that carbanion stabilization by α-sulfonyl groups is at least partially due to conjugation, which increases the *p* character of the carbanionic lone pair and thus decreases the *p* character of the cyclopropyl C–C bonds, thereby increasing ring strain.	(*BV&-75*)

V. Pentacovalency

A. The Pentavalent State in Organic Chemistry

The importance of the pentavalent state to organic chemistry may readily be seen by considering the broad scope of substitution reactions at carbon, silicon, phosphorus, and related elements. In addition to being involved in the transition state of the S_N2 reaction of all these elements, substitution at silicon and phosphorus may also involve pentavalent intermediates. In view of the importance of phosphate ester hydrolyses in life processes, it is not surprising that most of our knowledge of such intermediates is concerned with phosphoranes having two or more P—O bonds. The role of such oxyphosphoranes in phosphate hydrolysis and in particular the effects of phosphorane polytopal rearrangements (vide infra) on the stereochemistry of the overall reaction have been reviewed (*R-68, R-70, W-68, UR-72*).

Phosphorane bonding was originally considered in terms of $sp^3\,d$ hybridization at phosphorus, but more recent work (*M-69a*) has centered on the formation of a 3-center-4-electron bond involving the two apical ligands in the trigonal bipyramid. However, the assumption of the trigonal bipyramidal form may be explained in terms of electron repulsions without any reference to the bonding, as proposed by *Gillespie* (*G-60*).

Recent semi-empirical and a-priori MO studies (*HHM-72, RAM-72, SV-73*) on PH_5 and PF_5 have clarified the role of d-orbitals in phosphorane bonding. For both molecules inclusion of d-orbitals in the basis set stabilized the a_1 HOMO by σ overlap, which corresponds to the 3-center apical bond, but destabilized all the other orbitals, with which they did not overlap. A literature study by *Musher* (*M-72*) showed that this destabilization is quite general for all hypervalent molecules; thus, the effect of d-orbitals on the total energy is quite small, even when dp–π bonding is taken into account. However, the effect of d-orbitals on the charge distribution is not negligible; *ab initio* calculations (*SV-73*) showed a charge transfer to phosphorus of 0.27 electron units for PH_5 and 0.55 units for PF_5 upon d-orbital inclusion. As expected, dp–π bonding was found to be more effective for the basal than the apical ligands, but the preference of π electron donating ligands for the basal position was controlled primarily by destabilizing interactions with filled MOs and stabilizing interactions with empty MOs rather than by dp–π bonding, (*HHM-72*). Thus, (*SV-73*), the charge transfer from fluorine in PF_5, 0.11 electron units, was the same for the apical and basal fluorines.

The extent of *d*-orbital contributions of course varies with the ligands. Of all possible phosphoranes, PH_3F_2 probably comes closest in bonding to the *Rundle* three-center type (*R-63a*). *Ab-initio* SCF calculations with and without *d*-orbitals in the basis set (*KK-75*) show that *d*-orbitals contribute only 40 kcal/mole to the total 170 kcal/mole energy difference between the phosphorane and the dissociation products $PH_3 + 2F$. The population of the d_{z^2} orbital was only 22% of that expected on the basis of sp^3d hybridization. The back-bonding, i.e. $dp-\pi$, contribution to the stability was only 15 kcal/mole; the total population of the d_{xz} and d_{yz} orbitals was only 0.08 compared to 0.16 in PF_5.

Extended Hückel calculations have been made (*H-75*) to explain the increase in the difference between the apical and basal P—F bond lengths as the number of basal ligands of low electronegativity is increased, i.e. in going down the series PF_5, PF_4Me, PF_3Me_2, and PF_2Me_3. Obviously, as the number of methyl groups increases, the bonding becomes more and more of the Rundle three-center type, with less and less contribution from *d*-orbitals on phosphorus, but whether this affects the σ or π bonds more is not immediately obvious. In fact, the calculations show that when *d*-orbitals are included in the basis set, the apical σ overlap actually increases with increasing number of methyl groups. However, the effect is clearly more than a simple decrease in apical $dp-\pi$ bonding since increasing methyl substitution increases the electron density on phosphorus. An increased repulsive interaction of the apical F lone pairs with a combination of the basal σ bonds having π symmetry is also to be taken into account.

Phosphoranes can be quite stable when the ligands are highly electronegative, but compounds with five P—C bonds are rare and generally unstable. In particular, pentaalkylphosphoranes should spontaneously lose hydrogen to form an ylide and thus not be isolable.

The pentaarylphosphoranes have no α hydrogen and so cannot form an ylide. Thus, pentaphenylphosphorane has been known since 1949 (*WR-49*), when it was prepared from triphenylphosphane oxide and phenyllithium. It is not particularly stable, decomposing to triphenylphosphane and phenyl radical above its melting point at 124 °C, and being rapidly hydrolyzed to triphenylphosphane oxide. The instability is probably steric in origin, the size of the benzene ring and the shortness of the P—C bonds making for a very crowded structure; the stability can be increased by incorporating the phosphorus into one or more rings. Thus, I (*WK-64*) melts above 200 °C and is stable to hydrolysis. Similarly, II (*TK-71a*) is stable at room temperature but decomposes at higher temperature to trimethylphosphane and III, the half-live at 75 °C being 108 hr.

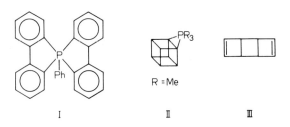

The stability of II is due to the fact that the small heterocyclic ring can be strain-free in the phosphorane but not in the corresponding phosphonium salt by spanning an apical and a basal position, giving a C–P–C angle of 90°.

Sulfuranes are generally much less stable than the corresponding phosphoranes; thus $(C_6F_5)_4S$, the only known sulfurane with four S–C bonds (*S-71a*), decomposes above 0 °C to $(C_6F_5)_2S$ and $(C_6F_5)_2$. Since the apical and basal positions are different, one would expect the ^{19}F NMR spectrum of this compound to show six signals, two for each ortho, meta, and para fluorine; instead only three signals are found, proving the apical and basal positions to be equivalent in the NMR. This equivalence is common to most pentacoordinate compounds.

Polytopal rearrangements (*UM&-70*) by which permutational isomers (those which differ only in the positions of the ligands on a framework) are interconverted, may occur by intermolecular exchange with bond breaking and reforming, or may, as is usual with the phosphoranes, occur by one of several intramolecular pathways. The intramolecular path most commonly referenced was first proposed by *Berry* (*B-60*) to explain the NMR equivalence of the five fluorines in PF_5. This process, pseudorotation, is actually vibrational rather than rotational, two apical and two basal ligands becoming equivalent by simultaneous bending vibrations, while the fifth ligand remains fixed; the transition state for this process is a tetragonal pyramid with the fixed ligand at the apex.

More recently another mechanism, turnstile rotation (TR) (*GH&-71*) has, been proposed. The Berry pseudorotation illustrated in IV is perceived as a (1 + 4) process in which one ligand remaining stationary pivots the motions of the other four. By contrast, the TR is a (2 + 3) process in which the motions of two ligands and those of the other three as respective groups are coordinated in bringing about a configurational change in the trigonal bipyramid; illustrated in V is a (14)(235) which is one of the possible processes giving the configurational change corresponding to the pseudorotation exhibited in IV. In terms of results, the two processes are equivalent in that any pseudorotation can be reproduced by any of four possible turnstile rotations. However, formation of the transition state for pseudorotation requires an increase in the basal-basal angle, whereas the transition state for turnstile rotation has a 90° basal-basal angle. This could explain the ^{19}F–NMR spectrum of VI which shows only a single peak above -90 °C, as well as the sharp ^{19}F–NMR singlet of VII which broadens only at -165 °C, thus showing an even lower barrier (≤ 5 kcal) (*RU-74*) for the configurational rearrangement involved. The structural representations in VI and VII show clearly that the TR process is best adapted to effect relatively strain-free configurational changes which are in consonance with the low activation barriers observed. The alternative (to VII for example), representation in VIII of the corresponding pseudorotational process illustrates the very considerable strain introduced in the adamantanoid moiety leading to the expectation of a much higher activation barrier than is observed. It is claimed (*GH&-71*) that rearrangements of cyclic phosphoranes, where the ring spans an apical and a basal position, are much more likely to involve a turnstile rotation than a pseudorotation.

Support for the occurrence of turnstile rotation in preference to pseudorotation is given by the much faster exchange with H_2O^{18} of IX than of X (*WK-74*); such exchange (see Sec. V.C) is known to involve phosphorane intermediates which undergo permutational isomerization when steric or electronic constraints prevent the incoming and departing ligands from being simultaneously apical. In this case, formation from IX of phosphorane XI in which the bridging carbon is apical, by attack of $^{18}OH^-$ on phosphorus, is favored by relief of strain, whereas there is no strain in X and, consequently, no incentive for phosphorane formation. However, exchange of the apical and basal ^{16}OH and ^{18}OH ligands in XI by pseudorotation or the equivalent turnstile rotation should be disfavored since it would give the strained structure XII. Therefore, one would have expected exchange to be faster for X than IX. This contradiction can be resolved by a $(TR)^2$ process (UR-72), i.e. a process equivalent to two successive turnstile rotations but without any isolable intermediate since it is merely a continuation of the same rotational process; this would result in exchange of the two hydroxy ligands in XI with no change in the positions of the other ligands. Thus, the $(TR)^2$ process can proceed with equal rapidity in both XI and the phosphorane from X giving the usual situation in which permutational isomerization is immeasurably fast in comparison to phosphorane formation or decomposition. Note that such a process is impossible with Berry pseudorotation, since two such successive pseudorotations would merely regenerate the original structure.

In general the activation energy for an intramolecular polytopal rearrangement depends much more on the ligands involved than upon the mechanism. Thus, any process which places an electropositive atom (e.g. carbon) in an apical position or causes a five-membered ring to span two basal positions will be disfavored but not necessarily prohibited. For example, the free energy barrier to pseudorotation in R_2PF_3 varies with R as follows:

R	ΔG ± or ΔE$_a$. Kcal / Mole		Reference
H	10.2	–	(GBC-74)
Cl	–	7.2	(MM-65)
Me$_2$N	19.6	–	(MD&-76)
Ph	18.7	–	(MDL-73)
Me	17.8	–	(MD&-76)

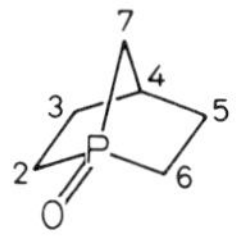

IX, 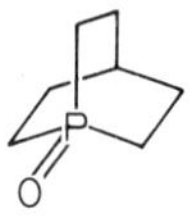 **X**, 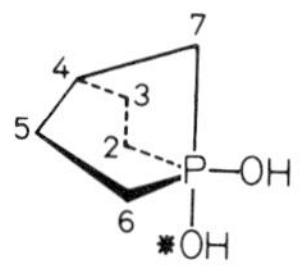 **XI**, 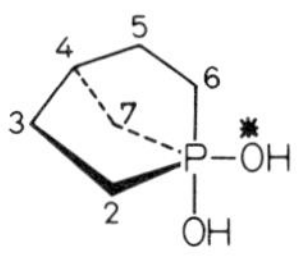 **XII**

The barriers for Cl, H and Me are as expected from the electronegativities and thus the apicophilicities of these groups. The high barrier for the dimethylamino group can be attributed to steric hindrance, the apical position being more sterically hindered, as well as the necessity for dp–π bonding which is naturally much stronger in the basal position. An earlier report (FDC-70) that the exchange in Me$_2$PF$_3$ is intermolecular has been shown to be in error resulting from the use of Pyrex NMR cells; there is no evidence for an intermolecular process in Teflon cells (MD&-76). Consequently, it is clear that electronegativity effects may raise the barrier to pseudorotation but cannot entirely prevent it.

The diastereomeric reaction products, XIII and XIV (R-68, R-70), of PhP(OMe)$_2$ and 3-benzylidene-2,4-pentanedione show spectroscopically equivalent methoxyl groups above about 0 °C and interconversion of the two diastereomers above 50 °C. The isomerization of XIII to XIV can be written as two successive pseudorotations (or turnstile rotations) through an intermediate with two apical P–C bonds, or, more likely, four pseudorotations through three intermediates with one apical P–C bond each. On the other hand, all the ligands in the XV are frozen in place below 127 °C, above which temperature decomposition into an open-chain structure occurs (R-68).

The synthesis and characterization of several oxyphosphoranes (phosphoranes with several P–O bonds) has recently been reviewed by *Ramierez* (R-68, R-70), who is responsible for a large proportion of the work in this area. Due to the shortness of the P–O bonds and the resultant crowding, only cyclic oxyphosphoranes are stable. However, acyclic oxyphosphoranes do exist. Pentaethoxyphosphorane is formed when a mixture of ethyl phosphite and diethyl peroxide is allowed to stand several days at room temperature (*DR-64*); it cannot be isolated but decomposes on standing to triethyl phosphate and diethyl ether. It is a powerful alkylating agent with respect to almost any acidic hydrogen (*DS-66*), e.g. converting diethyl malonate to diethyl ethylmalonate in 88% yield. Pentaphenoxyphosphorane is not so reactive but is rapidly converted by catechol to XVI and XVII, a clear indication of the preference for cyclic oxyphosphoranes (*RBS-68*).

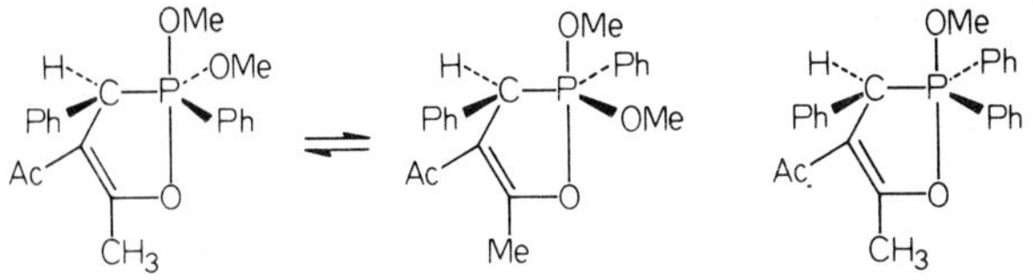

The crowdedness of acyclic oxyphosphoranes has been confirmed by X-ray diffraction of XVIII (*R-68*). The P–O alkyl bonds are notably shorter than the 1.71 Å calculated for a P–O single bond by the *Schomaker-Stevenson* rules (*SS-41*); apparently $dp-\pi$-bonding plays a considerable role in stabilizing phosphoranes. Thus, the apical P–F bond length in PF_5 is 1.58 Å (*BH-65*) compared to a Schomaker-Stevenson length of 1.65 Å. The bonds to the ring oxygens are somewhat longer since delocalization of the oxygen lone pairs through the aromatic system reduces $dp-\pi$ bonding. These short bond lengths produce rather short distances between methyl groups and nonbonded oxygens; for example the O_1-C_4 and O_3-C_5 distances are 2.63 and 2.70 Å respectively, compared to 3.4 Å for the sum of the van der Waal's radii of methyl and oxygen.

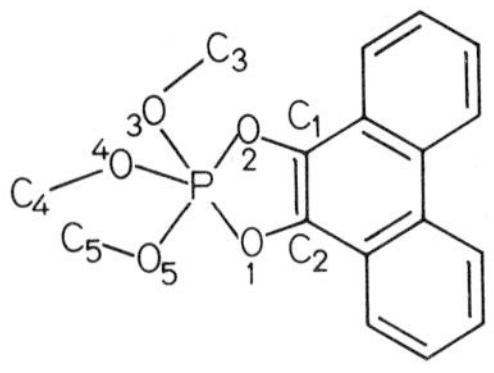

P – 0,1 = 1.753 Å
P – 0,2 = 1.641 Å
P – 0,3 = 1.649 Å
P – 0,4 = 1.601 Å
P – 0,5 = 1.586 Å

The crowding is even more extreme in pentaphenoxyphosphorane, (*SR&-76*) which assumes a nearly perfect trigonal bipyramidal configuration, with apical and basal P—O bond lengths of 1.66 and 1.60 Å, respectively. However, the C—O lengths are shorter in the apical than in the basal position, which can be attributed to reduced $dp-\pi$ bonding in the apical position. The benzene rings assume a modified propeller arrangement that minimizes steric interactions. Under these circumstances, the shortest distance between a carbon and a nonbonded oxygen is 2.34 Å. On the other hand, the distances between apical and basal oxygen atoms were as short as 2.28 Å, whereas those between two basal oxygens were 2.72–2.86 Å. This demonstrates the severe crowding that exists in the apical position, where each ligand is subject to steric interactions from three other ligands, and indicates the severe restrictions that probably apply to the polarity rule in real phosphoranes. For instance, the preference of chlorine for the apical position is tempered by the large size of this atom, a fact which also causes a countervailing preference for the basal position.

A number of spirobicyclic oxyphosphoranes have been found by X-ray diffraction to be nearly square pyramidal. In the ideal square pyramid (*H-74*), the four basal bonds forming the corners of the square should be of equal length and longer than the single apical bond; the basal-basal angle should be 88°; the apical-basal angle should be 105°; and the angle between nonadjacent basal ligands should be 150°. No compounds having exactly this ideal structure are known, but X-ray diffraction results for XIX (R = F or Me), (*WM&-74*), an analog of XIX with four sulfur atoms (*ESS-73*), and XX (*HRT-73*) have been interpreted by *Holmes* (*H-74*) in terms of modified square pyramids. Although each ring in XIX showed a long and a short P—O bond, even the long P—O bonds were shorter than expected by the Schomaker-Stevenson rules, and the OPO angle were those of a square pyramid rather than a trigonal bipyramid. *Holmes* (*H-74*) has also shown that many temperature-dependent NMR spectra previously interpreted in terms of pseudorotations or turnstile rotations can also be understood in terms of square pyramidal ground states. Thus, the occurrence of two methyl proton signals in XXI (*HW&-69*) at low temperature and one at high temperature (T_c = 37 °C) has been interpreted in terms of a fast pseudorotation at low temperature and a complex multistep process at high temperature. However, this can also be interpreted in terms of a square pyramid of fixed conformation at low temperature and a high temperature polytopal rearrangement involving a trigonal bipyramidal transition state. Similar interpretations apply to VII.

At first glance the X-ray diffraction results for XVII (*SRM-76*) are even closer to the dimensions of a square pyramid than those discussed above; the OPO angles are as expected, and the four endocyclic P—O bonds are of equal length at 1.65–1.66 Å and longer than the P—OPh bond at 1.60 Å. However, the dihedral angles between planes formed by three or more atoms show significant deviations from square pyramidal geometry, and *Ramirez* (*SRM-76*) has interpreted the structure as a 15° TR, i.e. a structure 1/4 of the way along the reaction coordinate of a trigonal bipyramid undergoing turnstile rotation. This interpretation is intriguing in view of a recent *ab-initio* MO study of pseudorotation and turnstile rotation in PH_5 (*AYC-76*). In agreement with all previous work of this kind, pseudorotation was found to be of much lower energy than turnstile rotation, a conclusion which is general for all acyclic phosphoranes. However, whereas pseudorotation was characterized as a simple concerted

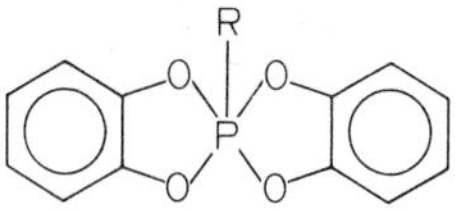

process with a 1.95 kcal/mole activation energy and a square pyramidal transition state, turnstile rotation was a two-step process. The first step had an activation energy of 10.1 kcal/mole and involved compression of one of the basal-basal angles to 90° and a 9° tilt of the apical-basal pair toward the other apical ligand. It will be noted that the resulting C_s structure while not a square pyramid, does have bond angles around phosphorus similar to those in a square pyramid. There are of course two structures of this type, both identical with and corresponding to distortions of the initial and final trigonal bipyramids. Interconversion of these two structures requires only torsional motion and is of very low activation energy, 0.3 kcal/mole. Thus, turnstile rotation should be of extraordinary rapidity in phosphoranes which are already distorted toward the C_s structure, as in XIX, and probably also in VII. *Yates* has also shown (*AYC-76*) that the transition state for turnstile rotation, i.e. the 30° TR, can be pro-

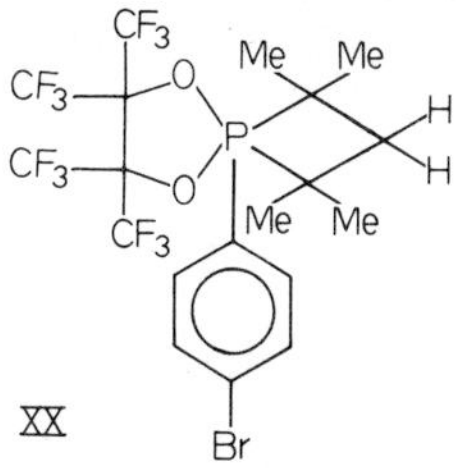

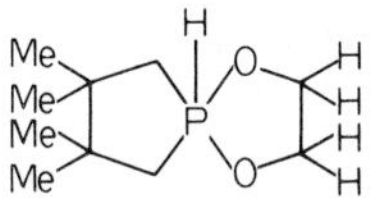

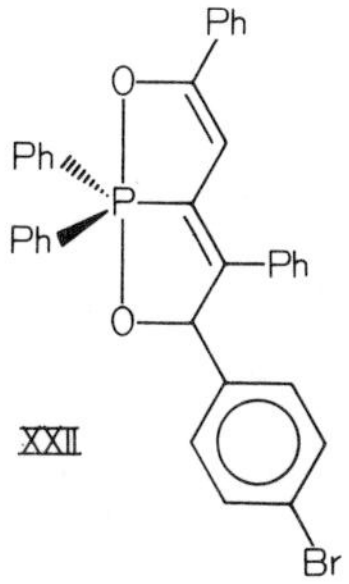

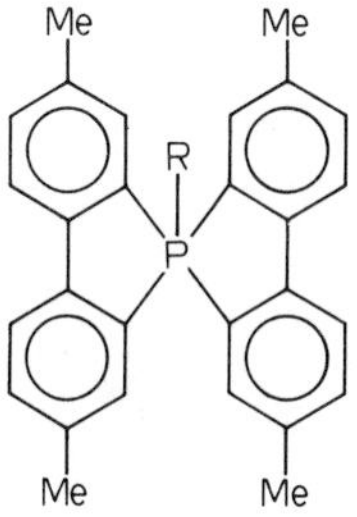

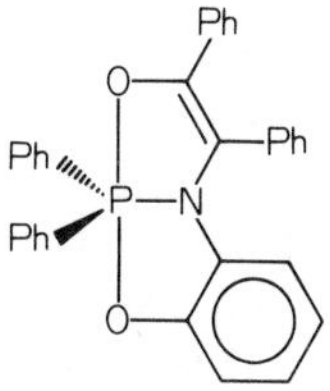

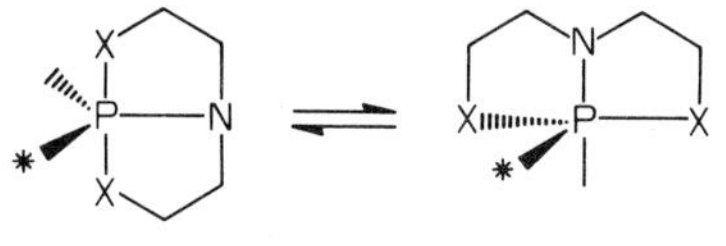

duced by simple vibrational excitations of a square pyramid; this implies that pseudorotation and turnstile rotation are indistinguishable experimentally.

It will be noted that although XVI is a trigonal bipyramid, the endocyclic P–O bonds are much shorter than the others. This implication of reduced P–O, $dp-\pi$ bonding when the oxygens are part of a small ring can explain the structures of XVII, XIX, and XX, since the resulting tendency for four long and one short bonds can only be satisfied by either a square pyramidal or a 15° TR structure, and the latter possibly minimizes steric interactions. Probably XIX and XX, previously interpreted as square pyramids, also assume the 15° TR, which in any case is nothing more than a distorted square pyramid. This explanation implies that deviations from the trigonal bipyramidal structure are limited to spirobicyclic phosphoranes with electronegative ligands. Indeed, X-ray diffraction of XXII (*SC&-71*) shows a trigonal bipyramid, and the NMR spectrum of XXIII (*WB-67*) can be interpreted only in terms of a trigonal bipyramid with pseudorotation above −60 °C (*H-74*).

The bicyclophosphoranes like XXIV can exist in two stable conformations depicted in XXV, one with nitrogen basal and the other with nitrogen apical, in pseudorotational equilibrium (* = pivot bond). X-ray diffraction of this phosphorane (*SSH-75*) shows it to be distorted towards the transition state for the pseudorotation equilibrium in XXV. The OPO angle is 171.6°, and the CPN angles are 130.6° and 116.5°, which amounts to a 33% distortion in the direction of the square pyramidal transition state for pseudorotation. The P–N bond length is 1.70 Å, exactly between the ranges for apical and basal P–N bonds.

The structure of a stable sulfurane, $Ph_2S(OC(CF_3)_2Ph)_2$ (*MA-71*), has also been determined by X-ray diffraction (*PMP-71*). This is one of a large number of stable cyclic and acyclic oxysulfuranes prepared by Martin; the subject has been recently reviewed, (*MP-76*). Here repulsions by the lone pair produce considerable distortion of the trigonal bipyramid, the apical O–S–O angle being 175° and the basal C–S–C angle 104°. By comparision, in XVIII the apical angle is 178°, and the three basal angles are 117°, 117°, and 125°. The S–O bond lengths are 1.889 and 1.916 Å, both considerably longer than the 1.69 Å Schomaker-Stevenson length. Apparently the S–O bonds are of the three-center type with little s or $dp-\pi$ contribution to the $p-\delta$ overlap; ideally this should of course result in an S–O bond order of 0.50. The difference in comparison to phosphoranes is understandable; phosphoranes unlike sulfuranes contain the third-row element in its highest oxidation state, a factor which enhances d-orbital overlap. It will be noted that $dp-\pi$ bonding is considerably stronger in sulfones than in sulfoxides.

Since the Group IV elements have no nonbonding pairs, they cannot become penta- or hexacoordinate without assuming positive or negative charge; therefore all structures in which carbon or silicon is formally pentavalent will be ionic. The notion of pentavalent carbon may seem strange, but the use of three-center-four-electron bonding permits, even if it does not encourage, the formation of hypervalent (*M-69a*) compounds by carbon or nitrogen. Actually, pentacoordinate carbon has been discussed for some time, even if not recognized as such, in the so-called "nonclassical carbonium ions" (*RM-51*). The parent pentacoordinate carbon cation, CH_5^+, has been known for some time in mass spectroscopy and is fairly stable in "magic acid", FSO_3H-SbF_5 (*OKS-69*). Experimental evidence and MO calculations indicate that the preferred conformation for this species is C_s (*O-72*), i.e. a tetrahedron in which the fourth valence is directed to the midpoint of the line connecting the fourth and fifth hydrogens. The species CCl_5^- has been observed (*MD-66*); this may however involve a linear C—Cl—Cl bond analogous to that in the polyhalide ions. X-ray diffraction analysis (*LG-76*) of tetraphenylphosphonium pentabromocarbonate, prepared by reaction of tetraphenylphosphonium bromide with tetrabromomethane in acetonitrile (*EP&-76*), showed tetrabromomethane tetrahedra bonded to bromide at the apices, probably analogous to the bonding in the Br_3^- anion. CNDO calculations (*GU-71*) indicate that the transitions states for the S_N2 reactions $CH_3OH + F^-$ and $CH_3CN + F^-$ are actually potential energy minima, making them formally intermediates. Strangely enough these calculations gave the greatest stability to conformers with hydrogens apical and the electronegative groups basal. It is suggested (*GU-71*) that this is due to the lack of $dp-\pi$ bonding, thereby hindering the transfer of negative charge from the ligands to carbon. CNDO calculations on PCl_2F_3 made without d-orbitals indicate that the most stable isomer has one chlorine apical (*GH&-71*). Also, CNDO calculations on PH_nF_{5-n} showed that inclusion of d-orbitals in the basis set increased the electron density on phosphorus by 15% (*GH&-71*).

On the basis of these calculations, which indicated very low energy barriers for pseudorotation and turnstile rotation, *Ugi* suggested (*GU-71*) that certain nucleophilic substitutions at carbon in small rings would proceed with retention of configuration. *Eschenmoser* (*TF&-70*) has shown that displacement reactions at carbon cannot proceed unless the entering and leaving groups can both achieve the apical orientation. Thus, *Ugi* would require for reaction with a steric course leading to retained configuration the intervention of a pentacoordinate carbon intermediate undergoing turnstile rotation. The reactions of *Eschenmoser*, XXVI and XXVII, apparently take place bimolecularly faster than the competing unimolecular processes of endocyclic displacement which require formation of the intramolecular trigonal bipyramid configuration, possibly with the energy expense of the turnstile rotation barrier.

XXVI

XXVII

There is, however, no verified experimental evidence at hand to support Ugi's predictions, which casts some doubt on the theory. The reaction of XXVIII with lithium bromide or sodium iodide in anhydrous acetone to give XXIX was believed (*ELU-75*) to take place with retention of configuration via the turnstile rotation pro-

105

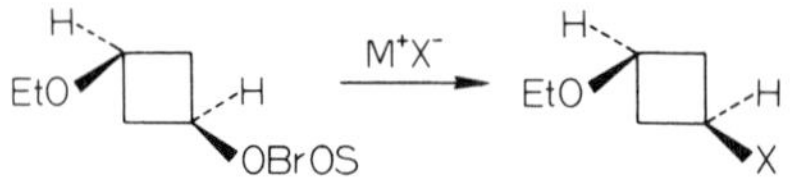

cess XXX. Later work by *Mislow* (*MOM-75*), however, has shown that the reaction actually occurs with inversion of configuration by an ordinary S_{N2} mechanism (i.e. no intermediate) followed by halide-catalyzed epimerization.

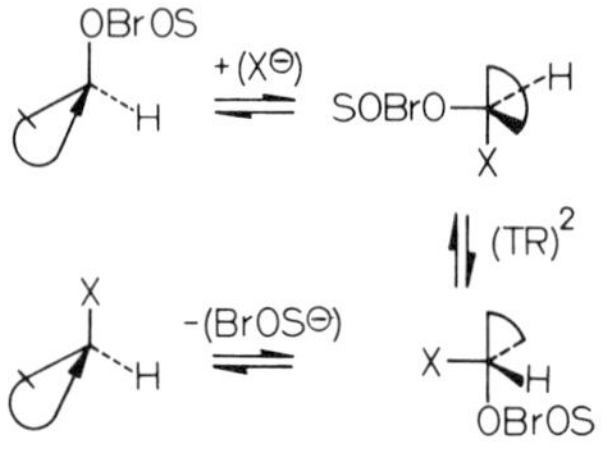

Recent X-ray structural determinations for the epimeric cyclopropyl substrates XXXI and XXXII, which react readily with tetraethylammonium acetate in acetone at room temperature (*YK&-76*), provide evidence of configuration retention in their bimolecular, nucleophilic, substitution reactions. However, extended Hückel calculations to develop a model for these S_{N2} reactions suggest that a fully-bonded, pentacoordinate carbon center is not realized here. Instead, a transition state corresponding to retention of configuration is attained by passage through a series of states evolved by the entering and leaving group motions, which, overall, constitute the pattern of turnstile rotations (*S-74, SS-76*). On the other hand XXXIII failed to react with tetraethylammonium acetate in acetone even after 48 hours at 120 °C in a sealed tube (*GC&-76*). This is in agreement with earlier work on cyclopropyl chloride and bromide (*RC-51*)

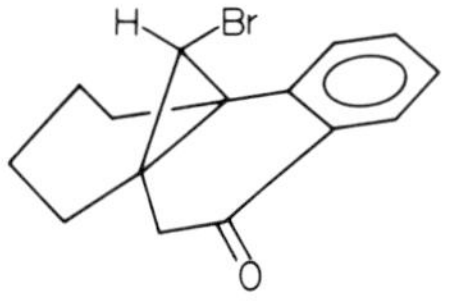

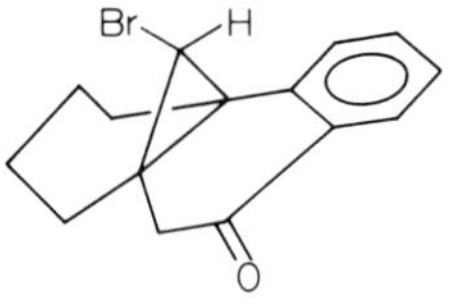

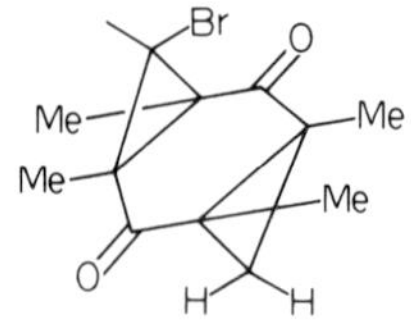

and suggests that the $S_{N}2$ reaction at carbon proceeds either with inversion of configuration or not at all. *Seebach (GC&-76)* has pointed out that the acetoxylation of XXXI and XXXII may well involve a stereospecific addition-elimination mechanism, with XXXIV and XXXV as highly strained olefinic intermediates. NMR spectra of the reaction mixture did show transient signals in the olefinic region, and XXXII does appear to acetoxylate faster than XXXI, as would be expected from the greater strain of XXXV compared to XXXIV.

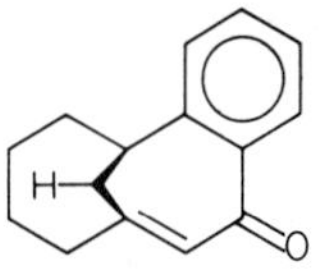

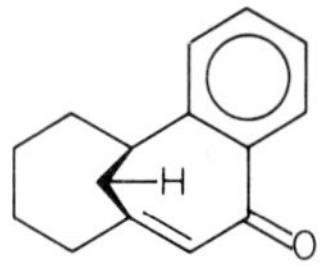

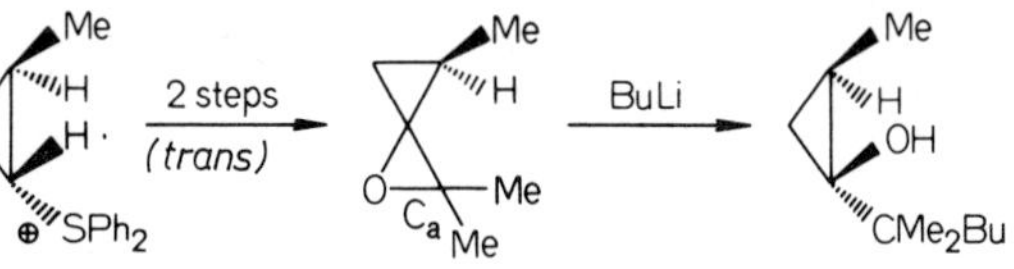

Trost (TB-73) has observed an intramolecular nucleophilic substitution at cyclopropyl carbon which appears to proceed with inversion of configuration. Reaction with acetone of the ylide generated from an 80% trans/20% cis mixture of XXXVI gave the expected spiro-epoxide XXXVII, which upon subsequent reaction with butyllithium gave a 78% cis/22% trans mixture of XXXVIII. As ring opening of the epoxide involves attack of butyl carbanion at C_a, this second step must proceed with retention of configuration at the cyclopropyl carbon. Since sulfonium ylides are known to retain the chirality of their parent sulfonium salts *(TM-75)*, the overall inversion of configuration observed implies that attack of $-O^-$ on the cyclopropyl carbon to form XXXVII proceeds with inversion of configuration. On the other hand, this is an intramolecular reaction, and it may well be that molecular constraints here enforce backside attack of $-O^-$ to give inversion of configuration even though cyclopropyl carbon could have an intrinsic preference for attack with retention as proposed by *Stohrer (SS-76)*; however it must be emphasized that there is at present no real evidence for this theory.

A number of tetravalent silicon compounds have been found to be pentacoordinate; for example, the so-called "silatranes" (*V-66*), which are tricyclic condensation products of trialkoxysilanes with trialkanolamines. The silatranes XXXIX and XL have been studied by X-ray diffraction *(TB-68, BTF-68)*; the Si—N distances are 2.19 in XXXIX and 2.34 Å in XL. For comparison the Si—N single bond length predicted by the Schomaker-Stevenson equation is 1.81 Å, and the sum of the van der Waals radii is 3.5 Å. Assuming a linear relationship between bond length and bond order, with 3.5 Å corresponding to zero bond order, these values correspond approximately to a 0.7 Si—N bond order. The configuration at silicon is close to a

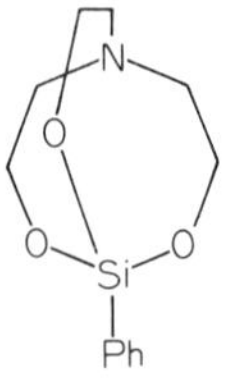

Pentacovalency

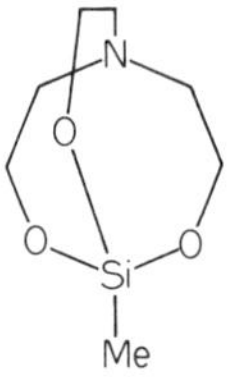

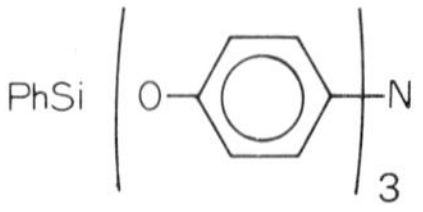

trigonal bipyramid with all three oxygens basal; in XXXIX the angles are: Ph–Si–N = 180°, Ph–Si–O = 97°, O–Si–O = 118°. Since the silatranes are formed by electron donation from nitrogen to silicon (which may or may not involve *d*-orbitals), the silicon is quite resistant to nucleophilic attack. As a consequence, EtOSi(OCH$_2$CH$_2$)$_3$N could be recovered in 66% yield after slowly boiling off the water from its aqueous solution at atmospheric pressures (*FVF-71*). Similar results are found for a phosphonium analog, HP(OCH$_2$CH$_2$)$_3$N (*CM&-76*) formed by reaction of P(OCH$_2$CH$_2$)$_3$N with a trialkyloxonium tetrafluoroborate. The fact that protonation rather than alkylation results with this strong alkylating agent is indicative of P–N coordination increasing the basicity, i.e., lone pair electron density on phosphorus. Such coordination is shown by X-ray diffraction of the phosphonium salt (*CM&-76*) which exhibits a nearly perfect trigonal bipyramidial configuration at phosphorus, with 120° OPO angles and an HPN angle of 172°. The P–N bond length is 1.986 Å; comparison with the Schomaker-Stevenson P–N bond length and the sum of the van der Waals radii results in a bond order about the same as in the analogous silatranes. The bridging phenylene groups of XL show alternating C–C bond lengths; apparently the close approach of nitrogen to silicon causes considerable distortion of the phenylene groups. This makes the existence of XL surprising, in view of the fact that a related compound, XLI (*FVF-71*), exhibited none of the characteristic properties of a silatrane.

An interesting example of pentavalent silicon is tetramethylammonium bis(-o-phenylenedioxy)phenylsiliconate, XLII prepared from catechol, a phenyltrialkoxysilane and a base (*F-64*). As expected, X-ray diffraction (*BTF-68*) shows the structure to be a trigonal bipyramid with the two heterocyclic rings spanning basal and apical positions, but this trigonal bipyramid is distorted in the direction of a tetragonal pyramid. The angle between the two apical oxygen is reduced from 180° to 167.7° and that between the two basal oxygens is increased from 120° to 127.9°. Were the structure a tetragonal pyramid these two angles would be equal and all four oxygens

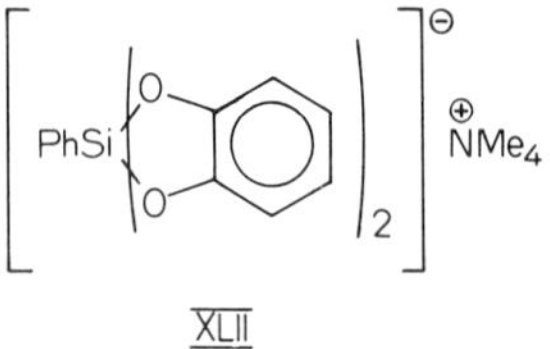

108

would be coplanar; as it is the deviation of the oxygens from this plane is about the same order as the *rms* amplitude of thermal vibration. It is to be noted that Berry pseudorotation occurs by passage through a tetragonal pyramid structure. The bond angle in the ring between basal and apical oxygens is 87.6°. The Si–O bond lengths are 1.794 Å to the apical oxygen and 1.700 Å to the basal oxygen; these rather long values are probably due to the negative charge on silicon.

The observation of pentacoordinate intermediates in solvolysis is based primarily on the occurrence of pseudorotation (or turnstile rotation); it is therefore of interest to compare the abilities of hypervalent phosphorus, sulfur, and silicon to undergo polytopal rearrangement. According to *Musher* (*M-69a*) the bonding in the covalent (basal) and hypervalent (apical) bonds of a trigonal bipyramid is completely different. It is reasonable to assume that the greater the *s* character of the apical bond, as indicated by a decrease in the length difference between the apical and basal bonds, the lesser should be the activation energy for a polytopal rearrangement which would make the apical and basal bonds equivalent. In PF_5 the ratio of the apical bond length to the difference between the basal and apical bond lengths is 37.8 (*BH-65*), indicating almost the same degree of *s* character in the apical as in the basal bonds. In the NMR all the fluorines are equivalent even at $-197\,°C$ (*M-70*). In XIII the apical and basal methoxy groups become equivalent above $0\,°C$; in a similar compound XVIII the apical-basal ratio, calculated as for PF_5, for the methoxy groups is 28.5. In SF_4 the apical-basal ratio is only 15.0 (Sec. I p. 8), indicating a major difference between the apical and basal bonds. Experimentally (*MP-59*) coalescence is observed at $-47°$ neat but at higher temperatures in dilute solution; this clearly indicates an intermolecular process, probably with an addition-elimination mechanism. The rapid exchange in $PhSF_3$ is also intermolecular (*S-62*), and one would suspect a similar explanation for the NMR spectrum of $(C_6F_5)_4S$, (*S-71a*).

Thus, the lone pair in sulfurane is apparently a major hindrance to pseudorotation. There is, however, one report of facile pseudorotation (*AM-76*) in which the ^{19}F NMR spectrum of XLIII at room temperature was best interpreted in terms of facile conversion to XLIV. The coalescence temperature for this process was $-100\,°C$, corresponding to a free energy barrier of 7.5 kcal/mole. Since the coalescence temperature was independent of concentration, the XLIII/XLIV conversion cannot be a bimolecular process involving a dimeric intermediate or transition state. Since the distinction between exo and endo trifluoromethyl groups is preserved in both compounds, the process cannot involve cleavage and reformation of the S–O bond, even though heterolytic S–O cleavage is almost certainly involved in the pyrolysis of XLIII at $250\,°C$ (*AM-76*). However, there seems to be some possibility that the process could be intermolecular with a nucleophilic catalyst, probably traces of chloride ion resulting from the preparation of XLIII from perfluoropinacol, SCl_2, and pyridine, although the lack of dependence of the NMR spectrum on solution concentration might seem to argue against this. Attack of chloride on XLIII would give a square pyramidal intermediate or transition state with equivalent S–O bonds, analogous to the structure of SF_5^- (*TCM-64*). Such nucleophilic catalysis is clearly observed in $C_6F_5SF_3$, for which the free energy barrier for permutational isomerization is 10.5 kcal in the presence of traces of HF and > 25 kcal in the presence of a scavenger for HF (*SO-72*).

On the other hand a free energy barrier of 10.3 kcal/mole has been reported (*LH-71*) for pseudorotation of SF_4 in the gas phase at 0.6 atm. Since this was determined by infrared at 228 cm^{-1}, which depends on the anharmonicity of the normal coordinate governing Berry pseudorotation, there can presumably be no confusion with the intermolecular process. If true, this would suggest that pseudorotation in phosphoranes or sulfuranes is greatly facilitated when all the ligands around the central atom are identical; this is certainly the case for XLIII. On the other hand, a rather high activation energy, 13.5 kcal/mole, has been observed for pseudorotation of XLV (*B-76*).

In $Ph_2SiF_3^-$ anion (*KM-68*) the fluorines became equivalent above about −20 °C. An intermolecular process might be expected here as well, since pseudorotation or turnstile rotation would require the phenyl groups to become apical.

XLIII ⇌ XLIV   XLV

In XLII the apical-basal ratio is 20.0, which should indicate that pentavalent silicon has a pseudorotation barrier which is intermediate between that in phosphoranes and sulfuranes. The ^{19}F NMR spectrum of $Pr_4N^+SiF_5^-$ showed a single peak at or below −60 °C, but above this temperature the peak broadened, which was attributed (*KM-68*) to the formation of bridged oligomers with hexacoordinate silicon. If such species are stable above −60 °C, there appears to be no reason why the related hexacoordinate silicon transition state for an intermolecular polytopal rearrangement could not occur below this temperature.

A recent NMR study of the spectrum of a highly pure sample of SF_4 showed no changes upon dilution with butene or addition of $Ph_3PNC_6H_4F$ as a fluoride scavenger (*KK&-75*). Comparison of the spectrum and its temperature dependence with spectra calculated on the basis of all possible intra- and intermolecular fluorine-exchange mechanisms showed good agreement for only one mechanism: Berry pseudorotation. Significant differences were observed between the spectra of pure and impure samples indicating that pseudorotation is masked by an intermolecular exchange in the presence of impurities. For the pure sample, coalescence at −47 °C (*MP-59*) and t = 0.001 sec (*KK&-75*) gave a calculated free energy of activation for pseudorotation of 10.3 kcal, exactly the same as in the gas phase (*LH-71*).

Survey of Relevant Articles on the Pentacovalent State in Organic Chemistry (Table V A)

System and Results	Conclusions	References
Phosphorane formation is 40 times faster for R=Ph than for R = ethoxy, but the product is more stable toward decomposition to pinacolone for R = ethoxy. 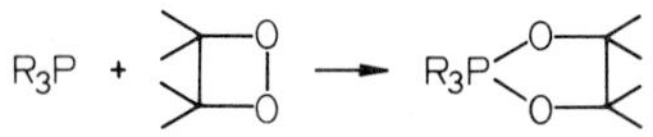 R_3P + [dioxetane] → R_3P[cyclic]	Since the ring must be apical-basal, one R must be apical; thus, the phosphorane is more stable when R is the electronegative ethoxy group. However, phosphorane formation requires nucleophilic attack of the trivalent phosphorus on the dioxetane and is thus hindered by electron withdrawing substituents.	(BB&-74)
In the NMR, the multiplet for the ring protons of A became a simple doublet at 172°; this did not occur for B and C. 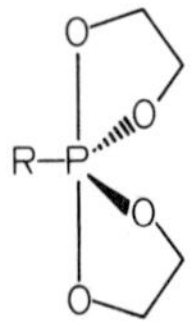 [structure] A, R = OEt; B, R = Ph; C, R = Me	At or above 172°, pseudorotation of A but not B or C may achieve a structure in which the rings are basal-basal.	(CC&-71)
Tritium content of pentaphenylphosphorane was very low even after 150 hr treatment with tritiated phenyllithium.	Exchange must occur via $LiPPh_6$, which is destabilized by the shortness of the P–C bonds making a very crowded structure.	(DP-68)
CNDO calculations on the S_N2 reaction systems $CH_3F + F^\ominus$, $CH_3OH + F^\ominus$, and $CH_3CN + F^\ominus$ indicate the pentavalent carbon structure to be an intermediate rather than a transition state, with greatest stability when the electronegative ligands occupy basal positions.	Authors predict that S_N2 could occur with retention of configuration at 3-. 4-, or 5-membered rings, the same as at phosphorus.	(GU-71)
$RLi + PCl_5 \rightarrow [R_2P+][R_3P]^- \xrightarrow{NaI} [R_2P]^+I^- + Na^+[R_3P]^-$ R 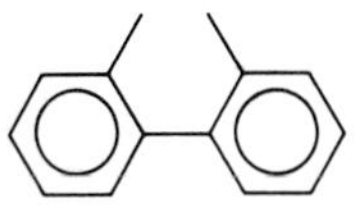 = [biphenyl structure]	Hexacoordinate phosphorus is stable when incorporated into one (or preferably more) heterocyclic rings, thus reducing crowding.	(H-65)
Treatment of optically active $LiPR_3$ (R = biphenyl) with aq HCl gave only racemic R_2P–RH; complete loss of optical activity.	Phosphorane R_2P–RH racemizes by pseudorotation; the tetragonal/pyramidal transition state for pseudorotation has a plane of symmetry.	(H-66)

Pentacovalency

System and Results	Conclusions	References

$R_3P^\ominus \xrightarrow{H^\oplus}$ [pentaphenyl phosphorane structure]

The phosphorane derived by hydrolysis of optically active (I) with aq HCl is optically active.

Here the tetragonal pyramidal transition state for pseudorotation no longer possesses a plane of symmetry. (H-66a)

[structures showing (I) with Li⁺ counterions and iodide intermediate]

$[R_2P^{32}]^\oplus [R_3P]^\ominus \xrightarrow{\Delta}_{\not\rightarrow} [R_2P]^\oplus [R_3P^{32}]^\ominus$

R = [dimethylbiphenyl structure]

X-ray diffraction
R = $(CF_3)_2CH-$; X = CF_3
apical O–O–O angle 165°, basal C–P–C angles = 113–127°, apical-basal = 75–96°, P–O = 1.71 and 1.79 Å.

The cyclic arylphosphoranes do not undergo any disproportionation equilibrium of the type $P(V) \rightleftharpoons P(IV) + P(VI)$, as does PCl_5, which exists in the solid state as $PCl_4^+ PCl_6^-$. (H-66b)

Trigonal bipyramid is distorted due to crowding of the ligands. No P–O, dp–π bonding. (HC&-71)

[phosphorane structure with RO, Ph, Ph, H, CH₃, O, X, X substituents]

Normal mode calculations based on a vibrational potential energy function give activation energies for pseudorotation of RPF_4 (R = F, Cl, Me) which increased with decreasing electronegativity of R, and thus with decreasing s character of the apical P–F bond. All calculated values were considerably higher than experimental.

Part of the deviation from experiment is due to the fact that the experimental NMR studies are in solution, whereas calculations always give the gas-phase activation energy. (HD-68)

The Pentavalent State in Organic Chemistry

System and Results	Conclusions	References
Ab initio SCF calculations on PF_5 with and without d-orbitals. Use of d-orbitals decreased the total positive charge on phosphorus by 0.8 electron units and increased basal and apical P–F overlap but stabilized the molecule by only 0.528% of the total energy.	d-Orbitals contribute to both the σ and π bonds, but the total change on adding d-orbitals is much less than that for formation of an dsp^3 hybrid.	(HVR-74)
Theoretical considerations on intramolecular rearrangements of phosphoranes show six modes, each leading to one or more of the 20 possible stereoisomers of a trigonal bipyramid with 5 different ligands. Pseudorotation and turnstile rotation are considered in the same mode since they lead to the same products by different mechanisms. Attempts to distinguish modes by NMR are discussed.	Experimental distinction between modes is difficult. Mechanisms within each mode can be determined only by MO calculations of the potential energy surface, which is of limited validity since even the mode may change with the ligand.	(M-72)
The pressure-composition isotherm for the system CCl_4 (A)/Bu_4NCl (B) shows the existence of specific compounds at mole ratio $A/B = 0.5$ and 1.0. The compound with $A/B = 0.5$ melts at 80° to a yellow liquid.	The compound with $A/B = 0.5$ is a mixed salt with equal amounts of Cl^- and CCl_5^--anions. Either a trigonal bipyramidal or a linear $[Cl_3C–Cl–Cl_2]$ structure could explain the yellow color.	(MD-66)

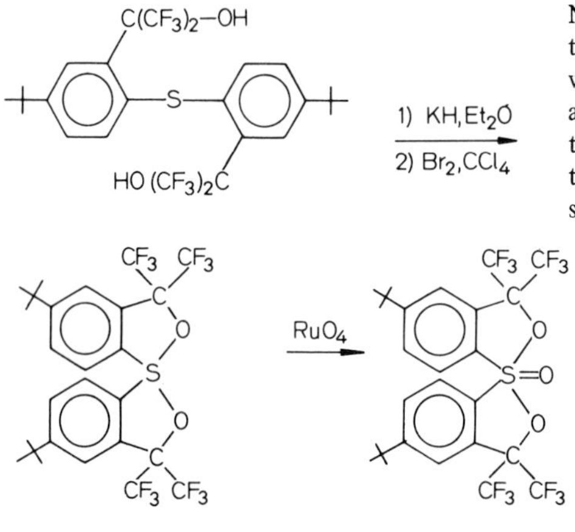

NMR consistent with expected trigonal bipyramidal structure with O (or lone pair) basal and alkoxy apical. The ring structure reduces crowding around the sulfur, thus stabilizing the sulfurane. (PM-72)

A   B

X-ray diffraction of Me_2NSiH_3 (mp 3.3°) shows a planar ten-membered ring with N at the corners of a pentagon. Si is trigonal bipyramidal with equal apical Si–N bonds of 1.95 Å. N is basic.	σ coordination at Si rather than $dp-\pi$ bonding.	(RH&-67)

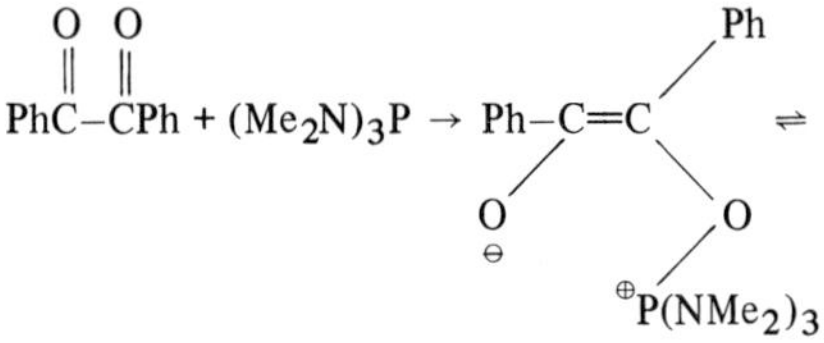

Steric hindrance of six methyl groups hinders formation of rigid cyclic phosphorane. (RP&-68)

113

System and Results 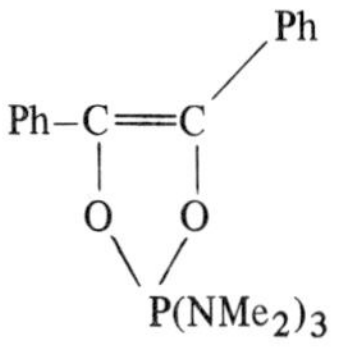	Conclusions	References

Ph–C=C(Ph) with two O's bridging to P(NMe$_2$)$_3$

Equilibrium varies with solvent; non polar solvents favor phosphorane.

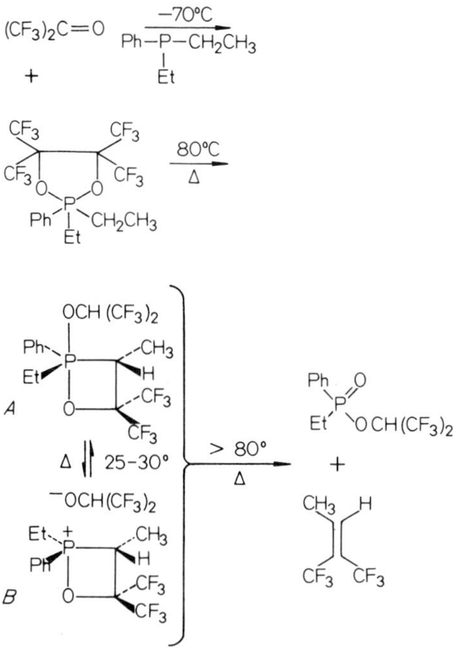

$(CF_3)_2C=O$ + Ph–P(Et)–CH$_2$CH$_3$ $\xrightarrow{-70°C}$

[4-membered ring: CF$_3$, CF$_3$, CF$_3$, CF$_3$ with O–P–O, Ph, CH$_2$CH$_3$, Et substituents] $\xrightarrow{80°C \; \Delta}$

The equilibrium between A and B is acid-catalyzed and involves bond cleavage (of the endocyclic P–O bond) and reformation. There is no combination of pseudorotations or turnstile rotations which will reproduce this equilibrium unless one allows intermediates in which the 4-membered ring is basal-basal.

(RP&-71)

A: [phosphorane with Ph, Et, OCH(CF$_3$)$_2$, CH$_3$, H, CF$_3$, CF$_3$]

$\Delta \updownarrow 25-30°$

B: [phosphorane isomer, Et, Ph, OCH(CF$_3$)$_2$, CH$_3$, H, CF$_3$, CF$_3$]

$\xrightarrow{>80° \; \Delta}$ Ph(Et)P(=O)OCH(CF$_3$)$_2$ + CH$_3$–CH(CF$_3$)$_2$

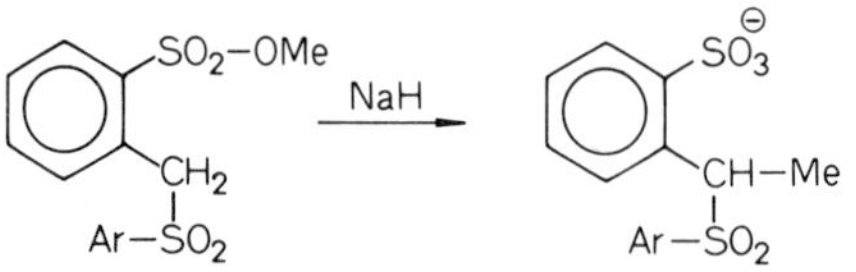

For intramolecular attack, the geometry would require either basal attack of the carbanion, basal departure of the oxygen, or a pseudorotating intermediate; all of these are apparently forbidden.

(TF&-70)

Deuterium tracer studies show this reaction to proceed intermolecularly.

Ar–SO$_2$–OMe with ortho CH$_2$–SO$_2$Ar $\xrightarrow{NaH}$ Ar–SO$_3^{\ominus}$ with ortho CH(Me)–SO$_2$Ar

The activation energy for pseudorotation in Ph$_2$P(OEt)$_3$ is about 12 kcal/mol.

The ease of pseudorotation compared to Ph$_2$PF$_3$ results from the lower electronegativity of the ethoxy group. See also (MDL-73).

(DD&-69)

In the temperature-dependent NMR spectra of MeP(OAr)$_4$ and Me$_2$P(OAr)$_3$, broadening of the methyl peak occurred at $-86°$ and $-80°$ respectively.

In general, phosphoranes with four P–O bonds undergo very rapid pseudorotation, too rapid to be observed by NMR, but in

(SKW-76)

System and Results	Conclusions	References
	MeP(OAr)$_4$ pseudorotation is slowed by steric hindrance. No such hindrance occurs in Me$_2$P(OAr)$_3$, for which the pseudorotation barrier is as expected. (See p. 100.)	
Temperature-dependent ^{31}P NMR spectra showed that free energy barriers for polytopal rearrangement were 8.8 kcal at $-85\,°$C for Me$_2$NPF$_4$ and ~ 4.2 kcal at $-177\,°$C for ClPF$_4$; the rearrangement involved simultaneous exchange of both apical fluorines with basal, and there was no detectable intermediate.	Either Berry pseudorotation or turnstile rotation would explain the results; there is no way to distinguish between them in this system. The relative pseudorotation rates for RPF$_4$ increased in the order Me$_2$N $<$ SR $<$ H $<$ Cl $<$ Me $<$ F, an order which cannot be rationalized by dp–π bonding alone and probably depends on a number of factors.	(EM&-74)
Proton NMR of A showed exchange of the 4 and 4' biphenylene methyl groups, as expected for Berry pseudorotation around Ar, with $\Delta G^* = 17.8$ kcal at 56°, but with a square pyramidal intermediate in which there is free rotation around the apical P–Ar bond. For B, exchange of the biphenylene methyl groups had $\Delta G^* = 26$ kcal at 202°, and involved a trigonal bipyramidal intermediate or transition state with Ar apical and basal-basal rings.	Permutational isomerization of cyclic organic phosphoranes is much more complex than can be simply accounted for by either Berry pseudorotation or turnstile rotation.	(WEB-74)

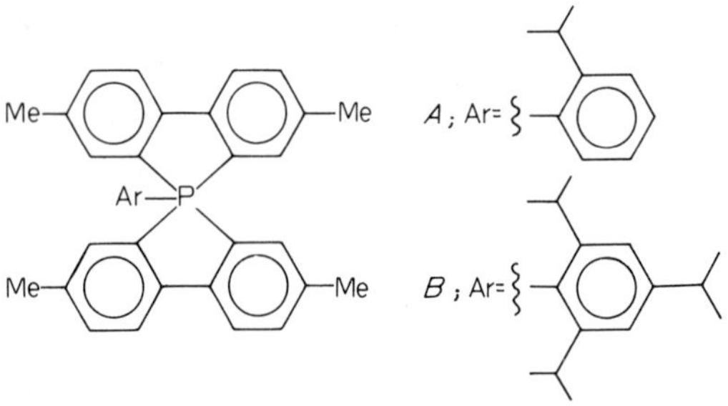

B. *dp*–π Bonding in Phosphate Hydrolysis

The pentacoordinate state is of primary interest in connection with its role as intermediate or transition state in nucleophilic substitution at phosphorus, sulfur, or silicon. More specifically, there are two points of interest in connection with nucleophilic substitution: the effect of *dp*–π bonding on reactivity, and the occurrence of pentacoordinate intermediates.

The effect of structural variations on rate must be considered in terms of the ground state and the transition state. Electron-withdrawing substituents should increase *dp*–π bonding in the ground state, thus stabilizing it and increasing the energy needed to reach the transition state; but they should also stabilize the pentacoordinate transition state. The reactivity at 68 °C of a series of phosphorus esters with PhMgBr was in the order

$$(EtO)_3PO < EtPO(OEt)_2 < Et_2POOEt \text{ and } (EtO)_3PO < PhPO(OEt)_2 < Ph_2POOEt,$$

which was also the order of increasing infrared stretching frequency of the P=O bond (*H-71*). Apparently, in this system at least, stabilization of the ground state by P=O *dp*–π bonding outweighs inductive stabilization of the transistion state, *dp*–π bonding probably being much less important in the transition state than in the ground state. This hypothesis can only be proven, however, by showing that the rate changes are due entirely to changes in the activation enthalpy, since electronic effects should not affect the entropy of activation to a significant extent.

Kinetic studies on the hydrolysis of phosphoryl and phosphonyl halides are of extreme practical importance in view of the inhibition of acetylcholinesterase by these compounds. This explains the use of several members of this class of compounds as insecticides and "nerve gases", since failure to remove acetylcholine results in rapid blocking of nerve impulses. Cholinesterase inhibition involves phosphorylation of a basic group in the esteratic site of the enzyme (*BB-66*, *EPN-67*). Regeneration of the free enzyme by reaction with OH⁻ or H_2O is a very slow process but can be speeded up by using stronger nucleophiles such as hydroxamic acids (cholinesterase reactivators).

Nucleophilic substitution at organic phosphorus halides by water or by cholinesterase reactivators has been reviewed by *Bruice* and *Benkovic* (*BB-66*). They explain the 15 fold increased rate of spontaneous (H_2O) hydrolysis of Et_2POCl at 0° in 5% aqueous acetone over that of Et(MeO)POCl and the fact that it is 860 times faster than $(MeO)_2POCl$ hydrolysis under the same conditions as being due to *dp*–π stabilization of the ground state. In this case determination of the Arrhenius parameters showed that the rate changes were due entirely to changes in the activation enthalpy.

This hypothesis has not been confirmed by newer work; in fact the variations in computed Arrhenius parameters are so great that meaningful conclusions about the mechanism are almost impossible to achieve, as may be seen from the following table. To judge from the work *Neimysheva* et al. (*NP&-67*) it would seem that substitution of methoxy for ethyl decreases the activation energy. Of course it must be remem-

Neutral hydrolysis 5% aq acetone	$k_0°$ min-l	$E_a \frac{kcal}{mole}$	log A	Ref.
Et_2POCl	1500	7.2	6.0	(HK-60)
Et_2POCl	551	7.5	5.8	(NP&-67)
MeO(Et)POCl	98	8.3	5.7	(HK-60)
MeO(Et)POCl	60	6.5	4.0	(NP&-67)
base hydrolysis	$k_{25°}$ l mole sec.			
Me\\P(O)OC$_6$H$_4$NO$_2$ /EtO	0.64	9.05	6.5	(LM&-66)
Me\\P(O)OC$_6$H$_4$NO$_2$ /EtO	0.71	10.0	7.2	(GB-66)
Me\\P(O)OC$_6$H$_4$NO$_2$ /EtO	0.04	12.0	7.4	(HK-56)
$(RtO)_2P(O)OC_6H_4NO_2$	0.011	12.2	7.0	(LM&-66)
$(EtO)_2P(O)OC_6H_4NO_2$	0.025	12.9	7.9	(GB-66)
$(EtO)_2P(O)OC_6H_4NO_2$	0.0087			(H-56)
$(EtO)_2P(=S)OC_6H_4NO_2$	0.0011	16.5	9.1	(LM&-66)

bered that often even relatively small errors in determining rate constants can lead to a large error in the Arrhenius parameters. In view of these results, published reports (*FCI-61, LSD-70*) of a correlation of rates or activation energy with the superdelocalizability, a measure of the positive charge on phosphorus or of the electron affinity of the lowest unoccupied orbital, can be viewed as coincidental.

The system $ClP(O)Et_n(OMe)_{2-n}$ was chosen to be studied (*HK-60, BB-66*) on the assumption that, since the ethyl and methoxyl groups have very similar steric factors, they should have similar entropies of activation. This may well be the case in the gas phase, but it is unlikely in solution, where the entropy of activation is primarily determined by solvation. For example, the activation energy and entropy for the neutral hydrolysis of Et_2POCl are, respectively, (*NS&-68*) 10.1 kcal/mole and −14.5 eu in isopropanol but 6.10 kcal/mole and −32.8 eu in acetonitrile. Apparently there is increased solvent interaction in the higher dielectric solvent, acetonitrile, which increases the degree of solvent striction in the transition state, making ΔS^* more negative. At the same time, however, such an increased solvation factor stabilizes the transition state, thereby lowering the activation energy; overall, this leads to a rate decrease at room temperature.

In base hydrolysis the S_{N2} transition state must be anionic. As the number of electronegative groups attached to phosphorus increases, the charge in the trigonal bipyramidal transition state becomes more evenly distributed, decreasing the degree of solvent orientation and making the entropy of activation less negative (AS-65). Clearly, any increase in activation energy due to a decrease in $dp-\pi$ bonding on going to the transition state can be compensated by an increase in pre-exponential factor, as may be seen from the following values for base hydrolysis (BP&-68). For each of the three pairs of compounds above there can be an apparent isokinetic temperature at which both members of the pair have the same rate. For $Et_2PO(OEt)$ and $EtPO(OEt)_2$ the apparent isokinetic temperature (BP&-68) is 27 °C; for the other two pairs it is well below 0 °C. Above the isokinetic temperature the increase in log A outweighs the increase in activation energy.

	$10^5 k_0°$ l/mole-sec.	$10^5 k_{80}°$ l/mole-sec.	$E_a \dfrac{kcal}{mole}$	log A
Et_2POOEt	2.39	132.	9.5	3.05
$EtPO(OEt)_2$	1.12	416.	14.0	6.35
$(EtO)_3PO$	6.18	470.	15.0	7.94

The rate decrease observed in going from $(EtO)_2P(O)OC_6H_4NO_2$ to $(EtO)_2P(S)OC_6H_4NO_2$ (LM&-66) is general, having also been observed (NS&-68) in the neutral H_2O hydrolysis of Et_2POCl and Et_2PSCl. As confirmed by the ^{31}P NMR chemical shifts and the dissociation constants of the corresponding acids (NS&-68), this occurs in spite of the fact that the electron density at phosphorus is greater for phosphoryl than for thiophosphoryl compounds, clearly indicating strong P=O, $dp-\pi$ bonding. Apparently, in this system stabilization of the transition state outweighs stabilization of the ground state, probably because the decrease in ground state $dp-\pi$ bonding is greater in going from $(RO)_3PO$ to $R_2P(O)OR$ than in going from $(RO)_3PO$ to $(RO)_3PS$. In both of these systems the rate decrease is due to an increase in activation energy; however in neutral hydrolysis (NS&-68) the increased solvation required in the less stable transition state for Et_2PSCl decreases the pre-exponential factor, whereas in base hydrolysis (LM&-66) placement of the negative charge on the more polarizable $P-S^-$ rather than $P-O^-$ decreases solvation in the transition state. On the other hand, substitution of RS for RO causes a large increase in the rate of base hydrolysis, e.g. a factor of 321 at 25 °C in going from Me(EtO)POF to Me(EtS)POF (L-57), which is due almost entirely to a change in activation enthalpy. A corresponding change is not observed in neutral hydrolysis; thus, Me(EtS)POF hydrolyzes 27 times faster than Et_2POF in base (L-57, LM&-66), but Et(MeS)POCl hydrolyzes 4.2 times slower than Et_2POCl (NP&-67) at 0 °C in 5% aqueous acetone. This is best attributed to the polarizability of sulfur, which permits the EtS group to accept negative charge in the transition state when the attacking nucleophile is an anion.

As has been mentioned, the importance of these kinetic studies lies in connection with cholinesterase inhibition. The changes in enthalpy and entropy of activation upon going from $(MeO)_2POF$ to $Et(MeO)POF$ to Et_2POF were parallel for attack by OH⁻ and by isonitrosoacetone (A-60), a known cholinesterase reactivator. For a series of para-substituted thioesters $(EtO)_2P(O)SAr$ (MH-68) good Hammett correlations were found for the hydrolysis rates and for the rate of fly brain cholinesterase inhibition in vitro; a similar plot could be drawn for the toxicity to houseflies, but here, where other factors such as diffusion to the nerve endings come into play, the scatter was considerably greater. Consequently, if $dp-\pi$ bonding effects hydrolysis rates, it must have similar effects on cholinesterase inhibition in vitro and in vivo, though these may be masked by such factors as changes in the rate of diffusion to nerve endings or in the strength of binding to the active site of the enzyme. However, solvation and polarizability factors, as well as the experimental errors appear to be far more important than $dp-\pi$-bonding.

Survey of Relevant Articles on $dp-\pi$ Bonding in Phosphate Hydrolysis (Table VB)

System and Results	Conclusions	References
Phosphorylation of ethanolamine by Sarin, $Me_2CHO(Me)POF$, was entirely at oxygen, whereas the chlorine analog was attacked by the nitrogen.	Since Cl is a good leaving group, the main factor is the nucleophilicity of the attacking species; (nitrogen is more nucleophilic than oxygen of alcohols). Cleavage of the P–F bond is slow, and here the main factor is not the leaving group facility but is the strength of the new bond, (P–O being stronger than P–N), in a purely addition-elimination mechanism.	(GW-67)
Order of reaction rates with PhMgBr at 68° $(EtO)_3PO < EtPO(OEt)_2 < Et_2PO(OEt)$ $(EtO)_3PO < PhPO(OEt)_2 < Ph_2PO(OEt)$	Stabilization of ground state by $dp-\pi$ bonding outweighs stabilization of transition state by electronegative alkoxy ligands.	(H-71)
Hydrolysis of $RPO(OEt)_2$ $R = CH_3 - \Delta H^* = 13.1$ kcal/mole; $\Delta S^* = -33.7$ eu $R = CH_2Cl - \Delta H^* = 12.0$ kcal/mole; $\Delta S^* = -31.9$ eu $R = CHCl_2 - \Delta H^* = 12.3$ kcal/mole; $\Delta S^* = -27.0$ eu	Increasing electronegativity of R stabilizes transition state but also increases P=O, $dp-\pi$ bonding in ground state, thus explaining the odd variation in ΔH^*. The steady increase in ΔS^* indicated increased stabilization of transition state, thus decreasing solvation.	(AS-65)
Logarithms of hydrolysis rates in neutral water of $RPO(OEt)_2$ and $RR'POOEt(R,R'$ = alkyl) vs sum of Taft σ^* values gave two parallel straight lines. Similar results for base hydrolysis but scatter was much greater. $(EtO)_3PO$ did not fit these plots.	Distance between the two lines is a measure of strength of EtO–P, $dp-\pi$ bonding. Deviation from linearity for phenyl- or vinylphosphinates or -phosphonates indicates some $dp-\pi$ bonding to these groups.	(BE-70)

System and Results	Conclusions	References
UV shows that base hydrolysis of MeP(O)(SEt)OPh occurs by loss of PhO−. Rate data for this compound fit LFE plot of (BE−70) for R_2POOEt.	This works only because solvolysis requirements of PhO− and ETO-leaving groups are similar. EtS−P, $dp-\pi$ bonding is negligible.	(BB&-70)
Review on nature and properties of acetylcholinesterase including its inhibition by the phosphorus "nerve gases".	The normal action of the enzyme involves acetylation of the active site by acetylcholine followed by hydrolytic cleavage of the acetyl group. R_2POX phosphorylates these same active sites, and hydrolysis of the product is very slow.	(EPN-67)
Linear relationship found for p-substituted $ArOPO(OEt)_2$ between superdelocalizability at P calculated by LCAO, MO and base hydrolysis rate or concentration needed to inhibit fly brain cholinesterase in vitro.	Since the total change in delocalizability in this system is too small, (see also LSD-70), the correlation is of dubious value.	(FCI-61)
Hammett plots for p-substituted ArO(Me)POCl vs σ^n (W-62a) values, reflecting inductive rather than resonance effects. base-catalyzed − hydrolysis: $\rho = 1$ base-catalyzed − reactivation of phosphorylated cholinesterase: $\rho = 1.56$ base-catalyzed − aging of phosphorylated cholinesterase: (ArO−P cleavage) $\rho = 1.1$.	Inductive effects of substituents on Ar are 3.6 times larger for reactivation than for hydrolysis due to the hydrophobic nature of the medium in the latter case, i.e., the parts of the enzyme surrounding the active site.	(HL-72)
Statistical analysis of new and some published kinetic data for base hydrolysis of R_2POF and $R_2P(O)OC_6H_4NO_2$ by Student's t test showed that the increase in ΔG^* on replacing R = alkyl by R = alkoxy is statistically significant.	Rate effect of $dp-\pi$ bonding in the ground state is real.	(LM&-66)
Linear relationships (found by least squares) between log k or E_a for base hydrolysis and superdelocalizability at P calculated by simple Hückel MO for RR'POF(R,R'-alkyl, alkoxy). $(MeO)_2POF$ and $(iPrO)_2POF$ did not fit these lines. Me(EtS)POF fit the lines only by assuming electron donation from P to empty d-orbitals on S in the MO calculation.	A linear relationship is expected with E_a but not log k, since these simple calculations cannot take entropic effects into account. $dp-\pi$ Bonding from P to S, as assumed, is highly unlikely and does not explain why the rate acceleration due to alkylthio occurs only in base.	(LSD-70)
Rate increase for H_2O hydrolysis in 5% aq. acetone upon going from MeClP(O)SEt to $MeClP(O)SCH_2CH_2Cl$. In the latter case E_a also increases, i.e., the rate increase is due entirely to entropy changes.	Since S is more polarizable than O it transmits electronic effects of substituents more readily.	(NEK-70)

C. Pentavalent Intermediates in Phosphate Hydrolyses

It has long been known that cyclic five-membered (but not six-membered) phosphate esters undergo base hydrolysis very much faster than their acyclic analogs (*W-68*). For example in acid or base ethylene phosphate hydrolyzes 10^8 times faster than dimethyl hydrogen phosphate, and methyl ethylene phosphate hydrolyzes 10^6 times faster than trimethyl phosphate (*W-68*). Only the rates of base hydrolysis can properly be compared since acyclic phosphates suffer acid hydrolysis by C—O cleavage (*B-70*). The rate of acceleration can be partially explained by ring strain in the cyclic five-membered esters (*SL-65, NCB-66*), but this alone does not explain the incorporation of ^{18}O into unreacted ethylene phosphate during the base hydrolysis of this diester in $H_2^{18}O$ (*HW-61*), nor does it explain the large amount of exocyclic cleavage (to give methanol and ethylene phosphate) in the hydrolysis of methyl ethylene phosphate (*CW-63*). These results suggest rapid equilibrium formation of a strain-free intermediate, almost certainly a phosphorane with the ring apical-basal, followed by decomposition of this intermediate in one of two ways. Rapid pseudorotation (or turnstile rotation) of intermediate I to II could be followed by cleavage of either apical P—O bond. This hypothesis is supported by the fact that methyl propylphostonate III hydrolyzes at the same rate as methyl ethylene phosphate but gives exclusively the product of ring opening (*DW-66a*). A stable phosphorane is easily formed from III, but pseudorotation to place the methoxy group apical would require formation of an apical P—C bond or of a basal-basal ring and is thus unlikely.

The rate of hydrolysis of methyl ethylene phosphate in acid solution increased with increasing acidity, but the yield of methanol, the product formed by pseudorotation of the intermediate phosphorane, went through a maximum at pH 2. This is attributed by *Westheimer* (*KC&-69*) to pseudorotation, which is uncatalyzed, becoming the rate-determining step. In acid solution the intermediate phosphoranes I and its pseudorotation product II, are neutral. Neither of these can undergo P—O cleavage to the product diester without prior protonation of the leaving oxygen. At pH 2 the concentration of I is not very high, and though pseudorotation is fast, protonation is relatively slow. However, in strong acid, protonation and, therefore, product formation can become faster than pseudorotation. At higher pH, increasing amounts of product are formed from the anion IV rather than through I. Since IV can form products directly without prior protonation, its lifetime is much shorter than that of I, and pseudorotation is unlikely to be observed. At pH 10, where all of the intermediate is IV, essentially all of the methyl ethylene phosphate hydrolyzes by ring cleavage.

Hydrolysis of this type is sometimes critically dependent on substituents. Thus, V hydrolyzes with O–P ring cleavage when R = Ph but with S–P ring cleavage when R = alkoxy (*GH-70*). If we assume that attack by hydroxide leads to the most stable phosphorane, i.e. VI, then pseudorotation or turnstile rotation to produce a phosphorane with apical P–S bond will also produce an apical P–O⁻ or P–R bond and thus be hindered when R = Ph. Similarly, hydrolysis of VII is faster than that of $(Me_2CH_2)_2POX$ when X = OEt or NMe_2. But, in acid solution the opposite result is obtained for X = NMe_2 or Cl. (*KH-73, CDT-70*). This is best explained by a change in mechanism. In base the reaction involves a trigonal-bipyramidal intermediate, formation of which is probably rate-determining for X = NMe_2 (*KH-73*), and decomposition for X = OEt (*CD&-73a*). Whichever step is rate-determining, relief of steric strain upon formation of a phosphorane with the 4-membered ring apical-basal should cause a rate acceleration. In acid (pH 1.8–3.1) phosphinamides hydrolyze 10^5 times faster than in base (*KH-73*) and phosphinamide VII reacts slower than its acyclic analog. Coupled with the large rate acceleration in acid of diphenylphosphinamide upon N-methylation, this suggests N-protonation, which converts the amino function from a poor to an excellent leaving group and enforces a direct $S_{N}2$ reaction with a transition state in which the four membered ring of VII is basal-basal. A similar situation probably applies to the phosphinoyl chloride since unprotonated chlorine should not have a greater electronegativity and thus a greater preference for the apical position than the ethoxy group (*KH-73*). The inability of the phosphinate ester to undergo direct reaction may be due to P–O, dp–π bonding, inherently much stronger than P–N or P–Cl, dp–π bonding, which would increase the preference of the alkoxy group for the basal position (*HHM-72*) and make it a poorer leaving group.

In this connection the NMR spectroscopic equivalence of the diastereotopic methylene hydrogens of the ethoxy groups of VIII above 30 °C has been attributed (*DWD-71*) to a concerted two-step pseudorotation or turnstile rotation involving a transition state in which the four-membered ring is basal-basal. SCF–MO calculations (*M-70a*) suggest that such a transition state would be fairly low energy when two of the exocyclic groups are strongly electronegative.

If the entering and leaving groups in a substitution reaction involving a pentacoordinate intermediate or transition state are identical or closely similar, then microscopic reversibility demands that both groups be either apical or basal, thus leading to inversion of configuration. If we assume *Westheimer's* rule (*W-68*) that entering and leaving groups must be apical (reasonable since apical bonds are weaker than basal),

V   VI   VII   VIII

then the same would apply generally. On this basis any nucleophilic substitution at phosphorus occurring with retention of configuration can be taken as proof of a phosphorane intermediate undergoing permutational isomerization, i.e. some kind of polytopal rearrangement of its bonds.

The steric course of nucleophilic displacements on tetrahedral phosphorus, which are completely predictable on the basis of the oxyphosphorane concept (*RU-74*), can be seen with the aid of the following table in which the possible results are summarized.

Table of Stereochemical Possibilities in Nucleophilic Displacements at Tetrahedral Phosphorus

Y = Nucleophile; X = Leaving Group; TR = Turnstile Rotation; BPR = Berry Pseudorotation.

$$L_3-\underset{X}{\overset{L_1}{P}}-L_2 + Y \rightleftarrows [\text{TRIGONAL BIPYRAMID}] \rightleftarrows \text{PRODUCT}$$

Mode of Entry of Y into Tetrahedron	Permutational Isomerization of Trigonal Bipyramid	Stereochemical Result at the P-center
Opposite X	None	Inversion
Opposite X	One TR or BPR	No reaction
Opposite X	$(TR)^2$ or two TR or BPR	Inversion
Opposite X	$(TR)^3$ or three TR or BPR	*Retention
Opposite L_1, L_2 or L_3	None	No reaction
Opposite L_1, L_2 or L_3	One TR or BPR	*Retention if Y can become equatorial
Opposite L_1, L_2 or L_3	$(TR)^2$ or two TR or BPR	Inversion
Opposite L_1, L_2 or L_3	$(TR)^3$ or three TR or BPR	*Retention

* Note that the $(TR)^3$-process leads to retention in both modes of entry of Y into the tetrahedron.

In view of these considerations, the conversion of *cis* IX to pure *cis* X and of *trans* IX to *trans* X for R = benzyl (*MBC-72*) is not surprising since inversion of configuration would require that the five-membered ring be basal-basal. On the other hand, when R = phenyl, hydrolysis of *cis* or *trans* IX gives the same 1 : 1 mixture of *cis* and *trans* X. This can be explained by the pseudorotation equilibrium *CE*; when R is a good leaving group (i.e. forms a fairly stable anion) the equilibrium will be driven towards XIII, resulting in retention of configuration. Hydrolysis of phospholanium salts IX also proceeds with stereomutation when R = methoxy (*M-75*). Since the ratio of *cis* to *trans* X differs in the product mixture derived from hydrolysis of *cis* and *trans* IX, and since methoxy is a far better leaving group than phenyl, this stereomutation cannot be due to the pseudorotation equilibrium. It has been attributed to competition between direct $S_{N}2$ reaction in a transition state with the five-membered ring basal-basal and formation of the phosphorane shown in *CE* with the phospholane ring apical-basal and the methoxy group basal (*M-75*). This implies that the energy released upon going from the strained phospholanium salt IX to the unstrained phosphorane in which the ring is apical-basal is similar in magnitude to the energy required to place the electronegative methoxy group in a basal position. When this electronic energy effect is greater than the ring strain energy, the reaction proceeds with inversion of configuration, probably by direct $S_{N}2$, as in the deoxygenation of X by Si_2Cl_6 (*EC&-70*), which involves the strongly nucleophilic attacking group $^{\ominus}SiCl_3$ and the excellent leaving group $-OSiCl_3$. However, even this reaction involves some stereomutation, which proves that the ring strain energy in a phospholanium system is considerable (*M-75*). Thus, deoxygenation of XII by Si_2Cl_6 proceeds with complete retention of configuration (*DZ&-69*), since the strain in the phosphetanium system is far greater than the energy preference of any possible substituent for the apical position. Nonetheless, when the ring strain energy is high and the leaving group is poor, stereomutation can also occur due to the pseudorotation equilibrium analogous to *CE*, as in the hydrolysis of IX when R = phenyl (*MBC-72*) and of XI when R = benzyl (*CCT-69, CST-69*).

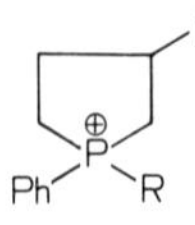

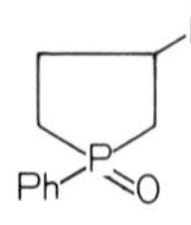

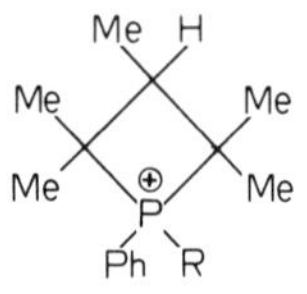

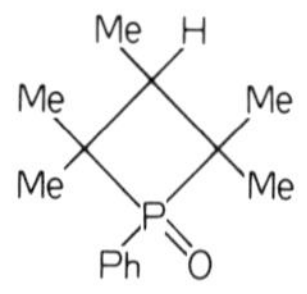

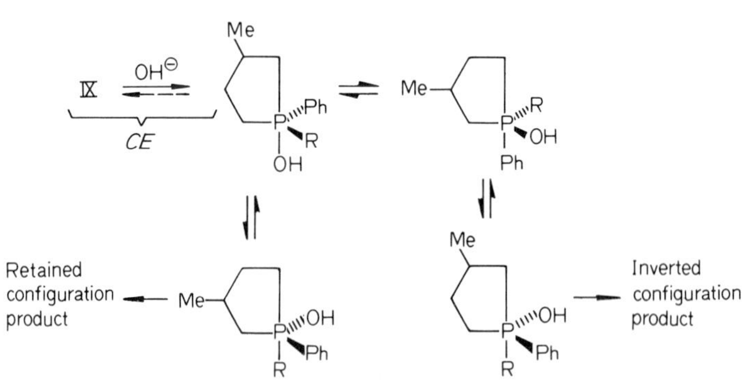

Hydrolysis of XIV occurs with complete inversion of configuration (*M-71*), proving that the seven-membered ring is the smallest which can assume the basal-basal configuration without strain. Hydrolysis of XV (*MC-70*) leads to mixtures of *cis* and *trans* XVI, the composition of the mixture depending on whether the 4-methyl group in XV was *cis* or *trans* to the leaving benzyl group. This is best explained by simultaneous formation of two phosphoranes, one with the ring basal-basal, leading to inversion of configuration, and the other with the ring apical-basal, leading *via* pseudorotation to the product of retained configuration. The former phosphorane, XVII, suffers methyl-benzyl repulsions when these two groups are *cis*, leading as observed to increased retention in the hydrolysis of *cis* XV. Note that the six-membered ring is completely unstrained when the phosphorus is tetrahedral; apparently, formation of both phosphoranes must involve an increase in ring strain, and both phosphoranes, once formed, should decompose rapidly to products.

To understand the hydrolytic stereomutation of XI but not IX when R = benzyl, it is necessary to look at the kinetics. Hydrolysis of cyclic or acyclic phosphonium salts is third-order, second-order in hydroxide ion. These circumstances are generally explained by a four-step mechanism (*MA&-65*) involving:

1) rapid equilibrium addition of hydroxide to give a hydroxyphosphorane,
2) rapid equilibrium deprotonation by a second hydroxide to give the anionic phosphorane R_4PO^-,
3) rate-determining cleavage of an R–P bond in R_4PO^-, and
4) rapid protonation of the resulting carbanion R^-.

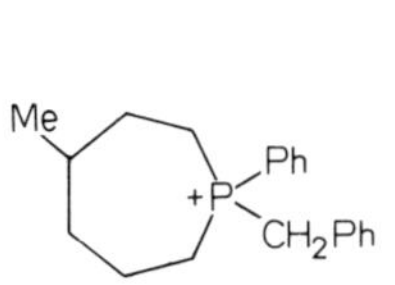

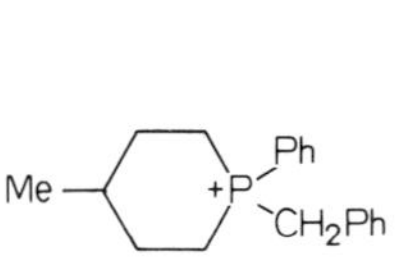

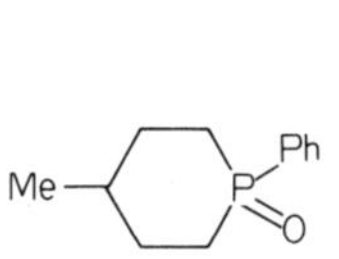

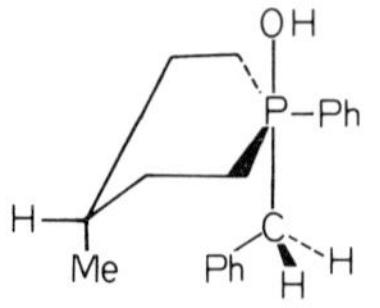

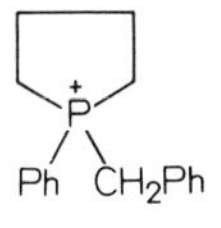

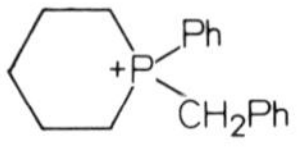

The overall rate constant of this process can be expressed as the product $K_1 K_2 k_3$. Hydrolysis of acyclic phosphonium salts generally occurs with inversion of configuration, suggesting that steps 2) and 3) are concerted or that R_4PO^- is short-lived enough to prevent a pseudorotation step which places the $-O^-$ group in a basal position. Although hydrolysis of XI is 650 times faster than that of XVIII at 25 °C (*CTW-71*), the activation energies are almost identical; the entropies of activation are + 11.0 eu for XI and – 1.6 eu for XVIII. If carbanion formation is fairly well developed in the transition state for a rate determining step 3), then the entropy of activation should be negative due to the high solvation required by carbanions. As a result the transition state for XI must closely resemble the phosphorane, with only a slight weakening of the P-benzyl bond. In all likelihood, the fast rate for XI is due primarily to the increases in K_1 and K_2. At the other extreme, the entropy of activation for hydrolysis of XIX is – 26.4 eu, indicating a nearly fully developed carbanion with the phosphorus tetrahedral in the transition state. This would explain the generally low activation energies, 14 kcal/mole for XVIII and 18 kcal/mole for XI. Evidently the transition state assumes whatever geometry minimizes ring strain.

Pseudorotation of oxyphosphorane intermediates may also be important in certain acyclic systems. When R = $-OCH_3$ hydrolysis of XX gives a 98% yield of acetoin and dimethyl phosphate and a 3% yield of methanol and XXI (*FU-67*). This is explained by the decomposition reactions 1 and 2 of pseudorotating oxyphosphorane intermediates. When R = CH_3, a very significant change in product distribution occurs wherein path 1, leading to methanol, accounts for 95% of the product. This is indicative of a 200-fold decrease in the rate of reaction 2. The inhibition of pseudorotation in the oxyphosphorane intermediate can only be explained on the assumption that the phosphoryl oxygen is constrained to occupy a basal position. This is not unexpected on theoretical grounds; the greater extent of $dp-\pi$ bonding in the phosphoryl group should lessen the electron density on this oxygen atom while decreasing its electronegativity relative to an alkoxy oxygen atom.

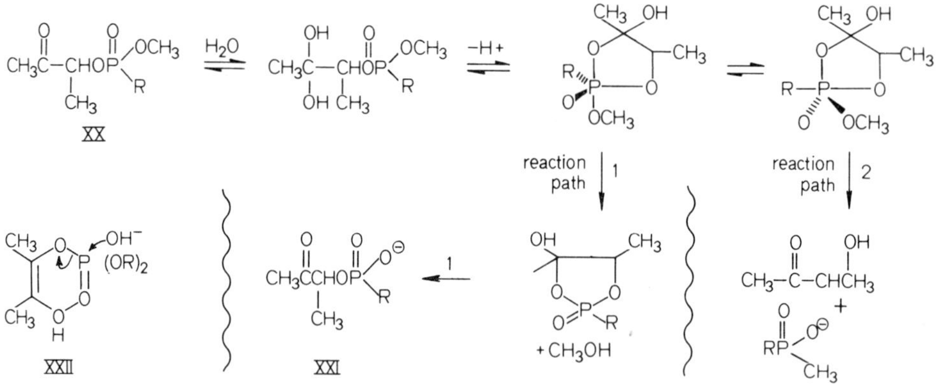

More recent work has established the preference of the phosphoryl oxygen for the basal position in the trigonal bipyramid intermediate on a quantitative basis. The CNDO calculations by *Gillespie* (*GH&-71*) on $^-OP(OH)_3Me$ with d-orbitals explicitly included showed an energy increase of 22 kcal/mole upon making Me apical vs. 32 kcal/mole for the apical oxyanion center (O^-). Experimentally (*HA-70*), in the hydrolysis of diphenylbenzylphosphane oxide in alkaline H_2O^{18} the recovered starting material contained no O^{18}, proving that pseudorotation is slow compared to the lifetime of the intermediate. This is understandable since exchange here would require an intermediate in which either the phosphoryl oxygen or both phenyl groups are apical. Interestingly enough, ab-initio MO calculations on H_4PO^- (*DA-76*) without d-orbitals showed the preference of the oxyanion center for the basal position to be 15.5 kcal; the major factor was hyperconjugation between O^- and an antibonding π MO on P, which is favored by the decreased symmetry of the species with O^- basal.

Apparently $dp-\pi$ bonding and ligand-ligand repulsion also contributed to the stability of the phosphorane with O^- basal but to a much lesser extent.

The recent work of *Brown* and *Frearson* (*BF-68*) lends further support to the hypothesis of pseudorotating oxyphosphorane intermediates in the hydrolysis of phosphoacetoin triesters. Second-order rate constants in 1 mol^{-1} sec^{-1} were as follows for the hydroxide catalyzed hydrolysis:

Phosphoacetoin R'OPO(OR)$_2$	Formation of (RO)$_2$PO$_2^-$	Formation of R'OPO(OR)–O$^-$	% Phosphorane Formation
MeCOCHMeOPO(OMe)$_2$	387	9	97
MeCOCHMeOPO(OEt)$_2$	64	–	–
MeCOCHMeOPO(OCHMe$_2$)$_2$	2.2	–	–
MeCOCHMeOPO(OPh)$_2$	–	820	–
MeCOCMe$_2$OPO(OMe)$_2$	60	14	77

All of these compounds hydrolyze 10^5–10^6 times faster than trimethyl phosphate. These rates are in the accepted order of RO$^-$ leaving group ability. Evidently the phenolate group is so electronegative that it is uncomfortable in the basal position, thus leading entirely to hydrolysis by path 1. The rapid hydrolysis of MeCOCMe$_2$OPO(OMe)$_2$ eliminates an enolic transition state such as XXII.

Since base hydrolysis of phosphonium salts is known to involve intermediates, pseudorotation could be important here even if cyclic phosphoranes are not involved. This is best illustrated by the hydrolysis of optically active dialkoxyphosphonium salts XXIII from which optically active phosphinate esters were formed as follows (*DP-72*); where (*R*) and (*S*), as usual, denote configuration.

A	B	(*R*) XXIV	(*R*) XXV	(*S*) XXIV	(*S*) XXV
methyl	ethyl	1	50	37	12 %
ethyl	menthyl	2	20	38	40 %

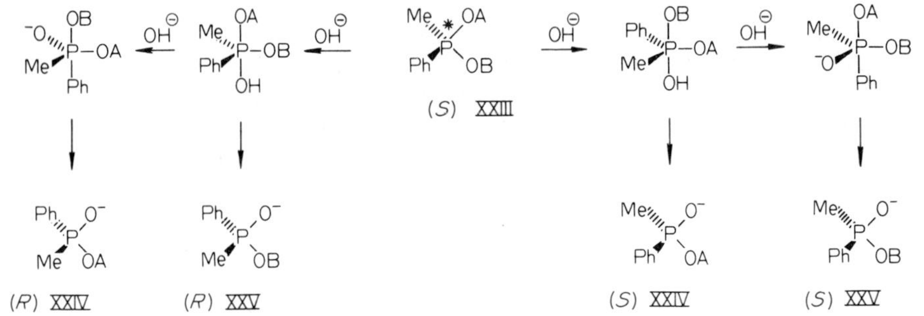

The ratio (*R*-XXIV + *R*-XXV)/(*S*-XXIV + *S*-XXV) is a measure of the preference for hydroxide attack opposite or adjacent to B; it will be seen that there is no real preference when A and B have the same steric factors (methyl and ethyl), but that attack adjacent to the bulky menthyl group is greatly disfavored. The ratios R-XXIV/R-XXV and S-XXIV/S-XXV measure the preference for direct decomposition of the intermediate phosphorane over pseudorotation. As might be expected, pseudorotation appears to be generally much less important than direct decomposition, since it would require the phenyl to become apical, the only exception being the phosphorane with apical menthoxy. This indicates that the preference for pseudorotation is determined mainly by the electronic factors which control the relative leaving group abilities of OA and OB, steric factors being generally unimportant.

Another example of acyclic pseudorotating phosphoranes acting as solvolytic intermediates involves the reaction with base of certain β-substituted triphenylvinylphosphonium salts, $Ph_3PCH=CHZ$ (*CCS-76*). Again it will be noted that the starting material is a phosphonium salt. This is not surprising since only with phosphonium salts do the kinetics (second order in hydroxide ion) actually require the presence of a phosphorane intermediate (with or without pseudorotation), which would otherwise not form in the absence of special stabilizing factors such as phosphorus being part of a strained ring system. In the above system, hydrolysis in NaOH/75% aqueous ethanol proceeds by loss of $CH_2=CHZ$ regardless of the nature of Z, but in benzene/-H_2O/NaOH suspension it proceeds by this path when Z = OPh, by loss of benzene when Z = OMe, and by phenyl migration to give $Ph_2P(O)CHPhCH_2Z$ when Z = SEt or SPh. These results can be explained by the tendency of the substituted vinyl group, which is more electronegative than phenyl, to assume the apical position in the intermediate phosphorane. However, when Z is an ether group, the carbanion formed by phosphorane decomposition, which proceeds simultaneously with deprotonation of the apical hydroxyl group, is destabilized by $pp-\pi$ electron donation from oxygen. Consequently, there is competition between direct decomposition of XXVI and pseudorotation to XXVII. This would, of course, expel a phenyl carbanion, whereas direct decomposition of the very unstable XXVI is favored by Z = OPh, a group which is marginally less electron donating than Z = OMe. When Z is SEt or SPh, the intermediate carbanion is stabilized by mechanisms discussed earlier (see Sect. IV D). This brings about deprotonation at the α carbon to form an ylide prior to hydroxide attack at phosphorus to form phosphorane XXVIII, which is ideally suited for phenyl migration. Ylide formation has also been demonstrated by the formation of $CD_2=CHZ$ in $NaOD/D_2O/PhH$ when Z = OPh. Clearly, in this case the ylide is unstable and reacts with the protic solvent medium before it can be transformed into phosphorane XXVII.

It must be remembered that phosphorane formation during phosphate ester hydrolysis has still not been proven to be general. For instance, although the rate of hydrolysis of XXIX in strong base to form the cyclic ester XXX is considerably less than the rate expected for deprotonation of the α OH group (*URO-70*), thus proving the existence of an intermediate, the rate at pH 8 is much less than that expected for protonation of XXXI, suggesting that the intermediate is most probably XXXI rather than phosphorane XXXII. Nucleophilic attack by the α–O^- of XXXI should be relatively slow due to the negative charge on the phosphate group, whereas loss of the phenoxide ion from XXXII should be quite fast.

It is well known that phosphates play a major role in biochemistry, and adenosine triphosphate (ATP) is one of the most important phosphates, deeply implicated in many energy-consuming reactions including acetylation of coenzyme A, an essential factor in sugar metabolism and protein synthesis. These reactions involve conversion of ATP with acetic acid or amino acids with loss of pyrophosphate to form adenyl carboxylate, which in turn undergoes a reaction with the appropriate enzyme resulting in transfer of the acyl group to coenzyme A, or in protein synthesis, to an RNA species. Buck (VB-74) has proposed that the intermediate adenyl carboxylate/enzyme complex is a phosphorane on the basis of the rapid esterification of acetic or trimethylbenzoic acid or O-alkylation of phenol by XXXV in hexane at room temperature. The same results were obtained with XXXVI but at a much slower rate, and the reaction of optically active XXXVII with acetic acid occurred with inversion of configuration at the chiral center of the 2-octyl substituent. This suggests a mechanism involving protonation at the ring oxygen, the basicity of which is greatly decreased by the acyl group in XXXVI, followed by simultaneous cleavage of the bond between phosphorus and the ring oxygen and displacement at the carbon of the

Pentacovalency

[Structures XXXV, XXXVI, XXXVII shown across top row]

XXXV: cyclic phosphorane with Me on ring, MeO, MeO, OMe substituents

XXXVI: similar cyclic phosphorane with -C(=O)-Me, H, Ph, OMe substituents

XXXVII: $(CH_3(CH_2)_5-CH-O)P$ cyclic structure with Me, COMe, H, Ph substituents

[Second row: structure with Nu:→CH₃ attacking, and product XXXVIII]

$(MeO)_2-P(=O)-CH_2-CH_2-C(=O)-CH_3$

XXXVIII

apical group, resulting in C–O cleavage and formation of a phosphoryl bond, as shown in XXXVIII.

The enzymic hydrolytic depolymerization of ribonucleic acid XXXIII is known to involve intermediate formation of a cyclic phosphate similar to XXX. It has been proposed (U-69) that this process involves a pseudorotating phosphorane intermediate, primarily because all previously suggested mechanisms required one histidine group at the active site of the enzyme to act first as a base toward the α OH, then as an acid toward the leaving group. This would require the entering and leaving groups to be adjacent to each other, which would mean that one or the other must be basal, or, if we assume the validity of Westheimer's rules, that pseudorotation occurs to switch the leaving group from basal to apical.

However, the availability of the nucleotide analog XXXIV (E-68) (E-70), which exists in two stereoisomeric forms, one crystalline with the sulfur endo to the bicyclic ring (SE-70), as shown, and the other noncrystalline with the sulfur exo, has enabled a definitive test of the mechanism to be made. This test was predicated on the results of enzymic reaction of XXXIV with two reagents;

1) reaction with cytidine (UEE-72) to generate a dinucleotide by a path which is the microsopic reverse of cyclic phosphate formation from ribonucleic acid, or
2) with H_2O^{18} (UR-70) to generate the monoester (the second step of ribonucleic acid hydrolysis).

In both cases reaction was followed by a second step of regeneration of XXXIV via a nonenzymic reaction known to occur by direct displacement. If the enzymic reaction involves an "adjacent" mechanism with pseudorotation, then this two-step process should switch the sulfur in XXXIV from *endo* to *exo* or *vice-versa*. This result was not observed, reaction of *endo* XXXIV with cytidine giving only *endo* XXXIV (UEE-72). Of course, free rotation around the CO–P bond in the monoester, impossible in the dinucleotide, means that reaction with $H_2^{18}O$ would be nonstereospecific; however, reaction of *endo* XXXIV gave *endo* XXXIV with no ^{18}O and *exo* XXXIV with complete retention of ^{18}O (UR-70). This proves that both steps of the enzymic reaction, cyclic phosphate formation and decomposition, must be "on-line", i.e. occur by direct displacement. This requires the simultaneous involvement of two histidine groups on the enzyme, one to act as base and the other as acid; this is supported by recent NMR and X-ray diffraction studies of the enzyme (*HU&-69*).

All of the mechanisms discussed above were derived on the basis of product analyses, isotope exchange studies, and the known behavior of analogous, stable oxyphosphoranes. Confirmation by comparison of observed and expected kinetics is hindered because of the complexity of the system of consecutive and parallel reactions resulting from the intermediacy of pseudorotating phosphoranes. This has been demonstrated (*GR&-73*) for the hydrolyses of trimethyl phosphate, methyl ethylene phosphate, and dimethyl acetoin phosphate in a critical review of all the evidence relating to oxyphosphorane intermediates. Further complications result from the presence of metal cations, which are essential in biological systems (*GR&-73*) to increase the susceptibility of phosphate mono- and diester anions to nucleophilic attack. Complexation or ionic bonding between the metal cation and phosphoryl oxygen could effect the preference of the phosphoryl group for the basal position as well as the stability of the oxyphosphorane, thereby changing both the kinetics and the stereochemistry (*W-73*). The occurrence of hexacoordinate phosphorus intermediates or transition states also complicated the kinetics. Such structures have been demonstrated in the hydrolysis of neutral oxyphosphoranes (*AW-73*). Similarly, the equilibrium constants for ionization of $R_nP(OPh)_{5-n}$ in dry acetonitrile, determined by conductivity measurements, could not be interpreted without assuming the presence of $R_nP(OPh)_{6-n}^{\ominus}$ and an equilibrium constant $< 10^{-3}$ for dissociation of this anion to the neutral oxyphosphorane (*LW-76*). This implies that addition of NaOPh to the oxyphosphorane solution would give at least 60% conversion to the anion, a fact verified by ^{31}P—NMR spectroscopy. At pH 12 or higher, hydrolysis of methyl ethylene phosphate is partially first- and partially second-order in hydroxide, and the amount of ring cleavage increases with basicity. This is best explained by hydroxide attack on anion IV to form one of two energetically equivalent hexacoordinate phosphorus species, one leading to ring cleavage and the other to methyl loss (*KC&-69*, *GR&-73*).

Survey of Relevant Articles on Pentavalent Intermediates in Phosphate Hydrolysis (Table V C)

System and Results	Conclusions	References
Tri(2-thienyl)methylphosphonium iodide hydrolyzes 680,000 times faster (at 40 °C) than triphenylmethyl phosphonium iodide.	The inductive-electron-withdrawing of the thienyl group far outweighs $dp-\pi$ bonding between the positively-charged phosphorus and the electron-rich thienyl substituent. Since the rate-determining step (see text) involves cleavage of an already-formed phosphorane, $dp-\pi$ bonding in the original phosphonium salt is unlikely to be of any importance.	(*A-70*)

System and Results						Conclusions	References
Hydrolysis of $[Ph_3\overset{\oplus}{P}R]Br^{\ominus}$ at 40 °C						$\delta\ ^{31}P$ values indicate that B, C and D are all electron-with-drawing relative to A, and thus should stabilize intermediate phosphorane. Entropies of activation indicate that the transition state has significant carbanionic character for A, B, and especially C, but not for D ($\Delta S^* < 0$ due to solvation). Here at least the most important factor affecting the rate is probably the carbanionic stability of the leaving group.	(AH-72)
	k, $1^2/-$ mol^{2}min	$\delta\ ^{31}P$ ppm	ΔH^* kcal	log A	ΔS^* eu		
A: R = benzyl	7.25	−19.9	18.2	12.4	−1.9		
B: R = 2-thienyl-methyl	183.	−18.2	16.8	12.3	−1.4		
C: R = 2-furyl-methyl	356.	−17.4	11.1	11.9	−4.2		
D: R = 3-furyl-methyl	2.95	−18.1	24.3	16.3	+1.6		

^{31}P NMR chemical shifts confirm the strong electron-withdrawing effect of the thiophene in tri(2-thienyl)methylphosphonium iodide. Furan has an even stronger electron-withdrawing effect, and tri(2-furyl)methylphosphonium iodide hydrolyzes 1 000 times faster than tri(2-thienyl)-methylphosphonium iodide (A-70), and has an equilibrium constant of 11.0 1/mole at −83° for phosphorane formation in NaOMe/MeOH.

The rapid hydrolysis of these phosphonium salts and cleavage of the phosphorus heterocyclic bond compared to triphenyl-methylphosphonium is due at least as much to stabilization of the intermediate phosphoranes as to the greater stability of 2-furyl and 2-thienyl compared to phenyl carbanions. (AHM-72)

$$[NP(OCH_2CF_3)_2]_x \xrightarrow[H_2O]{OH^{\ominus}} \begin{bmatrix} OCH_2CF_3 \\ | \\ [P-NH] \\ | \\ O \end{bmatrix}_x$$

	x = 3	x = 4
$k_{80°}$, (M^{-1} sec^{-1})	0.100	0.915
ΔH^*, (Kcal/mole)	16.6	20.5
ΔS^*, (eu)	−11.7	−3.7

Hydrolysis is simple S_N2. Tetramer reacts faster because of greater conformational freedom of eight-membered ring. (AW-72)

Base hydrolysis in 25 % aqueous diglyme of hexakis (aryloxy)-cyclotriphosphazenes; Second-order kinetics; No ArO^{18}H is formed in H$_2$O^{18}.
$[NP(OPh)_6]_3 : \Delta H^{\ddagger} = 12.55$ kcal/mole
$[NP(O-Ph-NO_2)_6]_3 : \Delta H^{\ddagger} = 10.25$ kcal/mole

: $\Delta H^{\ddagger} = 6.38$ kcal/mole

Simple S_N2 reaction with P–O cleavage. Results very similar to those for phosphate esters; in particular the fast hydrolysis of the spiro compound is due to relief of strain on formation of trigonal bipyramidal transition state. The required 120° angle N–P–N enforces a basal-basal configuration for the phosphazene ring and prevents initial P–N cleavage. (AW-72a)

Survey of Relevant Articles on Pentavalent Intermediates in Phosphate Hydrolyses (Table VC)

System and Results	Conclusions	References
Hydrolysis of (ArO)$_5$P in 25% aqueous dimethoxyethane is 100 times faster for Ar = p-methylphenyl than for the ortho isomer. Hydrolysis of Ar = o-chlorophenyl is slower than that of Ar = phenyl, but hydrolysis of Ar–p-chlorophenyl is 2000 times faster.	Such steric hindrance proves that the transition state is more crowded than the starting oxyphosphorane. Thus, the hydrolysis must involve a hexacoordinate phosphorus anionic species as intermediate or transition state.	(AW-73)
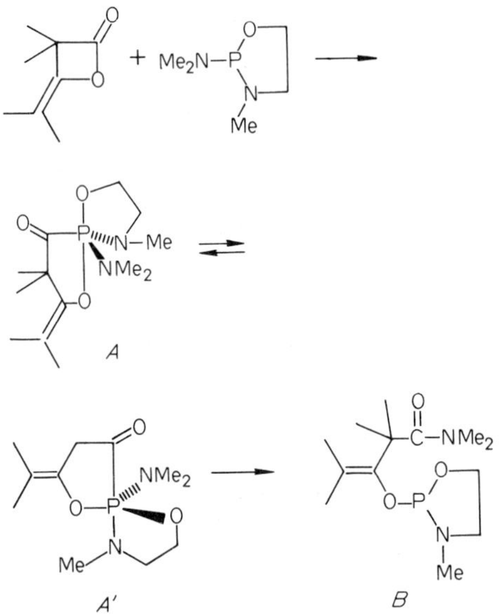 [Structures showing reaction of cyclic phosphate with Me$_2$N–P heterocycle, intermediates A, A′, and product B]	Formation of B probably involves pseudorotation of A to A′, followed by stepwise cleavage of the apical P–C and basal P–N bonds.	(BJK-72)
Epoxides are converted to episulfides by phosphine sulfides under acid catalysis with complete retention of configuration at carbon and phosphorus 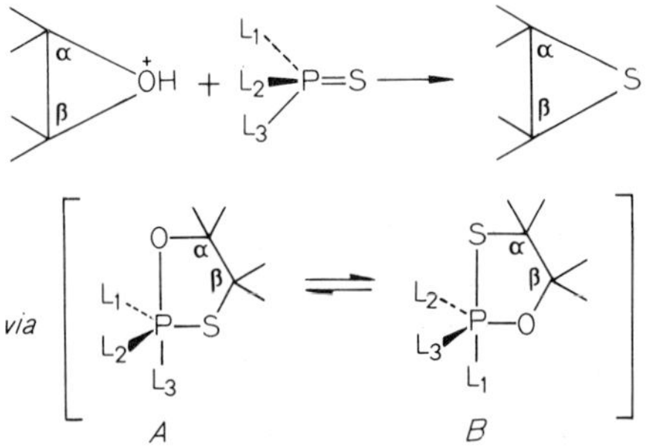 [structures showing epoxide + L$_2$–P=S → episulfide, via pentacoordinate intermediates A and B]	Protonation at O and C–O cleavage are followed by formation of A with inversion of configuration at Cα; but decomposition of the pseudorotation product B involves inversion of configuration at Cα, thus giving overall retention of configuration. The occurrence of pseudorotation guarantees retention at phosphorus.	(CF-72)
Hydrolysis of methyl propylphostonate is as fast as that of methyl ethylene phosphate but gives exclusively the product of ring opening.	Hydrolysis of the intermediate phosphorane here to put –OMe apical would also produce apical	(DW-66a)

Pentacovalency

System and Results	Conclusions	References
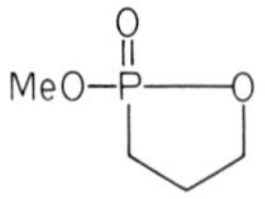 MeO–P(=O)–O (cyclic)	P–C bond and so occurs too slowly to be noticed during the lifetime of the intermediate. See text.	
Cyclic and acyclic phosphine oxides are reduced to the corresponding phosphanes by PhSiH$_3$ with complete retention of configuration. 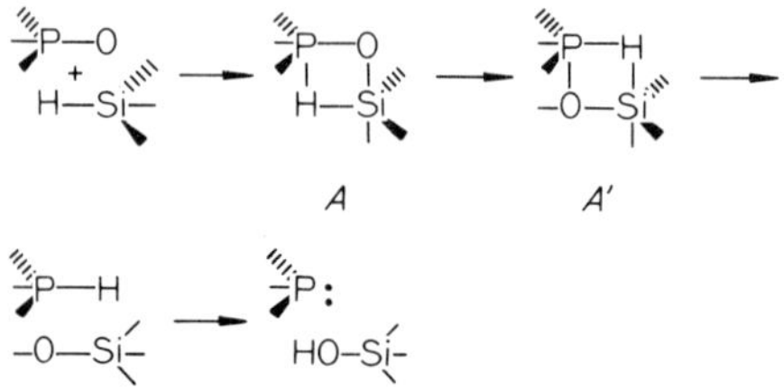 [scheme showing P–O + H–Si → A → A' → P–H/–O–Si → P: / HO–Si]	Pseudorotation or turnstile rotation of A results in retention.	(M-74)
Treatment of phosphine oxide A with HSiCl$_3$ and triethylamine in refluxing benzene for two hours gives a mixture of the *cis* and *trans* phosphanes. The composition of the mixture varied slightly depending on whether *cis* or *trans* A was used initially. 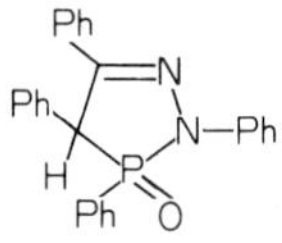 [structure A: Ph, Ph, H, Ph, N=N, N–Ph, P=O]	Oxidation products of HSiCl$_3$ (e.g. hexachlorodisiloxane) cannot reduce A but can add to it to form phosphoranes which survive for a sufficient interval to undergo pseudo (or turnstile) rotation, thus explaining the stereomutation result. Since the phosphoranes formed from *cis* and *trans* A differ in stability, the same equilibrium mixture does not result from both isomers.	(BT-75)
A, B, and C hydrolyze no faster than their acyclic analogs D, E, and F. 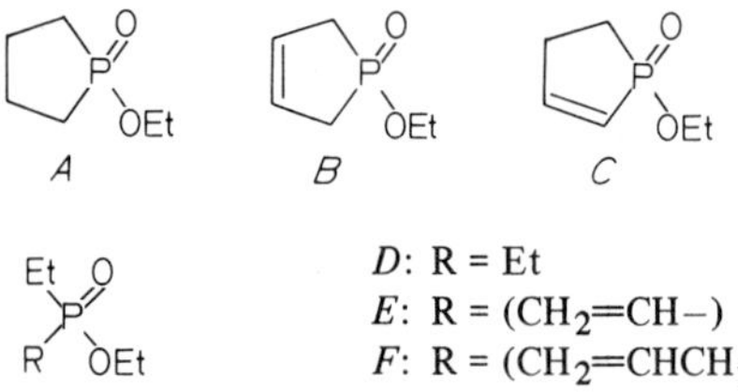 [structures A, B, C each with OEt; Et–O–P(=O)–OEt with R; D: R = Et; E: R = (CH$_2$=CH–); F: R = (CH$_2$=CHCH$_2$–)]	Here the intermediate phosphoranes must have basal-basal five-membered rings or apical P–C bonds and so are high energy structures which form slowly.	(DW-66)
NMR shows methylene hydrogens (diastereotopic) of the ethoxy groups in A to be equivalent (isochronous) above 30°. 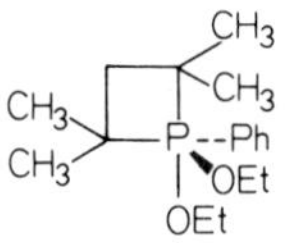 [structure A: CH$_3$, CH$_3$, CH$_3$, CH$_3$, P–Ph, OEt, OEt]	A must pass by successive Berry pseudorotations or turnstile rotations through a structure in which the four-membered ring is basal-basal. (See also M-70.)	(DWD-71)

134

System and Results	Conclusions	References
Hydrolysis of A by KOH in 70% aqueous dimethyl sulfoxide at 60 °C is 4.2 times faster than that of the acyclic analog B, yet proceeds with 74.6% ring retention, whereas hydrolysis of C is 95% by ring cleavage. 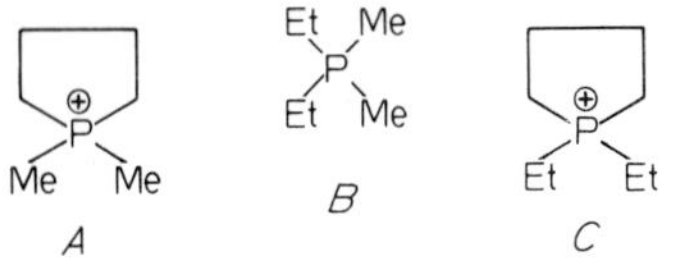	Steric acceleration due to relief of ring strain observed for A proves that formation of a pseudorotating phosphorane intermediate is rate-determining. Pseudorotation is favored for A but not for C because methyl is a much better carbanionic leaving group than ethyl and higher alkyl groups, which have equal leaving group abilities.	(MO-73)
In aqueous dioxane, acid hydrolysis of A proceeds primarily (85%) by loss of benzyl alcohol. In the presence of hydroxylamine the hydrolysis rate increases by a factor of about 30, and the only products are dibenzyl phosphate and the oxime of pyruvic acid. 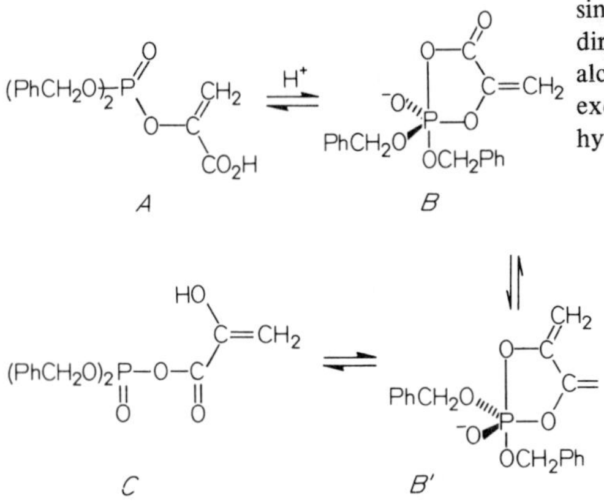	Hydrolysis of A involves attack of the carboxyl group on P to form phosphorane B; pseudorotation of B to C is 30 times faster than direct decomposition by loss of the apical benzyloxy group. However, since C is formed reversibly, direct decomposition to benzyl alcohol is the observed reaction except when C is trapped by hydroxylamine.	(SB-71)
Inversion of configuration in this reaction is favored by polar solvents or by added NaCl. The effect of added NaCl is greater in nonpolar than polar solvents. Added tetramethylammonium chloride gave even more inversion than sodium chloride. 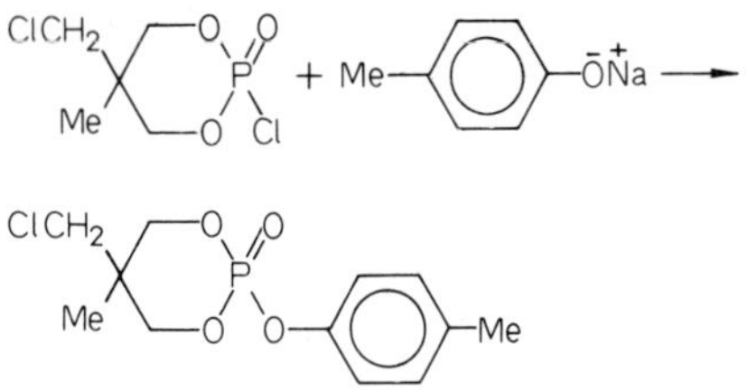	Complexation of sodium ion with the phosphoryl oxygen reduces P–O, dp–π bonding and destabilizes the intermediate phosphorane, thus reducing pseudorotation and resulting in retention of configuration. Polar solvents increase the solubility of byproduct NaCl. Added Me_4NCl precipitates NaCl and forms Me_4NOAr, which because of the decreased solvation of $Me_4N^{\oplus}$ is a more powerful nucleophile than NaOAr.	(W-73)

System and Results	Conclusions	References
Exchange of H_2O^{18} is faster with strained, e.g. A, than unstrained, e.g. B, phosphine oxides in neutral or basic solution, but slower in strongly acidic solution. 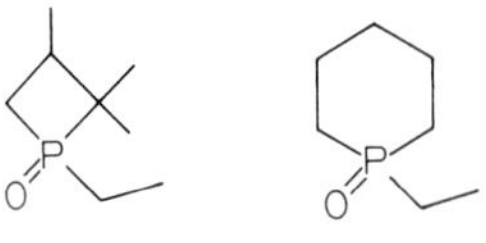	In strong acid the incoming and departing ligands are highly solvated and thus highly electronegative, and the resulting preference for the apical position is enough to outweigh relief of strain on forming a phosphorane with apical-basal ring and apical-basal H_2O^+ ligands.	(WK-74)
[Structures A and B shown]		
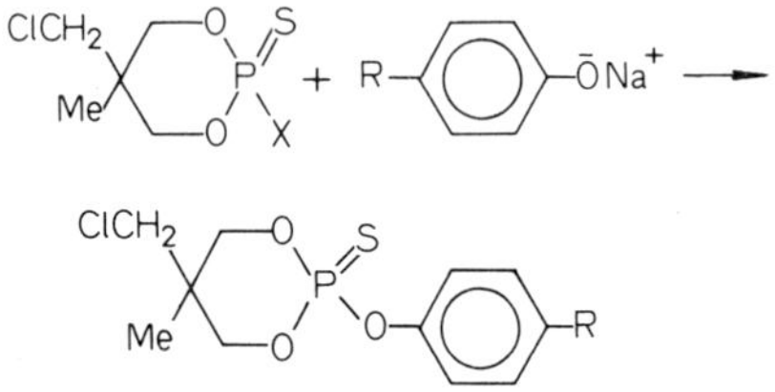 [Reaction scheme with ClCH$_2$ substrate + R—⌬—ŌNa⁺]	As observed earlier (W-73), inversion of configuration is favored by solvent polarity, added salt, greater basicity of attacking nucleophile (i.e. greater electronegativity of R), and greater leaving group ability of X.	(WT-74)
Attack of dimethylamino radical on A to displace Y proceeds with predominant inversion of configuration. 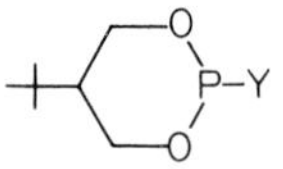 [Structure A with P–Y, Y = t–Bu⁻, PhCH$_2$⁻, Me$_2$N⁻]	The entering and leaving groups are apical and the 6-membered ring is basal-basal; thus, substituents in phosphoranyl radicals show the same preferences for apical and basal positions as in ordinary phosphoranes.	(BK&-74)
In both A and B hydrolysis is much faster at the bridgehead than at the other phosphorus, and the variation of kinetics with pH is similar to that for methyl ethylene phosphate. 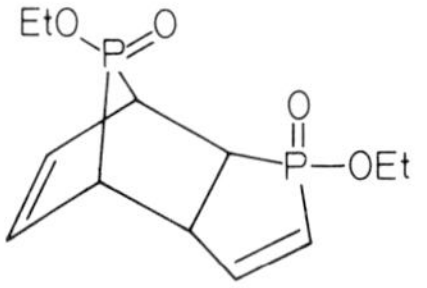 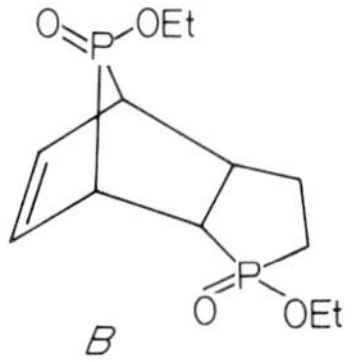 [Structures A and B shown]	Hydrolysis occurs by the same mechanism as with methyl ethylene phosphate, including pseudorotation becoming rate-determining in very strong acid (KC&-69), because the ring strain at the bridgehead is so great as to overcome the energy disadvantage of an apical alkyl in the intermediate phosphorane.	(KW-69)
Oxidation of R_3PS to R_3PO and S by H_2O_2 proceeds with retention of configuration for A and B but with racemization (slight excess of inversion over retention) for C. 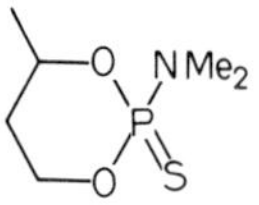 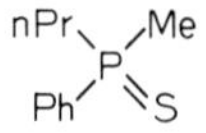 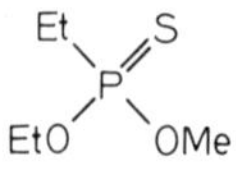 [Structures A, B, C shown]	The stereospecificity-determining step is the attack of OH⁻ on the R_3PSOH. For A and B this occurs at phosphorus along a line of centers with the most apicophilic ligand, which here is the alkoxy, leading to retention of configuration by pseudorotation or turnstile rotation of the resulting phosphorane intermediate. In C,	(SOM-76)

System and Results	Conclusions	References
	however, the only apicophilic ligand is the electronegative −SOH group; as a result pseudorotation of the intermediate phosphorane leads only to racemization.	
Reaction proceeds with 91% inversion of configuration in acetonitrile, and with even more inversion when $KClO_4$ or $Mg(ClO_4)_2$ is added, but proceeds with retention in the presence of $LiClO_4$. 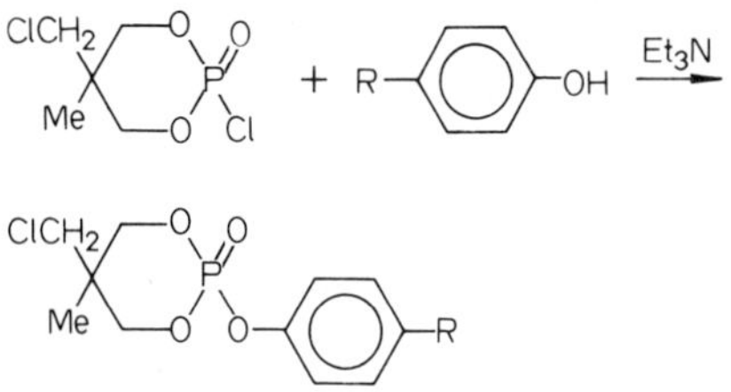 [Structures: cyclic phosphorane with ClCH₂, Me, O, P, Cl groups + R—C₆H₄—OH, Et₃N → cyclic phosphorane product with ClCH₂, Me, O, P, O—C₆H₄—R]	K and Mg act as electrophiles toward Cl and aid its departure, probably by a direct S_N2 process. Lithium, forming less ionic bonds has a greater affinity for O than Cl. Lithiation of the normally basal phosphoryl oxygen increases its apicophilicity and so favors formation of a pseudorotating phosphorane intermediate. (See also W-73 and WT-74).	(WW-76)

D. Pentacoordinate Sulfur Intermediates

In view of the small number of stable organic sulfuranes and the lack of demonstrated pseudorotation (as distinct from intermolecular polytopal rearrangements) in such cases, one would not expect to find clear examples of pentacoordinate intermediates in nucleophilic substitution at sulfur. For example, although I and II hydrolyze respectively 2×10^7 and 7×10^5 times faster than their acyclic analogs, diphenyl sulfate and phenyl α-toluenesulfonate, reaction in $H_2^{18}O$ gave unreacted ester with no incorporated ^{18}O. These results clearly indicate that, if there is an intermediate, it does not pseudorotate (KKW-65). It is to be noted that exactly the opposite result was found for ethylene phosphate (HW-61). This failure of pseudorotation to occur can be explained by the

$>S=O$ and $>\overset{\oplus}{S}-\overset{\ominus}{O}$ bonds

in the pentavalent structure. Just as the phosphoryl group in the oxyphosphoranes derived from cyclic phosphates prefer the basal position, so these groups appear to prefer the basal position. The presence of two groups preferring to be basal plus a ring spanning an apical and a basal position completely specifies the pentavalent structure, and, consequently, raises the activation energy (and the half-life) for pseudorotation to a value greater than the lifetime of the pentavalent intermediate. Actually, since there is no pseudorotation there is no need to assume the existence of an actual intermediate, but, on the other hand, consideration of the pseudorotation phenomenon is demanded for those circumstances in which the appropriate intermediate is established.

Five-membered cyclic sulfites also show considerable rate enhancement over their acyclic counterparts, though the magnitude is not as great as in the case of the sulfates. For instance, ethylene sulfite undergoes alkaline hydrolysis 360 times faster than dimethyl sulfite (*KPW-63*), and catechol cyclic sulfite hydrolyzes more than 1500 times faster than diphenyl sulfite (*DTV-62*), this despite the fact that the five-membered ring in sulfites is completely free of strain. The recently determined activation parameters (*BTW-68*) for this hydrolysis provide an explanation for the rate enhancement.

ethylene sulfite $\Delta AH^{\neq} = 10.2$ kcal/mole $\Delta S^{\neq} = -14.7$ esu
dimethyl sulfite $\Delta AH^{\neq} = 11.2$ kcal/mole $\Delta S^{\neq} = -23.9$ esu
catechol sulfite $\Delta AH^{\neq} = 5.6$ kcal/mole $\Delta S^{\neq} = -6.2$ esu
diphenyl sulfite $\Delta AH^{\neq} = 6.6$ kcal/mole $\Delta S^{\neq} = -17.4$ esu

The five-membered cyclic sulfites, unlike their acyclic counterparts, are relatively rigid structures and are held in a position well suited for the formation for a trigonal bipyramidal intermediate or transition state with one of the basal positions occupied by a nonbonding electron pair. The presence of the nonbonding electron pair and the S^+–O^- bond, both of which prefer to be basal, would be expected to inhibit pseudorotation; thus, there is again no necessity to assume an actual intermediate.

Sulfites and sulfoxides possess nonbonding electron pairs, which are also present in the sulfuranes derived from them. This results in severe distortion of the trigonal bipyramid. For example in $Ph_2S(OC(CF_3)_2Ph)_2$ (*PMP-71*) the basal C–S–C angle is only 104° instead of 120°, and the apical O–S–O angle is 175°. In SF_4 the basal F–S–F angle is 101° (*HB-72*). This factor in turn reduces the hydrolysis rate difference between cyclic and acyclic esters, since the reduction in ring strain achieved by formation of a trigonal bipyramid with the ring spanning apical and basal positions is not as great (*HB-72*). It also increases the ability of small rings to span two basal positions, thus allowing direct $S_{N}2$ reaction with inversion of configuration, as illustrated in the base hydrolysis of *cis* III to *trans* IV (*TM-69*).

Another example of the effect of the low basal-basal angle in sulfuranes is found in the stable sulfurane (*MP-76*) prepared by reaction of IV-A with RfO^-K^+ and bromine; ^{19}F NMR spectroscopy showed the two ring-bonded trifluoromethyl groups to be equivalent even at $-90\,°C$, which is indicative of structure IV-B rather than IV-C for this sulfurane.

IV-A   IV-B   IV-C   where $R_f = -C(CF_3)_2Ph$

Almost all known examples of nucleophilic substitution at chiral sulfur proceed with inversion of configuration, i.e. simple S_N2 (*CF-68*). Thus, all the nucleophilic substitutions at optically active compounds of tetrahedral hexavalent sulfur, which have been studied up to the present, e.g. sulfinates or sulfinamides (*JC-74*), proceed with inversion of configuration. Inversion is also common at pyramidal, tetravalent sulfur and occurs even where it would not be expected, as in the conversion of ArSOMe (Ar = *p*-tolyl) to ArMeS=NTos with TosN=S=NTos in pyridine (*YR&-73*). *Sheppard* and *Wudl* (*SW-69*) had predicted that the related reaction of Ar_2SO with TosN=C=O would give Ar_2S=NTos with retention of configuration by pseudorotation of sulfurane V. Since the observed reaction is second order in TosN=S=NTos with a highly negative entropy of activation (*CD&-70a*), which strongly suggests a cyclic transition state, the mechanism is best formulated as involving intermediate VI, which forms *via* the solvent-coordinated species VII and VIII. In any case, an acyclic transition state with one TosN=S=NTos attacking sulfinyl sulfur and the other attacking oxygen is impossible since it would involve simultaneous oxidation and reduction to TosN=S(O)=NTos and TosN=S. In nonpolar solvents, e.g. benzene or methylene chloride (*YR&-73*), the reaction proceeds with retention of configuration, possibly *via* conversion of the pyramidal species IX into sulfurane X, which pseudorotates prior to S—O cleavage. Alternative reaction routes may also be invoked to accommodate these facts. On the other hand a similar mechanism may be invoked to explain the inversion of configuration in the reaction of sulfoxides with isocyanates

(*GJC-73*). *Sheppard* and *Wudl* (*SW-69*) have considered the possibility that the thermal racemization of *t*-butylethylmethylsulfonium perchlorate involves reaction with solvent to form a pseudorotating tetravalent sulfur intermediate rather than the umbrella type inversion originally proposed (*DT-66*). True to the expected nucleophilic order the racemization is faster in acetic acid containing sodium acetate than in water, and faster still in ethanol, the rate in ethanol being 1.66 times that in water (*DT-66*). That this is the expected nucleophilic order is shown by the fact that solvolysis, which is known to involve cleavage at one of the α carbons, and which occurs at about one fifteenth the rate of racemization, shows the same solvent effects, being 2.71 times faster in ethanol than in water. Since S_N2 reactions at sulfinyl sulfur show the same nucleophilic order as at tetrahedral carbon (*K-68*), the same is probably true here. On the other hand the racemizations of XI and XII (perchlorate salts) at 100 °C were about 2500 times slower than that of methylethyladamantylsulfonium perchlorate (*GCF-70*). Racemization by pyramidal inversion is expected to be slow for XI and XII due to the 120° endocyclic bond angle in the transition state, whereas racemization *via* an intermediate sulfurane capable of pseudorotation should be faster for cyclic than for acyclic sulfonium salts.

Sulfurane intermediates are probably involved in the HCl catalyzed cleavage and racemization of sulfoxides. Since in HCl-dioxane containing $H_2^{18}O$, racemization of optically active sulfoxides occurs at the same rate as ^{18}O exchange (*MS&-64*), the loss of optical activity cannot be due to pseudorotation of ArRS(OH)Cl. Displacement of H_2O from protonated $R_2S(OH)Cl$ by chloride ion to give a symmetrical dichlorosulfurane and/or the corresponding chlorosulfonium chloride ion pair is mechanistically in agreement with the observed kinetic dependencies, i.e. $k_{rac} = k_{exchange} = k'[H]^2$. However, it has been shown (*KO-71*) that formation of $ArSCl_2$ in a medium producing racemization invariably results in solvolysis forming $R^\oplus$ and ArRSCl. The only reasonable achiral intermediate in the stereomutation of sulfoxides is, therefore, $ArRS(OH)_2$. The stereomutation of aryl benzyl sulfoxides occurs rapidly in HCl-dioxane containing 33% water, at which high concentration of water the nucleophilicity of chloride is quite low. If $ArRS(OH)_2$ is involved, it must be a covalent sulfurane since the change from chloride as the stronger nucleophile in forming $R_2S(OH)Cl$ from the protonated sulfoxide (i.e. hydroxysulfonium chloride) to water as the stronger nucleophile in isotopic oxygen exchange of $R_2S(OH)Cl$ must involve a change of valence from 3 to 4 at sulfur (*K-68*). At low water concentration ($\leqslant 2\%$), chloride attacks the $R_2S(OH)Cl$ giving dichlorosulfurane in equilibrium with chlorosulfonium chloride and carbonium ion cleavage results (*KO-71*). Halosulfonium cations must also be involved in the halogenation of sulfides, which may result either in carbonium ion cleavage or in α-substitution by halogen (*WH-70*) depending on substitution and solvent medium factors.

$$ArRS(OH)_2 + H_3O^{18} \rightleftharpoons \underset{XIII}{\text{[intermediate]}} \rightleftharpoons HO-S-^{18}OH + H_3O$$

XIII

However, the logic which suggests that ArRS(OH)$_2$ is the actual intermediate formed in the rate determining step of the stereomutation reaction must also explain the kinetic observation that racemization and exchange occur at identical rates. It has been argued (*KO-71*) that this kinetic imperative is consistent only with the assumption that ArRS(OH)$_2$ experiences displacement in H$_2$O^{18} at extremely rapid rates. An attractive way of explaining this is represented in XIII by H-bonding exchange in the Berry-pseudorotation, tetragonal pyramid structure of the intermediate.

All of the above examples involve sulfurane intermediates in nucleophilic substitution at tricoordinate sulfur (sulfoxides, sulfites, sulfonium salts, etc); this subject has been recently reviewed in detail (*T-76a*) and requires no further discussion here.

The thermal thio-allylic rearrangement (*KJ-70*) is another example of a sulfurane intermediate; this reaction, exemplified by XIV (R=R' = deuterium or methyl) occurs readily in solution at 140°–180°C. The kinetics show competing unimolecular and bimolecular processes; the bimolecular process has a slightly lower activation energy but a much more negative entropy of activation, especially in polar solvents, and is thus the slower of the two processes. The unimolecular process does not occur for aryl(oxy) ethers, nor for sulfides which do not contain aryl groups. The activation energy was decreased by electron-withdrawing substituents on the aryl group, but in a linear free energy plot against σ_p^-, ρ was only 1.15 compared to 2.58 for ionization of ArSH. Nonetheless, the magnitude of ρ does not reflect the extremely large degree of negative charge resident on sulfur in the rearrangement transition state as does the fact that $\Delta(\Delta H^{\ddagger})$ is nearly 6 Kcal/Mol over the full reaction series (p–CH$_3$ to p–NO$_2$), while the compensating factor $\Delta(\Delta S^{\ddagger})$ is more than 17 e.u. The activation energy was also decreased (by nearly 4.5 Kcal) by β methyl substitution on the allyl group but not significantly by α or γ methyl substitution. Conceivably, a full negative charge is developed on sulfur and positive charge on the β-allylic carbon, consistent with the formation of the zwitterionic sulfurane intermediates, XV and XVI. An ion-pair mechanism is eliminated since an allylic carbonium ion should be stabilized more by α or γ than by β methyl substitution. A radical pair mechanism is unlikely since at these temperatures some diffusion out of the solvent cage is inevitable, yet no disulfides or other radical coupling products were found.

The most direct and compelling evidence (*KS-76*) for intermediate sulfurane formation in this process is established by the appropriate heavy atom isotope (S$_{32}$/S$_{34}$) experiment. The results eliminate unequivocally a dissociative ion or radical pair mechanism, (maximum calculated isotope effect), as well as a concerted transition state with no change in the degree of bonding to sulfur, (so called no-isotope effect expected by theory), and are in full agreement with the observed minimum isotope effect calculated on the basis of a full additional bond to sulfur

[Structures XIV, XV, XVI with "Pseudorotation barrier" label]

in the activated complex. An equilibrium, heavy atom isotope effect is also observed in consonance with the expectation that the heavier isotope is favored in bonding to carbon (by about 4 cals/mol) over the light.

Clearly there must be two sulfurane intermediates in this reversible rearrangement. Microscopic reversibility therefore demands that their interconversion can take place only over an energy barrier, which in this case is easily identified as a pseudorotation process. Even more facile but related rearrangements involving the higher valence sulfur of aryl allyl sulfoxides and sulfones have also been characterized (KSB-76). Analogous results and conclusions regarding the occurrence of zwitterionic, sulfurane intermediates, interconverted by pseudorotation in the rate determining step, have been obtained in these cases. Thus, even with an activation energy for pseudorotation as high as 12 Kcal/mole, pseudorotation can be quite facile at 140° (B-60). By comparison the overall activation energy for unimolecular isomerization of $PhSCH_2CH=CD_2$ was 31.5 kcal/mole in o-dichlorobenzene and 32.9 kcal/mole in nitrobenzene; the entropies of activation in these solvents were − 11.4 and 1.3 eu, respectively (KJ-77).

Pseudorotation in sulfuranes formed as reactive intermediates should not be considered as surprising or totally unprecedented constructs, (OY&-68). Pseudorotation in relatively stable sulfuranes has been previously cited (AM-76). Moreover, the analogously structured intermediates occurring in the rearrangement of β-ketosilanes to siloxyalkenes have been shown (see sect. V E) through stereochemical studies with chiral silicon (BMB-75) and the application of kinetic isotope criteria to involve necessarily a pseudorotation barrier (KB-77). These reports also provide precedence in kinetic details for a course of reaction similar to the thioallylic rearrangements (KS-76).

A similar anionically-charged sulfurane can be invoked in the bimolecular ligand exchange of thione bis-ylides $R_2C=S=CR'_2$. Such species are, of course, unstable with respect to cyclization to the corresponding episulfide unless R and R' are electron-donating and electron-withdrawing, respectively. As an example we may cite the ligand exchange of XVII and XVIII at 30 °C in methylene chloride (AB-76), which occurs via XX. This species almost certainly arises in a two step process involving attack of an ylide carbon bonded to the sulfur of a second bis-ylide molecule to form intermediate XIX on the way to XX. The similarity of this mechanism to the reaction of ArMeSO with TosNSNTos via IX and X with retention of configuration (*YR&-73*, p. 139) will be noted.

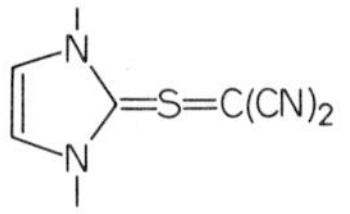

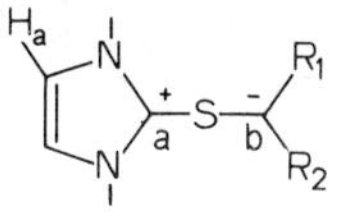

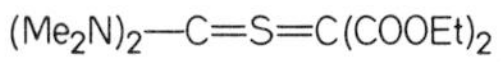

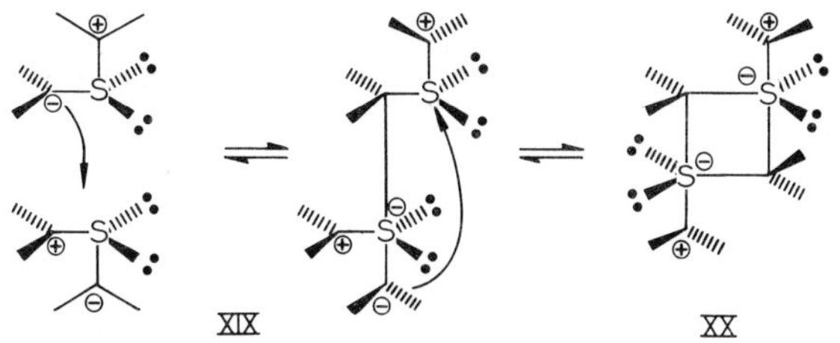

The ease of reaction here is possibly due to the fact that the sulfur is already hypervalent, as required by the through-conjugative stabilization of the bis-ylide illustrated in XVIIa. The efficiency of this through-conjugation, which is most easily predicated upon the involvement of sulfur d-orbitals, is seen by NMR in the correlation of the ring current induced diamagnetic anisotropic deshielding of H_a with the degree of total shielding of C_b for three compounds of the type XVIIa. As expected, the chemical shift of C_a varied little through the series (*AB-76*).

As an example of a sulfinyl-allylic rearrangement we may cite XXI (*RSL-76*), in which all fluorines are equivalent in the ^{19}F–NMR at temperatures above $-124\,°C$, thus indicating a degenerate isomerization process with a free energy of activation of 6.8 kcal/mole. *Lemal* (*RSL-76*) prefers to interpret this in terms of an unprecedented pseudopericyclic rearrangement, involving overlap of the sulfur lone pair with the electron-poor double bond illustrated in the conversion to XXIa. An allylic rearrangement stemming from biphilic (oxidative) addition (*PGB-60*) to sulfur is at least as possible. This may seem surprising in view of the electron-withdrawing nature of the trifluoromethyl group, which in fact makes XXI an excellent dienophile. There are, however, several considerations which account for the great facility in this rearrangement of XXIa. First, it has been noted that the corresponding sulfide XXIb undergoes rearrangement analogously but with considerably higher activation entropy and much lower activation entropy; ($\Delta H^{\ddagger} = 18.8$ kcal/mol and $\Delta S^{\ddagger} = -7.7$ eu for the sulfide vs. $\Delta H^{\ddagger} = 6.6$ and $\Delta S^{\ddagger} = 0.5$ for the sulfoxide), (*BRL-77*). This of course is quite in keeping with the greater facility of the sulfinyl-allylic compared to the thiaallylic rearrangement (*KJ-70, KS-76*) arising from the greater ability of the higher valency sulfur to accomodate octet expansion (*KSB-76*).

The geometry of XXI is a second factor favoring rearrangement *via* a sulfurane structure. The intermediate (or transition state) XXII formed in this way has either a very distorted trigonal bipyramid structure, or could be regarded as well advanced toward the tetragonal pyramid configuration through which pseudorotation could proceed.

Pentacovalency

[Structures XXI, XXIa, XXIb, XXII, XXIII, XXIV, XXV, XXVI, XXVIa]

The formation of XXV from XXIII on oxidation with peroxytrifluoroacetic acid occurs without isolation of the presumed intermediate even at −60 °C, which implies a free energy of activation of about 10 kcal/mole for the bond-switching rearrangement involved. This reaction, first observed by *Lautenschlaeger* (*L-69*) and subsequently analyzed by *Lemal* et. at. (*RSL-76*), can be formulated with a sulfurane intermediate XXVI possessing a somwhat strained trigonal bipyramid structure. Pseudorotation via the square pyramid structure XXVIa may also be considered as a viable pathway for rearrangement.

An even more intriguing example of sulfoxidation (*L-69*), which can be diagnosed as a homosulfinyl-allylic rearrangement, is the transformation of XXVII to XXIX with m-chloroperbenzoic acid (*RSL-76*). In this case the intermediate sulfoxide XXVIII could be isolated at low temperatures; it had a half live of 0.5 hr. at 55 °C (*RSL-76*). If the presumed sulfurane intermediate were structured as a trigonal bipyramid XXX it would be severly strained by the incorporation of two apical bonds in a 5-membered ring since the oxyanion ligand and the lone pair could only occupy basal positions. Again the tetragonal pyramid structure XXXa represents both relief of this strain and a pathway for permutational isomerism leading to the stable product XXIX.

An interesting illustration of the effect of ring strain on the sulfinyl-allylic rearrangement is the epimerization of XXXI to XXXII (*AWC-75*). Apparently this is not simply pyramidal inversion at sulfur as indicated by the activation parameters, 27.8 kcal/mole and −4 eu in benzene at 110°–140 °C, compared to 43 kcal/mole and

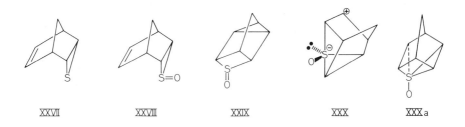

XXVII   XXVIII   XXIX   XXX   XXXa

+ 3 eu for inversion of *p*-tolyl methyl sulfoxide at 240° (*RGM-68*). This conclusion was established by the scrambling of a bridgehead deuterium label in XXXII, accompanying migration of the bridging C–S bonds. That this migration occurs only into the large (four-carbon) bridge is proven by the fact that epimerization is found to take place under these conditions for XXXIII but not XXXIV. This observation is explained by the tilt of the sulfur toward the large bridge as revealed by molecular (*Dreiding*) models (*AWC-75*), but, given this direction of migration, it is not possible to draw an intermediate sulfurane in this system which is free of severe strains since it would involve a one- or two-carbon bridge of the two apical C–S bonds. However, as in the conversion of XXI → XXIa via XXII, a series of tetragonal pyramid intermediates, as expressed in XXXV in a conceivable associative mechanism, (*KS-76*), may accommodate the steric requirements of this change. *Anastassiou*, (*AWC-75*) on the other hand, has tentatively recommended a dissociation — recombination mechanism of biradical nature, basing this conclusion largely on the observation that XXXI rearranges 2.5 times more slowly in acetonitrile than benzene at about 140°. Nonetheless, as pointed out earlier in connection with such antipolar solvent effects (*KJ-70*, *KB-77*), a decision between these alternative mechanistic proposals cannot be regarded as unequivocal until the activation parameters pertinent to the solvent influence have been determined.

A sulfurane intermediate is also implicated in the anchimerically assisted decomposition of XXXVI, which decomposes 24,500 times faster than *t*-butyl perbenzoate itself at 60 °C in chlorobenzene (*LM-76*). NMR analysis of the reaction mixture showed the presence of sulfurane XXXVII, which was identified by comparison with an authentic sample prepared by reaction of XXXVIII with *t*-butyl hypochlorite. The inference of a sulfurane structure for XXXVII is suggested by the fact that XXXIX shows no anchimeric assistance in the peroxide decomposition (*TBM-63*); here the trigonality of the carbonyl carbon requires the C(O)O–S grouping to be coplanar with the dibenzothiophene ring system, which is incompatible with trigonal bipyramidal sulfurane geometry. Decomposition of XXXVII under the reaction conditions gave XL, XLI, and isobutylene; XXXVI also gave these products but gave *t*-butanol and acetone as well, which can only arise from the *t*-butoxy radical. Formation of radical-derived products has been explained (*TBM-63*) by the intermediacy of XLII, which has both biradical and bipolar character, as proven by the rate dependence on solvent polarity. Thus, either XXXVII and XLII form independently in parallel paths, or XLII decomposes independently to radicals and to XXXVII, which in turn would decompose by S–O cleavage to XLIII (*BM-62*), followed by loss of *t*-butyl carbonium ion. The rate dependence on solvent polarity makes the latter mechanism seem more likely (*LM-76*). A similar sulfurane intermediate XLIV has been observed in the decomposition of XLV (*MC-74*).

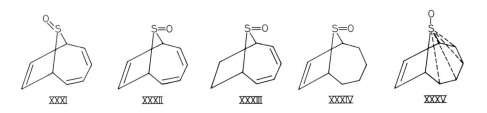

XXXI   XXXII   XXXIII   XXXIV   XXXV

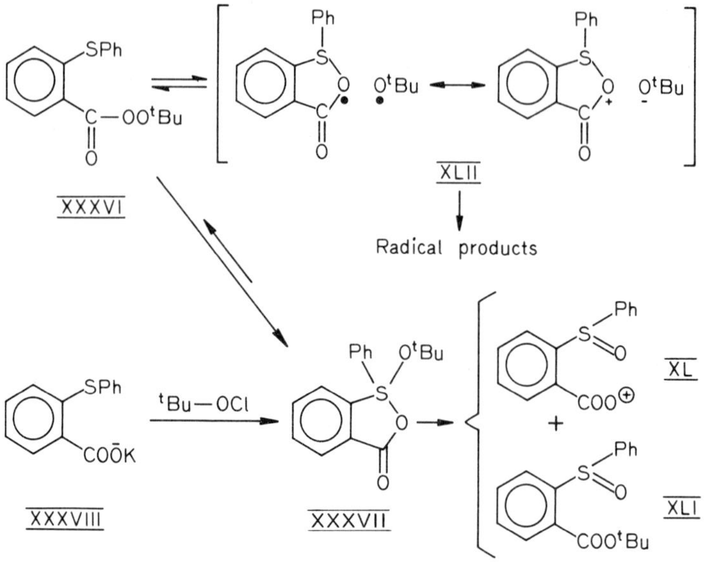

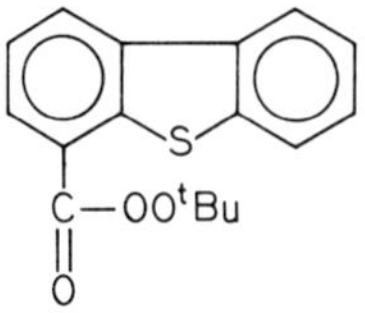

Pentacovalency

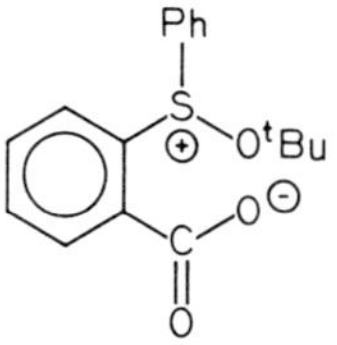

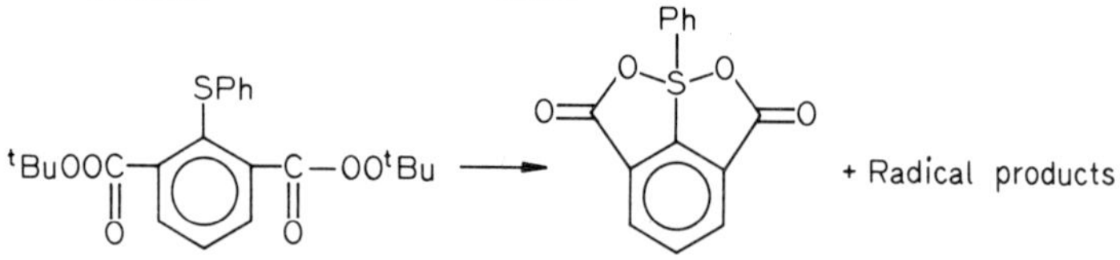

Survey of Relevant Articles on Pentacoordinate Sulfur Intermediates (Table VD)

System and Results	Conclusions	References
X-ray diffraction of the crystalline adduct of Br_2 and tetrahydrothiophene showed it to be a charge transfer complex with pyramidal sulfur and the S–Br–Br group linear with S–Br = 2.32 Å and Br–Br = 2.72 Å.	The Br–Br length is 0.43 Å longer than that in Br_2, suggesting that the species is best formulated as a bromosulfonium bromide. In this case sulfurane formation is prevented by the electron-donating inductive effect of the aliphatic ring carbons and possibly also by the slightly lower electronegativity of bromine compared to chlorine. (See *BB&-69* below).	(AW&-70)

System and Results	Conclusions	References
Linear free energy plot of rates of hydrolysis of A showed that although reaction was favored by electron-withdrawing substituents, resonance interactions were almost identical in the ground and transition state except for X = MeO, for which resonance was greater in the ground state. Reaction in H_2O^{18} gave no O^{18} in recovered amide. 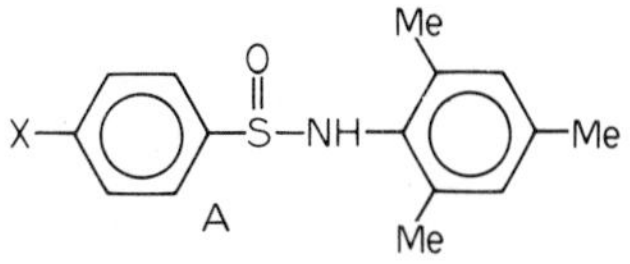 X―⌬―S(=O)―NH―⌬―Me (with Me, Me substituents) A	Direct $S_{N}2$ reaction with no intermediates.	(BA-71)
X-ray diffraction at 143 °K of the crystalline adduct of Cl_2 and bis(p-chlorophenyl) sulfide showed it to be a distorted trigonal bipyramid with apical S―Cl lengths of 2.259 Å, a basal C―S―C angle of 109° and a Cl―S―Cl angle of 174°.	(See AW&-70)	(BB&-69)
Kinetics of nucleophilic attack by aniline, butylamine, or hydroxide on benzenesulfonyl halides. Rates very similar for Cl, Br and I; k_{Cl}/k_F = 4.6 for OH^-, 4,200 for $BuNH_2$, and 1.65×10^5 for aniline.	This suggests a two-step mechanism with an expanded octet intermediate. For Cl, Br, and I bond formation is rate determining, so rate is not affected by S―X bond strength, whereas for F bond breaking is rate determining.	(CSI-72)
The ^{19}F―NMR spectrum of the reaction mixture shows two doublets at $-80°$, which collapse to two broad mounds at above $-40°$ in a solvent consisting primarily of $FCCl_3$. 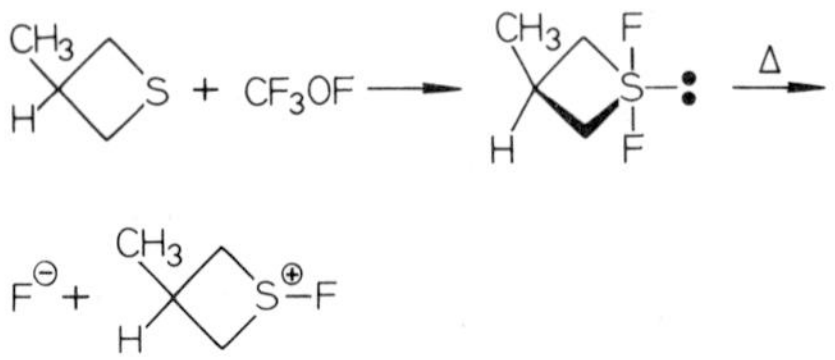 [cyclobutane with CH_3, H]―S + CF_3OF → [cyclobutane with CH_3, H]―S(F)(F): ―Δ→ $F^\ominus$ + [cyclobutane with CH_3, H]―$S^\oplus$―F	The methyl group makes the two fluorines always nonequivalent. If apical and basal fluorines were undergoing fast exchange by pseudorotation the spectrum should consist of two singlets. Thus the occurence of F―F coupling proves the 4-membered ring in this sulfurane must be basal-basal, with the CSC angle reduced by repulsion from the basal lone pair, and no pseudorotation. Collapse at $-40°$ suggests ionization to F^- and a fluorosulfonium ion, which would be favored by the polar solvent.	(DDH-73)
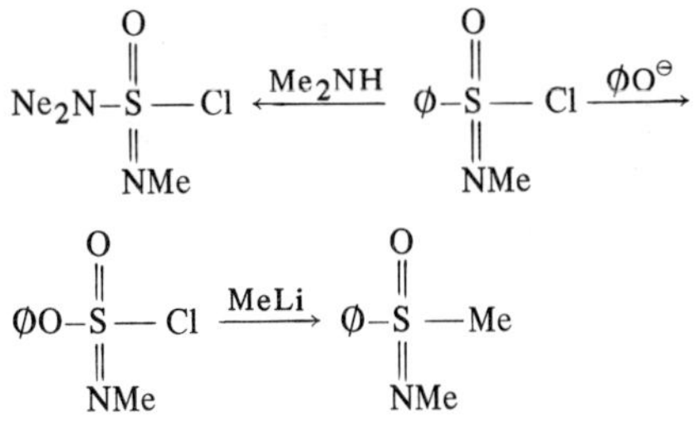 Ne_2N―S(=O)(NMe)―Cl $\xleftarrow{Me_2NH}$ ∅―S(=O)(NMe)―Cl $\xrightarrow{\phi O^\ominus}$ ∅O―S(=O)(NMe)―Cl $\xrightarrow{MeLi}$ ∅―S(=O)(NMe)―Me	All reactions proceed with inversion of configuration.	(JJ-71)

Pentacovalency

System and Results	Conclusions	References
Kinetics of acetate-catalyzed exchange of methyl p-toluenesulfinate with CD_3OH involves two terms, one nucleophilic attack by CD_3O^- and the other general base catalysis by acetate ion. The ratio of the rate constants for the latter process in MeOH and MeOD is only 1.4–1.5.	In general, microscopic reversibility demands that exchange reactions involving general base catalysis must involve an intermediate, in this case a sulfurane. However, the kinetic isotope effect suggests combined general acid and base catalysis, so no intermediate is required.	(KW-72)
Reduction of p-tolylmethyl sulfoxide to the sulfide by HI in aq $HClO_4$ was second order in [HI], first order in sulfoxide, and first order in [I^-], exactly the same kinetics as for racemization by HBr or HCl.	Reduction and racemization must have the same mechanism up to the rate-determining step, which may be written as loss of water from a protonated iodohydroxysulfurane. The resulting iodosulfonium ion is attacked by iodide to give I_2, whereas the other halosulfonium ions reform sulfuranes; both reactions very fast.	(LM&-70)
O^{18} exchange of labelled p-tolyl or p-chlorophenyl benzyl sulfoxide with dimethyl sulfoxide proceeds with retention of configuration.	Since sulfoxides are known to be dimeric neat or in nonpolar solvents, exchange probably proceeds via formation and pseudorotation of A.	(OY&-68)
X-ray diffraction of A gives apical S–O lengths as 1.780 and 1.777 Å, basal S–O as 1.439 Å, apical O–S–O angle as 172.3°, basal C–S–C angle as 117.7°.	Since there is no lone pair, the structure is quite close to an ideal trigonal bipyramid, as indicated by the much larger basal C–S–C angle in A than in B (104°) (PMP-71), in spite of the fact that the sulfur is formally hexavalent. However, the higher positive charge on sulfur does result in greater d-orbital contributions, as indicated by the basal S–O, which is much closer in length to a sulfone than a sulfoxide, and by the apical S–O length, which is much shorter than for B.	(PMP-74)
The apparent second-order dependence on hydroxide concentration of base hydrolysis rates of A, B, and C in strong base was shown to be due to the inappropriateness of this parameter. The dependence on $a_w 10^{H-}$, where H− is the Hammett acidity function, was strictly first-order even in 6M NaOH.	There is no need to assume the existence of a sulfurane intermediate which is deprotonated in strong base.	(SIK-72)

System and Results	Conclusions	References

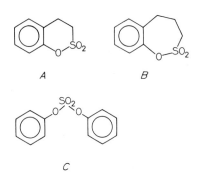

A

B

C

| At concentrations (equimolar) below 0.3 M in CCl$_4$ or CDCl$_3$ of substrate and halogenating agent (chlorine, bromine, chlorosuccinimide, bromosuccinimide), the ratio of yields of cleavage products to benzyl α-halobenzyl sulfide in the halogenation of dibenzyl sulfide was independent of concentration. | Both cleavage and α-substitution products involve an intermediate halosulfonium cation. | (WH-70) |

E. Pentacoordinate Silicon Intermediates

Although retention of configuration during nucleophilic substitution at optically active silicon is quite common, the occurrence of pseudorotation of pentacoordinate intermediates is doubtful in many cases. Most of the work in this field is that of *Sommer* and coworkers (*S-65*), whose conclusions are based primarily on the stereochemistry of nucleophilic substitution on optically active derivates of naphthylphenylmethylsilane. These reactions generally proceed with very high stereospecificity, often 100%, and are almost always bimolecular.

For substitution reactions at silicon proceeding with retention of configuration, *Sommer* (*S-65*) proposes what he calls an S_Ni–Si mechanism, which involves a cyclic four or six-membered transition state, III or IIIa, with basal attack by the nucleophile or basal departure of the leaving group. Such a mechanism explains the fact that retention of configuration only occurs in nonpolar solvents (*S-65*). Inversion reactions, which operate by the S_N2–Si mechanism I, involve charge separation and so are favored by polar solvents of high dielectric constant, by good leaving groups, and by attacking groups which can exist as stable anions. This is illustrated by reaction II, where R_3Si^* refers to the optically active naphthylphenylmethylsilyl group. For X = Cl all these reactions proceed with inversion of configuration regardless of the nature of R', clearly an illustration of the fact that chloride is an extremely good leaving group. For F, MeO, or H as leaving group, if we take a series of organolithium compounds in

Pentacovalency

$$X-\underset{B\ \ C}{\overset{A}{\underset{|}{\overset{|}{C}}}}-R' \quad X-\underset{Li\overset{\cdot\cdot\cdot}{-}R'\ B}{\overset{A}{\underset{|}{\overset{|}{C}}}}-C \quad X-\underset{\underset{R'--Li}{\overset{Li}{|}}}{\overset{A}{\underset{|}{\overset{|}{C}}}}\underset{R'\ \ B}{-}C$$

$$\text{I} \qquad\qquad \text{III} \qquad\qquad \text{III a}$$

$$R_3Si^*X + R'Li \longrightarrow R_3Si^*R' + LiX$$
$$\text{II}$$

which R' gradually increases in stability as a carbanion, then at a certain point in the series the stereochemistry will change from retention to inversion, this point being different for each leaving group. With Ph_2CHLi even R_3Si^*H reacts with inversion of configuration in spite of the extremely poor leaving group ability of H. The nature of the cation also affects the stereochemistry. In the exchange of R_3Si^*OMe with n-butyl alcohol and lithium, sodium, or potassium n-butoxide (*SF-68*), potassium gives the greatest amount of inversion and lithium the greatest amount of retention at a given concentration of alcohol. This clearly is due to the ionic character of the O—M bond, the less ionic bonds favoring the 4-center S_Ni mechanism.

Given a sufficiently powerful nucleophile even poor leaving groups can react by S_N2; thus, $AgBF_4$ reacts with R_3Si^*H in ether to give R_3Si^*F with inversion of configuration (*SKF-72*). On the other hand diisobutylaluminum hydride in xylene reacts even with R_3Si^*Cl by S_Ni (to give R_3Si^*H) (*SMG-72*). This reaction however depends greatly on solvent; in ether the crossover from retention to inversion occurs at R_3Si^*OMe. Similarly R_3Si^*H reacts with solid KOH in xylene giving 92% retention of configuration; in isopropanol the stereospecificity drops to 60%. Racemization occurs in methanol (*SKF-72*), apparently due to increasing importance of an S_N2 pathway in the more polar media.

It could of course be assumed that the S_Ni reaction involves apical attack to form an intermediate with the leaving group basal, followed by pseudorotation or turnstile rotation to put the leaving group apical, but there is evidence against this. In the first place the rates of reaction and preference for inversion of R_3Si^*X are proportional to the leaving group ability of X rather than to its electronegativity, which implies that bond breaking rather than formation of a pentacoordinate intermediate is rate-determining. If we assume that formation of the intermediate is fast relative to its decomposition, then it should survive long enough to undergo several pseudorotations, leading to extensive racemization, which is generally not observed. If we shorten the assumed lifetime of the intermediate so as to allow only one or two successive pseudorotations, then the rates of formation and decomposition of the intermediate become approximately equal, which should lead to a more complex dependence of rate upon X than is actually observed. Furthermore, the methanolysis of Me_3SiCH_2Ar followed a Hammett relationship with $\rho = 4.88$ (*S-65*); this high ρ has been used as argument for considerable Si—C cleavage in the transition state and against formation of an intermediate having any effect on the rate, since the electronic effect of Ar on the stability of any pentacoordinate intermediate should be small due to the intervening CH_2-group.

The evidence at hand indicates, therefore, that although the S_Ni reaction involves a 90° angle between the entering and leaving groups, it probably does not involve a pseudorotating pentacoordinate intermediate. However, the situation may be altered in structures where the silicon is a bridgehead reaction center. Thus, substitution at IV, V, and VI (X = H or Cl) also involves this type of S_Ni reaction geometry, with the constraint being configurational rather than electronic. Models (*S-65*) indicate that the C—Si—C angles in these systems are 90–100°, thereby affording considerable relief of strain upon formation of a trigonal bipyramid with the rings spanning apical and basal positions.

All three (IV, V, and VI) chlorides are far more reactive toward water or $LiAlH_4$ than are the corresponding acyclic silanes. This is completely analogous to the rapid hydrolysis of ethylene phosphate (see Sec. III). On the other hand, the corresponding carbon compounds, e.g. VII, are extremely inert since the entering and leaving groups cannot here both be apical. The positions of the infrared Si—H stretching frequency of IV and V in the Smith-Angelotti correlation (*SA-59*, p. 29) correspond to increased s character in the Si—H bond, as if there were electron-withdrawing groups on silicon. This results in acceleration of nucleophilic substitution, in agreement with the high ρ values observed (*S-65*). Since electrophilic assistance to the leaving group, as in the S_Ni mechanism, is hardly necessary here, *Sommer* (*S-65*) proposes that all these solvolyses involve S_N2 with frontside (basal) attack or departure. Although the concerted MO model developed by *Stohrer* (*SS-76*) for S_N2 with retention at carbon in small rings could perhaps be extended to silicon, the known stability of pentacoordinate trigonal bipyramidal silicon compounds renders an actual intermediate undergoing turnstile rotation considerably more likely than for the analogous carbon compounds. In particular, the observed inertness of VII to nucleophilic substitution is in agreement with Stohrer's model and is the complete opposite of the behavior of IV.

More recent work renders the existence of pentacoordinate silicon intermediates in certain nucleophilic substitutions quite likely. A good example can be seen in the methanolysis of R_3SiH (*EJ-74*), which involves nucleophilic attack of methoxide on silicon and electrophilic attack of methanol on the hydridic hydrogen to generate H_2. The solvent isotope effect in this reaction, k_{MeOH}/k_{MeOD}, ranged from 2.3 to 5.03, decreasing with increasing electron donation by R, whereas k_{Si-H}/k_{Si-D} was much smaller and showed little dependence on R. The small silane isotope effect implies a bent transition state VIII, in which the vibration lost in going from Si—H to Si—D is a low-frequency bending rather than a high-frequency stretching vibration (*CH-69*). The dependence of the two isotope effects on R implies that the nucleophilic precedes the electrophilic attack so that there is a distinct pentacoordinate silicon anion intermediate. This is also implied by the values of the *Broensted* coefficient (*OB&-72*). If the reaction were concerted, the observed rate increase with increasing electron

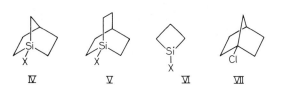

IV   V   VI   VII

acceptance would imply (according to Hammond's postulate) that the transition state increasingly resembles the starting materials as electron acceptance by R increases. As a consequence, less cleavage of the H–O bond should be expected in transition state VIII, and therefore the solvent isotope effect should increase with increasing electron donation by R, which is exactly the opposite of what is observed.

According to precedence in phosphorus chemistry, the best chance for observation of pentacoordinate silicon intermediates undergoing turnstile rotation should involve nucleophilic substitution at small ring silanes under conditions favoring S_N2 rather than S_Ni. However, reaction of *cis* or *trans* IX (*CFM-75*) with KOH in aqueous methanol gave a complex mixture of products, mostly methoxysilanes or disiloxanes, all formed by ring-opening cleavage of the CH_2–Si bond. On the other hand, gas chromatography/mass spectroscopy did suggest the presence of X in the early stages of the reaction, decomposing to XI. Hydrogen evolution was 3,500 times faster from *cis* IX than from XII, one of the reaction products, but the rate of hydrogen evolution from IX decreased greatly after about 30% reaction. It should be noted that the larger atomic radius of silicon compared to phosphorus and the negative charge conferred on pentacoordinate silicon (which does not occur in attack on phosphonium ions) would tend to greatly destabilize the pentacoordinate silicon species. It is not surprising, therefore, to find ring cleavage competing with and apparently slightly faster than turnstile rotation leading to hydride displacement and retention of the ring. Of course, the results could be interpreted in terms of competition between S_N2 and an S_Ni process involving basal attack or departure, but if an intermediate of trigonal bipyramidal structure is involved in nucleophilic attack on an acyclic silicon hydride in methanol, then it is almost certainly involved in similar attack on a silacyclobutane.

The only reaction for which *Sommer* proposed an actual pentacoordinate intermediate is the racemization of R_3Si^*F in pentane or t-butyl alcohol containing methanol. The high energy of the Si–F bond makes an S_N1 (ionization) mechanism extremely unlikely here, and exchange with traces of HF is unlikely since added HF inhibits the reaction, and R_3Si^*OMe is inert under these conditions (*SB-69*). The racemization of optically active chlorosilanes in polar aprotic solvents such as nitrobenzene, nitromethane, and acetonitrile, for which *Sommer* originally (*S-65*) proposed an ionization mechanism, probably also involves complex formation with solvent. However, the racemization also is observed in nucleophilic solvents of low dielectric constant such as dimethoxyethane or ethyl benzoate; it is nearly instantaneous in ethyl acetoacetate (*CLM-68*). It occurs too in less nucleophilic solvents such as dioxane or tetrahydrofuran, given electrophilic assistance to polarization of the Si—Cl bond, by, e.g., $SbCl_3$, $SnCl_4$, or lithium salts. Racemization in CCl_4 containing hexamethylphosphoric triamide, dimethylformamide, or a similar nucleophile was second order in the nucleophile, indicating a symmetrical octahedral intermediate with hexacoordinate silicon (*CL-71*). This is in agreement with the octahedral structure of silicon acetylacetonates (*S-65*). It also explains how the rather crowded conditions resulting from addition of two molecules of CH_3CN solvent produce the 565-fold rate decrease observed on going from XIII to XIV (*CCL-70*).

On the other hand racemization of chlorosilanes by chloride ion has been shown to involve S_N2 and/or S_Ni exchange. Racemization of XV in chloroform (*SH&-71*) or of XIII in acetone-dioxane (*GP-69*) by cyclohexylammonium chloride — ^{36}Cl was exactly twice as fast as exchange, suggesting that each act of exchange proceeds by S_N2, to give a racemic mixture as soon as half of the original chlorosilane has exchanged. For XIII in chloroform, the ratio of racemization to exchange was 1.3, indicating that some of the exchange proceeds by S_Ni (*SH&-71*); in the nonpolar solvent benzene exchange of XIII was 40—80 times faster than racemization, indicating that, as might be expected, almost all of the exchange is by S_Ni (*GP-69*). Were the reaction to involve equilibrium formation of a pentacoordinate or hexacoordinate complex, one should expect $k_{rac}/k_{ex} \gg 2$.

These same conclusions were noted in connection with a linear free energy correlation between the hydrolysis rates in aqueous methanol of (2-chloroethyl)- and (2-hydroxyethyl)aryldimethylsilanes and Hammett σ constants, (*VH&-73*). Had a siliconium ion been involved, a correlation with σ^+ values should have been expected. The high ρ values, which are observed in all media except those of high water content, imply the formation of a carbonium ion stabilized by silicon, as proposed by *Eaborn* (*CEW-70*). In media of high water content the linear free energy plot is curved because compounds in the reaction series with electron attracting substituents on the aryl group possess low ρ values. This characteristic is generally found in solvent-assisted processes.

Allylsilanes undergo a 1,3 migration analogous to the thiallylic rearrangement (*KJ-77*); the reaction proceeds quite readily at 500 °C in the gas phase using a flow system with very short contact time or at 275° in solution (*KS-72*). However, the negligible variation of activation energies and frequency factors on going from Me$_3$SiCHMeCH=CH$_2$ to Me$_2$PhSiCHMeCH=CH$_2$ indicates that *d*-orbitals are not involved (*KS-72*). A suprafacial sigmatropic rearrangement (*WH-69*) involving the Si–3*p*-orbital is suggested and confirmed by the observation that every act of rearrangement is accompanied by inversion of configuration of a chiral reactant (*SK-73*). The carbon analog does not react as readily, probably because its 2*p* orbital is too small to bridge the allylic orbital framework easily. The greater ability of silicon compared to carbon for sigmatropic rearrangements is fairly general; thus isomerization of 1-methylcyclopentadiene to the 2- and 3-methyl isomers due to hydrogen shifting is detectable only above 25 °C (*MSE-63*), whereas scrambling of the olefinic hydrogens in trimethylsilylcyclopentadiene due to silicon migration is rapid on the NMR time scale at this temperature and is even faster for the corresponding germanium and tin compounds (*DR-70*).

Cyanosilanes should be considered here since they exist as an equilibrium mixture of R$_3$SiCN and R$_3$SiNC. The normal form dominates the mixture at room temperature, but the less stable isocyanide form prevails at high temperature. For Me$_3$SiCN, IR data (*BF-70*) indicate 0.13–0.16% iso at room temperature, corresponding to ~4 kcal energy difference between the two forms, but the equilibrium is very strongly temperature-dependent, with the iso form favored at 50 °C (*DG&-73*). As might be expected from the $^+$N≡C$^-$ polarity of isocyanide, the iso form is favored by electron donating substituents on silicon and hindered by electron withdrawing substituents (*ST-76*). Kinetic studies (*ST-76*) at room temperature on the conversion of Me$_3$SiNC, formed at 225 °C and rapidly quenched, to the normal form showed that the process is intramolecular in 0.2 M chloronaphthalene solution. Predictably, organic isocyanides also isomerize to the corresponding cyanides (*SR-63*), but only at high temperature. The much greater ease of reaction for silicon could fit either an S$_N$i type of reaction involving octet expansion of the silicon or a sigmatropic rearrangement on the vehicle of a 3*p* silicon orbital (*SK-73*). Since atoms of differing electronegativity are involved, the former is preferred, even though no base is needed to initiate the reaction. At any rate, it is clear that reactions involving a three-membered ring at silicon as part of a trigonal bipyramidal intermediate or transition state are common and proceed readily.

Because of its ability to form trigonal bipyramidal complexes in which small rings can be accommodated without strain by bridging apical and basal positions, silicon shows a notable migratory aptitude under the influence of nucleophiles. Thus, deprotonation by amine bases of the hydroxyl group in R$_3$SiCR$_2'$OH gives R$_3$SiOCR$_2'$H, almost certainly by attack of –O$^\ominus$ on silicon. As expected for such an intramolecular nucleophilic substitution, the reaction is 13 times faster for R = Ph than for R = Me, but variations in R' have a much greater effect (*BLM-67*). For a series of *p*-substituted Ph$_3$SiMeArOH, a Hammett plot gave ρ = 4.60 at 36° in dimethyl sulfoxide with triethylamine as base, indicating that the Si–C bond is very weak in the transition state, giving it a high degree of carbanion character (*BLM-67*). For optically active silylcarbinols, the rearrangement has been reported to proceed with 100% retention of configuration at silicon (*BWL-67*), suggesting an S$_N$i type transition state rather than a rapidly pseudo-

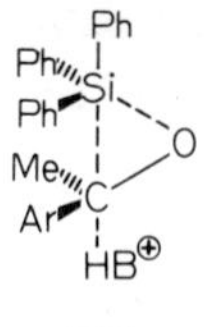

XVI 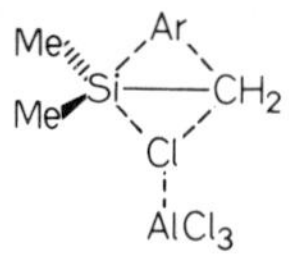   **XVII** 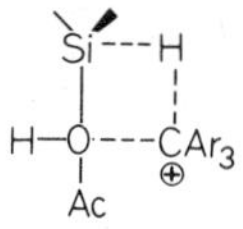   **XVIII**

rotating intermediate. However, further studies using chiral carbon substrates (*BM-71a, BP-71*) provided proof of inversion at carbon and indicated that reprotonation occurs through backside attack on carbon by the protonated base, BH⁺ as in XVI. If there is a free carbanion formed during the course of displacement its lifetime is too short to permit even one rotation around the C–O bond.

It should be noted that this conversion of silylcarbinols to alkoxysilanes is only one example of a quite general type of rearrangement in which silicon migrates between atoms of differing electronegativity; see also (*ET-73*). Other examples include the rearrangement of aminomethylsilanes to methylaminosilanes (*BD-74*); the silylhydrazine rearrangement (*WB-72*); the rearrangement of silylcarboxylic acids to silanols (*B-74*); the thermal decomposition of β-ketosilanes to siloxyalkenes (*B-74*); and the photolysis in alcohols of acylsilanes to siloxyacetals, which involves an intermediate siloxycarbene (*DB-73*). The evidence suggests that siloxycarbenes are formed only in polar solvents; the reaction in carbon tetrachloride has generally been interpreted in terms of silyl radicals, but a more recent study (*PI-74*) suggests that it involves formation of a pentacoordinate silicon adduct with solvent followed by decomposition to chlorosilane and acyl and trichloromethyl radicals. These rearrangements have been reviewed extensively by *West* (*W-69*), and rearrangements involving Si–O bonds have been reviewed by *Brook* (*B-74*).

All of these examples involve a migration from a lesser to a more electronegative atom. Such reactions, when they occur in the presence of catalytic quantities of strong base, can be reversed by excess base. Examples include the conversion of N,O- to N,N-bis(organosilyl(hydroxylamines (*WB-73*) by excess alkyllithium, and the conversion of benzyloxysilanes to the corresponding carbinols (*WL&-71*). Apparently, the chain reaction in the presence of catalytic quantities of base is controlled by the strength of the bond formed, whereas the equilibrium in excess base is controlled by the relative stabilities of the two anions. Thus, benzhydryloxysilanes are not converted to carbinols since the carbanion is so much more stable than the isomeric (*WL&-71*) oxyanion resulting from silicon migration, XIX.

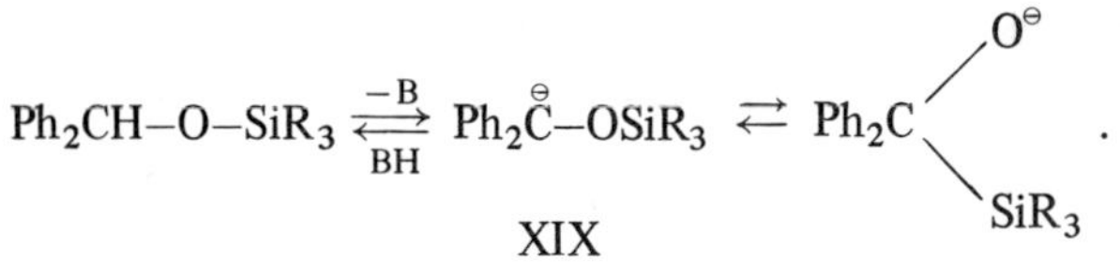

$$Ph_2CH-O-SiR_3 \underset{BH}{\overset{-B}{\rightleftharpoons}} Ph_2\overset{\ominus}{C}-OSiR_3 \rightleftharpoons Ph_2C\underset{SiR_3}{\overset{O^\ominus}{\diagup}}$$

XIX

Brook (*B-74*) has argued that the sila-allylic rearrangement discovered by *Kwart* and *Slutsky* (*KS-72*) should also involve a trigonal bipyramidal intermediate or transition state formed by use of Si–3 *d*-orbitals, and thus proceed with retention of configuration. However, experiments with optically active α-naphthylphenylmethylallylsilane (*SK-73*) showed that deuterium scrambling (silicon migration) and inversion proceed at the same rate as expected for a 1,3-sigmatropic rearrangement (*WH-69*) involving both lobes of the Si–3 p-orbital. The large size of the Si–3 p-orbital makes the transition state for this reaction fairly easy to attain, as indicated by the entropy of activation of -6.2 eu for α-methylallyltrimethylsilane and -4.0 eu for the exactly analogous conversion of propargyl- to allenyltrimethylsilane (*SK-73*) compared to $\Delta S^{\neq} = -34.2$ eu for siloxane formation from $Ph_3SiC(OH)MePh$ (*B-74*). A similar entropy of activation has been found for the sila-Cope rearrangement of *cis-* and *trans-*propenylallyldimethylsilane, XX (*SK-73a*), which involves an intermediate having an Si=C double bond. The activation energy for this process is only 3 kcal/mole higher than for the corresponding carbon system, 1,5-heptadiene (*FS-68*), which indicates that the large size of the Si–3 p-orbital, although it destabilizes the $(3p-2p)\pi$ bond, greatly facilitates sigmatropic rearrangements.

There is a clear distinction between these sigmatropic reactions and the rearrangements discussed by *Brook* (*B-74*) and *West* (*W-69*), all of which involve migration between atoms of different electronegativity, and most of which involve a 1,2 migration of silicon. The presence of electronegative ligands and the formation of three- and four-membered rings would both be expected to stabilize pentacoordinate silicon species formed by use of d-orbitals, whereas no such stabilization exists in the case of allylic ligands.

XX

Recent work has considered the mechanism of the thermal rearrangement of β-ketosilanes to siloxyalkenes, which occurs above 100 °C. The cleanly first-order kinetics and the entropies of activation, which range from -4.2 eu for Ph_3SiCH_2COMe to -16.9 eu for Me_3SiCH_2COPh (*BMB-75*), are consistent with a unimolecular reaction involving a cyclic intermediate or transition state. This is also in keeping with the observation of the complete retention of configuration at silicon. A concerted mechanism has been independently proposed by *Brook* (*BMB-75*) and *Larson* (*LF-75*) on the basis of the Hammett ρ values, which were -0.78 for the series Ph_3SiCH_2COAr (*BMB-75*) and $+0.54$ for $ArMe_2SiCH_2COMe$ (*LF-75*), and on the basis of the Arrhenius activation parameters for Ph_3SiCH_2COPh, which varied negligibly on changing the solvent from toluene to decalin (*B-74*). *Brook* (*BMB-75*) has indicated how a suprafacial migration of silicon can occur with retention of configuration, a possible violation of the Woodward-Hoffmann rules, by use of d-orbitals on silicon to overlap the p orbitals of the termini. However, toluene and decalin are similar in polarity, so that little change in activation parameters would be expected. Moreover, as *Kwart* (*KB-77*) has pointed out, the ρ values were measured at temperatures close to the isokinetic, at which temperature all compounds in the series would react at identical rates. Rate measurements (*KB-77*) for Me$_3$SiCH$_2$COPh show a change in activation energy from 30.6 to 22.9 kcal and in entropy of activation from + 2,7 to − 14.9 eu on changing the solvent composition from pure benzene to largely nitrobenzene; this is clear evidence for an intermediate with charge separation. Final proof of a pentacoordinate silicon intermediate comes from the heavy atom isotope effect (^{28}Si/^{29}Si), which was much smaller than calculated for a dissociative mechanism but well within the range for an associative mechanism (*KB-77*). On this basis it is probably also reasonable to assume a pentacoordinate silicon intermediate in the thermal isomerization of epoxysilanes to siloxyalkenes (*BB&-75, EG-76*).

Thermal rearrangements involving migration of two triorganosilyl groups are also known. *West* observed (*NW-76*) reversible migration between nitrogen and oxygen in tris(triorganosilyl)hydroxylamines at 139°–162 °C, with activation parameters of 29.8 kcal and − 3.3 eu for conversion of (Me$_3$Si)$_2$NOSiMe$_2$H to HMe$_2$Si(Me$_3$Si)NOSiMe$_3$, and 33.0 kcal and − 11.2 eu for conversion of (Me$_3$Si)$_2$NOSiMe$_2$Ph. For the first of these rearrangements the rates were essentially identical in decalin, benzonitrile, or neat. Similar activation parameters, 30.3 kcal and − 9.6 eu at 140°–175 °C, were found for conversion of XXa to XXb, but the relative rates here were not quite independent of solvent, being 1.0 in benzene, 1.8 in o-dichlorobenzene, and 3.5 in propylene carbonate. Finally, in the chirality test 95% retention of configuration was observed at silicon (*RKP-74*). Regretably the dependence of the activation parameters on solvent composition, which is the most reliable probe of the occurrence of dipolar intermediates or transition states, was not determined.

Rather similar activation parameters, 37 kcal and − 1.3 eu at 164°–220°, were reported for the conversion of Me$_3$SiCBrPh$_2$ to BrMe$_2$SiCMePh$_2$. Nonetheless, *Brook* (*BB&-75a*) has argued that the concerted dyotropic mechanism proposed by *West* (*NW-76*) and *Reetz* (*RKP-74*) for their rearrangements, which mechanism engenders a highly ordered transition state such as XXc, should require a much more highly negative entropy of activation than is actually observed. The suprafacial dyotropic process needed to achieve retention of configuration (*R-73*) is of course thermally forbidden, but *Reetz* (*RKP-74*) has claimed that it is allowed when the migrating groups have low-lying unoccupied orbitals. *Brook* proposed (*BB&-75a*) a two-step process involving rate-determining migration of bromine (or any other nucleophilic group) from carbon to silicon to form the "inverse ylid" XXd, followed by methyl migration to form the product. For the *Reetz* (*RKP-74*) rearrangement, the initial migration would have to be from carbon to oxygen, and the intermediate would be either an oxonium ylid or a singly-bridged species with pentacoordinate silicon (probably undergoing rapid turnstile rotation), and with the negative charge distributed between silicon and carbon. Either way the resulting charge separation would demand clear dependence of the activation parameters on solvent as found in related cases involving dipolar intermediates with pentacoordinate silicon (*KB-77*).

For the hydroxylamine rearrangement, one would expect on the basis of electronegativity considerations that the initial and rate-determining migration would be from oxygen to nitrogen, but the reverse might also be possible if the silyl group attached to oxygen has a particularly low migratory aptitude. Thus (*NW-76*) no interconversion of (Me$_3$Si)$_2$NOSiMe$_2$tBu and tBuMe$_2$Si(Me$_3$Si)NOSiMe$_3$ has been achieved;

steric hindrance and electron-donation by the tert.-butyl group make tBuMe$_2$Si- incapable of migration.

This reversible hydroxylamine rearrangement is complicated by an irreversible isomerization above about 180 °C to give R$_3$SiN(R)SiR$_2$OSiR$_3$ (*WNB-76*). A study of this reaction for cases in which the three silyl groups in the initial hydroxylamine are different, particularly cases in which the reversible rearrangement is impossible, e.g. (tBuMe$_2$Si)$_2$NOSiMe$_3$, showed that the group which inserts into the N–O bond is one of those originally bound to nitrogen. Moreover, while the products depend strongly on the nature of the silyl groups, the kinetics do not. While one could postulate silyl migration from nitrogen to oxygen to form a singly-bridged species which decomposes by N–O cleavage, such a mechanism seems unlikely in view of the ionic character of the N–O bond in the bridged species, which should hinder cleavage. Furthermore, the lack of rate dependence on solvent and on the nature of the silyl groups is not compatible with this mechanism. *West* (*WNB-76*) proposes a more likely mechanism: rate-determining homolytic N–O cleavage followed by reversible attack of the siloxy radical within the solvent cage on one of the silicon atoms of the disilylamino radical to form a betaine XXe with pentacoordinate silicon. Decomposition of XXe by migration of R would explain the observed dependence of product on the structure of the starting material, as for example the exclusive formation of Me$_3$SiNHSiMe$_2$OSiMe$_3$ from (Me$_3$Si)$_2$NOSiMe$_2$H or HMe$_2$Si(Me$_3$Si)NOSiMe$_3$ (which are of course in equilibrium) since hydride migration should occur much more readily than methyl migration.

Silicon may participate in reactions at adjacent carbon even where there is no apparent nucleophile. Thus, the conversion of PhSiMe$_2$CH$_2$Cl to PhCH$_2$SiMe$_2$Cl by aluminum chloride, a silicon analog of the Wagner-Meerwein rearrangement for which *Sommer* suggested a siliconium ion mechanism (*WSG-47*), has been shown not to involve C–Cl cleavage as the rate-determining step. *Eaborn* (*BEM-65*) has urged XVII, a "push-pull" mechanism with simultaneous phenyl migration and attack by chlorine on silicon as an alternative to electron deficient silicon. Even the attack of a silyl hydride upon a carbonium ion in strong acid has been suggested to involve a four-center transition state (*CH-69*) such as XVIII. These circumstances occur in the attack of triarylsilyl hydrides upon tris(2,6-dimethoxyphenyl)methyl carbonium ion in acetic acid, a medium in which this carbonium ion is stable. A Hammett plot of this reaction gives $\rho = -1.87$ compared to $\rho = -3.97$ for ionization of triarylmethyl chlorides in liquid SO_2 (*G-59, CH-69*).

Although siliconium ions do not exist, silicon is quite capable of stabilizing adjacent carbonium ions. Solvolysis of XXI in aqueous methanol with loss of ethylene gave recovered bromide containing XXI and XXII in a 2 : 1 ratio (*CEW-70*). Since XXIII was not detected, there could be a free, bridged carbonium ion (XXIV). The rate of bromide loss, about equal to the rate of ionization of tert-butyl chloride (*CEW-70*), was considerably greater than that expected from the inductive effect of the Me_3SiCH_2-group. The accelerating effect may be due to silicon bridging in the transition state, i.e. anchimeric assistance, which is formally analogous to a siliconium ion and, therefore, conceptually difficult. On the other hand, this may be another example of a Si–3p orbital participating in a 1,2 sigmatropic migration, and conceptually analogous to the symmetrical transition state of the sila-allylic rearrangement (*SK-73*). Silicon-carbon hyperconjugation (*P-70*) has also been discussed as a possibly significant factor in the observed solvolytic rate acceleration.

$Me_3SiCH_2CD_2Br$ $Me_3SiCD_2CH_2Br$ $Me_3SiCHDCHDBr$
XXI   XXII   XXIII

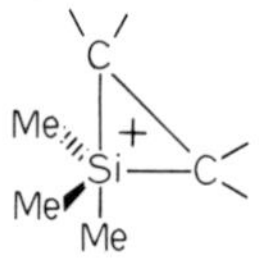

XXIV

For similar reactions of Me_3SiCMe_2Cl (*HO-71*) or $Me_3Cl_2Si_2CH_2Cl$ (*TK-71*), where more stable carbonium ions are to be expected as a result of ionization at a tertiary carbon (*HO-71*), or of strong electron donation by the disilyl group (*TK-71*), the kinetics suggest a two-step process involving formation of the silyl- or disilyl-stabilized carbonium ion, followed by silicon migration. Either of these steps may be rate-determining, depending on the stability of the intermediate carbonium ions; that is to say, the more stable the carbonium ion, the greater the tendency for silicon migration to be rate-determining.

Survey of Relevant Articles on Pentacoordinate Silicon Intermediates (Table V E)

System and Results	Conclusions	References
Aminomethylsilanes are converted to methylaminosilanes by BuLi. The reaction is faster for Ph_3SiCH_2NHR than for Me_3SiCH_2NHR and is faster for R = Me than for R = benzyl. For R = Ph the reaction does not occur.	Initial deprotonation at nitrogen is followed by rate-determining silicon migration in a mechanism similar to that for the silylcarbinol-siloxyal-	(*BD-74*)

System and Results	Conclusions	References
	kane rearrangement; (see text p. 154ff and *BLM-67*). Thus, the reaction is favored by species which increase the positive charge on silicon and the negative charge on nitrogen.	
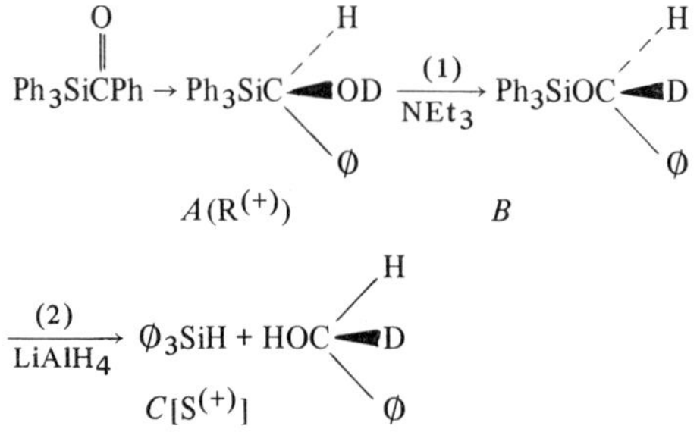 $Ph_3SiCPh \rightarrow Ph_3SiC{-}OD \xrightarrow[NEt_3]{(1)} Ph_3SiOC{-}D$ (with O, H, ø substituents) $A(R^{(+)})$ ... B $\xrightarrow[LiAlH_4]{(2)} \emptyset_3SiH + HOC{-}D$ $C[S^{(+)}]$	Since the configurations of A and C are fixed by X-ray diffraction, and since step (2) must proceed with retention, the Brooks rearrangement of A to B must proceed with inversion at carbon. Brook's assumption of retention at carbon was based upon application of Cram's rule, which apparently does not apply to reduction of α-silyl ketones.	(*BP-71*)
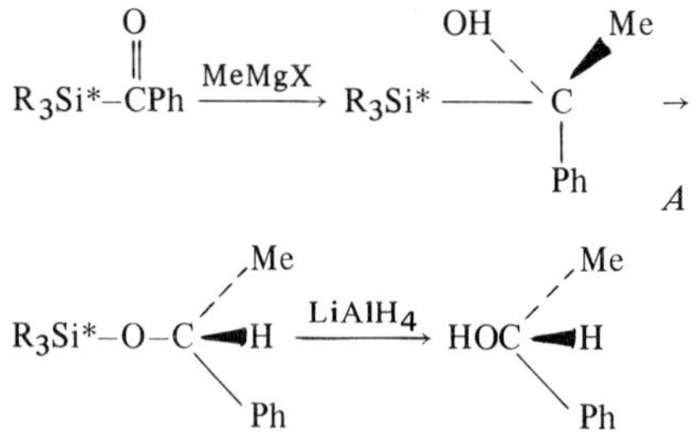 $R_3Si^*{-}CPh \xrightarrow{MeMgX} R_3Si^*{-}C(OH)(Me)(Ph)$ A $R_3Si^*{-}O{-}C(Me)(H)(Ph) \xrightarrow{LiAlH_4} HOC(Me)(H)(Ph)$ R_3Si^* = α-naphthylphenylmethylsilyl X-ray diffraction of A (as the *p*-bromobenzoate) showed it to have the $R^{(+)}$ configuration.	See (*BP-71*) (above)	(*BM-71a*)
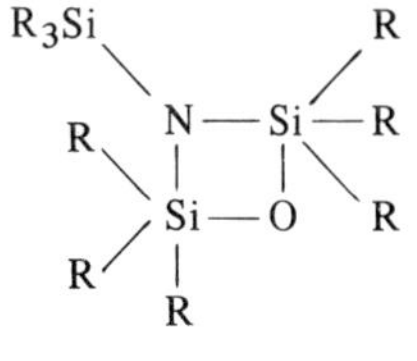 Structure A: cyclic with R_3Si, N—Si—R, Si—O bonds, R substituents $(R_3Si)_2NOSiR_3 \xrightarrow{200°/20\text{ hr}} R_3SiNR{-}SiR_2{-}OSiR_3$; R = Me or Et. Mixing the two homologs gave no cross products.	Intramolecular mechanism involving formation of intermediate or transition state A, followed by Si–N or Si–O cleavage to form an acyclic intermediate with one anionic, pentacoordinate silicon, from which R migrates to N.	(*BW-71a*)
$Me_3SiCH_2CH_2Br + H_2O \xrightarrow{80\% \text{ aq EtOH}} Me_3SiOH + C_2H_4$ $Me_3Si{-}CH_2CD_2Br$; $k_H/k_D = 1.12$ $Me_3Si{-}CD_2CH_2Br$; $k_H/k_D = 1.02$	Lack of expected β isotope effect suggests anchimeric assistance by Si to carbonium ion formation.	(*JHT-70*)

System and Results	Conclusions	References
In the *cis*-triorganosilyl acetylacetonates NMR shows a temperature-dependent equivalence of the two methyl groups of the acetylacetonate. Rates increase with increasing polarity of the organo groups, i.e., Et < Me < Ph < CF$_3$CH$_2$CH$_2$. For Me the Arrhenius activation energy is 13.05 kcal/mole, and the logarithmic frequency factor is 13.8.	Equilibration occurs via a pentacoordinate intermediate or transition state. 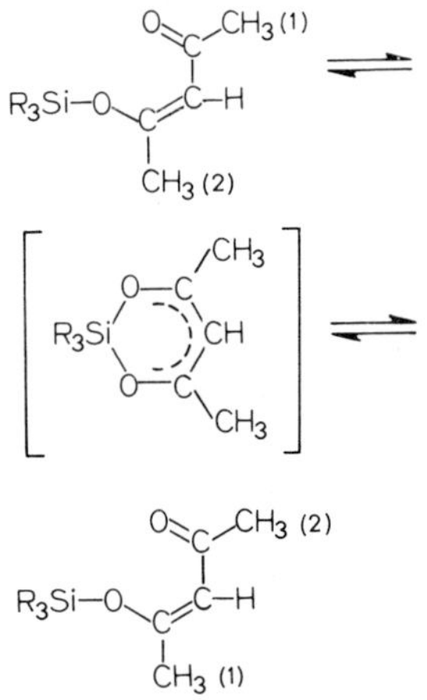 R$_3$Si–O–C(=O)CH$_3$(1) / C(CH$_3$)(2)=C–H ⇌ [R$_3$Si(O–C(CH$_3$))$_2$CH] ⇌ R$_3$Si–O–C(=O)CH$_3$(2) / C(CH$_3$)(1)=C–H	(PCH-70)
A: (aryl) with –N(R)SiMe$_2$R″ and –NHSiR′Me$_2$ substituents; 1) MeLi/Et$_2$O 2) pyrrole B: (aryl) with –NHR and –N(SiMeR′)–Si–Me$_2$R″ substituents	See (W-69) (below) and (BW-71a) (above)	(SKW-70)

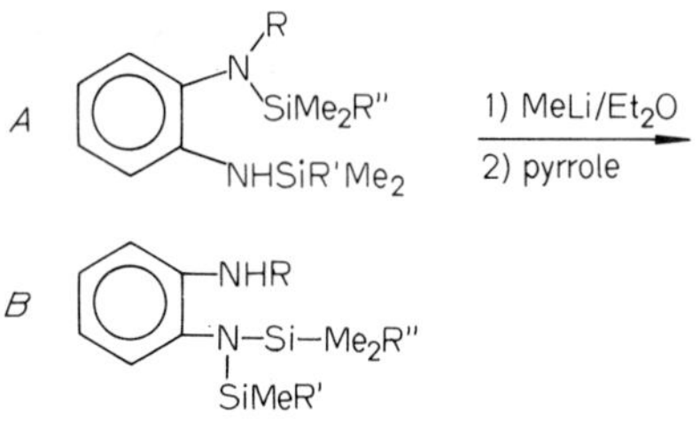

R	R′	R″	% B at equilibrium
Ph	Me	Me	85
Ph	Me	tBu	12
H	Me	Me	17
Me	Me	Me	8

Silanes with Si–H bonds are converted cleanly to silanols by ozone with retention of configuration. In the Taft linear free energy correlation, $\rho = -1.25$, and both R$_3$SiH and RSiH$_3$ fit the same correlation line.	Fast initial attack of ozone on silicon to form A is followed by rate-determining hydrogen abstraction by the terminal oxygen of A, thus explaining the ρ value; since this step is intramolecular the statistical factor is eliminated. The lifetime of A is short enough to prevent extensive pseudorotation, thus explaining the retention of configuration; but the product on standing at room temperature in presence of	(SP&-71)

System and Results	Conclusions	References
	ozone does racemize consistent with reversible complexation to pentacoordinate species undergoing pseudorotation. 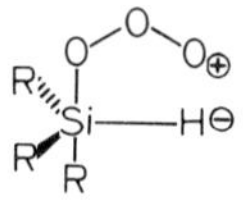	
Hydrolysis of Ph_3SiOR in 40% aqueous acetonitrile up to 0.005 M in hydroxide ion was second order in hydroxide for R = Me, but first order at high hydroxide concentration and second order at low for R = $MeOC_6H_4^-$. Methanolysis of the aryloxysilane in formic acid/sodium formate buffer showed simultaneous general acid/general base catalysis.	Simultaneous acid and base catalyis would seem to exclude any intermediate; thus the methanolysis is simple S_N2. However, MeO^- is a poorer leaving group than HO^-, and water is a much weaker acid than formic acid, and cannot catalyze fast ArO^- loss. Thus, there could possibly be fast reversible formation of a pentacoordinate silicon intermediate, which decomposes, apparently *via* a hexacoordinate transition state, upon attack of a second hydroxide.	(SP&-74)
$R_3Si^*H + PhCO_3H \rightarrow R_3Si^*OH$: Retention of configuration. $R_3Si^*H + :CX_2 \rightarrow R_3Si^*CHX_2$: Retention of configuration. Carbene generated from $PhHgCX_3$: X = Cl or Br R_3Si^* = α-naphthylphenylmethylsilyl.	Both reactions are formulated to proceed by S_Ni, with a cyclic three-membered transition state or so-called insertion mechanism.	(SUP-72)
Review of rearrangements of the type $(Me_3Si)NH$~$N(R)Si-Me_3 \rightarrow (Me_3Si)_2N$~$NHR$ brought about by catalytic quantities of RLi. Equilibrium between the two isomers is very rapid for hydrazines, ethylenediamines and o-phenylenediamines. R % $(Me_3Si)_2NNHR$ in equilibrium mixture H 51 Ph 91	Rearrangement is favored by formation of more stable anion; α-silyl anions are destabilized by electron-donating inductive effect of Si.	(W-69)

$Me_3SiNHOSiMe_3 \underset{H^+}{\overset{RLi}{\rightleftarrows}} Me_3SiNOSiMe_3^\ominus +$
                              A
$(Me_3Si)_2NO^\ominus \xrightarrow{MeI} (Me_3Si)_2NOMe$.
     B
IR spectroscopy shows that the equilibrium between anions A and B favors the latter and that the Si–N bonds are quite strong.
	See (W-69). B is the more stable anion because the charge is on the more electronegative element. In addition, repulsion by this negative charge makes the nitrogen lone pair more available for $dp-\pi$ bonding with silicon.	(WB-73a)

System and Results	Conclusions	References
PhCH$_2$OSiEt$_3$ $\underset{\text{Na-K in Et}_2\text{O}}{\overset{\text{excess tBuLi in C}_5\text{H}_{12}}{\rightleftarrows}}$ Et$_3$SiCHOH$\underset{\phi}{|}$	The Brooks rearrangement (see text), which requires catalytic quantities of base, can be reversed by using > 1 equivalent of base. In the former case the substrate is predominantly neutral, and the driving force is formation of a strong Si–O bond. In the latter case the substrate is entirely anionic, and the driving force is conversion of a carbanion to a much more stable alkoxide ion.	(WL&-71)
CNDO calculations on SiH$_5^-$ show it to be 16.9 kcal/mole more stable than SiH$_4$H + H−. In SiH$_4$X where X is a hydrogen-like atom with a Pauling electronegativity of 2.9, the preference of X for the apical position is 3.49 kcal/mole. Pseudorotation barriers were calculated as 2.87 kcal in SiH$_5^-$ and 2.94 kcal in SiX$_5^-$.	These pseudorotation barriers compare to 3.9 kcal for PH$_5$ (RAM-72) and slightly higher for PF$_5$. Apparently there is little charge rearrangement on formation of the square-pyramidal transition state for pseudorotation of SiH$_5^-$, possibly because of the overall negative charge.	(SS-73)
Electrolytic oxidation of Me$_3$SiNHOSiMe$_3$ results in hydrogen abstraction with rearrangement to form the nitroxide radical (Me$_3$Si)$_2$NO, for which the nitrogen hyperfine coupling constant is much lower than for most nitroxide radicals, even those with strongly electron-withdrawing groups, e.g. (CF$_3$)$_2$NO.	The rearrangement is favored by the much greater stability of the nitroxide radical over Me$_3$SiNOSiMe$_3$. The low A_N proves strong Si–N, dp–π bonding, which increases the separation between the π and π^* orbitals of the nitroxide N–O bond, stabilizing the π orbital which contains the N lone pair and destabilizing the π^* orbital which contains the unpaired electron, and, in effect, increasing the spin density on oxygen.	(WB-73)
Rates of reaction of AX and BX with diisobutyl-aluminium hydride are in order BCl > ACl but AOMe > BOMe. $AX =$ adamantane-like cage structure with Me groups on three Si vertices, Me-Si, Me-Si, Si-X BX = (Me$_3$SiCH$_2$)Si–X X = Cl, OMe.	Departure of –OMe by $S_N i$ involves a cyclic transition state, which is hindered in the acyclic system B. Departure of Cl is by $S_N 2$, even though in the adamantane system A it must proceed with retention of configuration (apical-basal entry and departure).	(HS-73)

F. Synthetic Applications of Pentacoordinate Intermediates

Synthetic applications of pentacoordinate sulfur or phosphorus intermediates are numerous, but most of them involve ylides rather than sulfuranes or phosphoranes. Applications involving sulfur or phosphorus ylides have been comprehensively discussed in earlier books (*J-66, TM-75*) and will not be treated in such detail here.

Trost (*TM-75*) has proposed that ylides be named as π-sulfuranes or π-phosphoranes in contradistinction to σ-sulfuranes or -phosphoranes, which are bonded to four or five ligands, respectively. This is convenient for ease of nomenclature, but it may obscure the real nature of the bonding in these species. Trost earlier has urged that the short P=C or S=C bond lengths in ylides (see Section III.A) cannot be taken as evidence for extensive $dp-\pi$ contributions to this bond. A recent MO study (*WWB-75*) showed that the C–X bond lengths in the CH_3X (X = O or S) radicals, anions, and cations could not be associated with overlap populations, but instead correlated linearly with ionic bond order. Thus, the short bond length in ylides is more an indication of strong ionic than of $dp-\pi$ character. In fact, most of the important reactions of ylides depend upon the carbanionic character of the ylide carbon.

The importance of the Wittig reaction as a synthetic tool is so well known that no examples need to be given. Although the reaction has been extensively studied and reviewed in full detail (See e.g. *J-66*), there is still some question as to the nature of the intermediate, generally having been formulated as a betaine. For stabilized ylides, where the negative charge localized on the ylide carbon is low, no intermediate is isolable because its formation by attack of the carbonyl carbon on the ylide carbon is slower than its decomposition. Nevertheless, in cases of unstabilized ylides, a crystalline betaine is sometimes isolable (*WWS-61, J-66*). In fact, although $Ph_3P=CH_2$ reacts with benzophenone to give the expected 1,1-diphenylethylene, $Me_3P=CH_2$ gives only a crystalline betaine from which the carbinol $[Me_3PCH_2C(OH)Ph_2]^+$ is formed upon acidification. It has been pointed out, however, (*AS-76*) that the isolation of a crystalline betaine during reaction with unstabilized ylides may be due simply to the presence of lithium, which is always present in complexation with ylides prepared by reaction of a phosphonium salt with alkyllithium. More recent work suggests that the intermediate may be a phosphorane rather than a betaine. For instance, in the reactions of $Ph_3P=CHCOOEt$ with benzaldehyde (*RPE-67*) and of I with *p*-nitrobenzaldehyde, (*F-72*), the rates increased with decreasing solvent polarity, and the entropy of activation was less than -30 eu, which is hardly the behavior expected of a reaction involving rate-determining formation of a zwitterion. Thus, the initial step in the Wittig reaction is perhaps best formulated as a four-center cycloaddition II, with simultaneous formation of a C–C and a P–O bond to give a neutral phosphorane. For I this would explain the rate increase in going from R = Ph to R = Et as a decrease in steric hindrance to attack of the carbonyl oxygen on phosphorus. Similarly, in the reaction of $Ph_3P=CHCH_3$ with various aldehydes and ketones (*VS-73*), the ^{31}P chemical shifts of the intermediates ranged from $+61.9$ to $+66.5$ vs H_3PO_4, which is characteristic of neutral pentavalent phosphorus rather than a tetrahedral phosphonium salt.

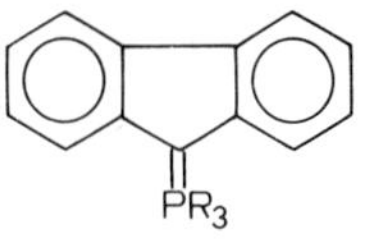

The phosphorane once formed would contain an apical P–O bond which is probably quite weak. This is illustrated by the one known X-ray diffraction study of a Wittig intermediate, i.e. III (CD-68). The P–O length here was 2.01 Å, which, although quite the longest P–O bond known, is still considerably shorter than the sum of the van der Waals radii. The four-membered ring was planar, and the C–O bond length was 1.39 Å, characteristic of a single bond. In this case, however, the phosphorane is destabilized by the tendency for the positive charge on the exocyclic phosphonium group to be delocalized over the P–C–P system, as shown by the identical lengths, 1.76 Å, of the endo- and exocyclic P–C bonds. The phosphorane would also be destabilized by electron-donating substituents on phosphorus. This would explain the tendency of the adduct of $Me_3P=CH_2$ and benzaldehyde (WWS-61) to decompose by P–O cleavage to give the betaine, whereas that from $Ph_3P=CH_2$ survives long enough to undergo pseudorotation (or more likely, turnstile rotation) and P–C cleavage to form the olefin and triphenylphosphane oxide. Note, too, that this sequence of reactions is much more consistent with the generally high specificity (yielding cis olefins in the Wittig reaction) than could be correlated with a betaine intermediate in which there would presumably be free rotation around the P–C bond.

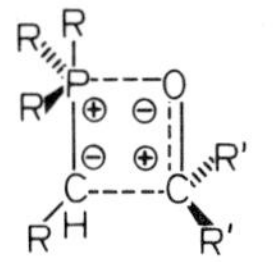

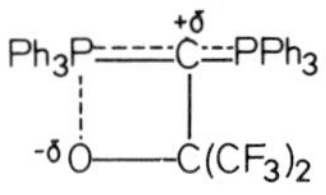

Ylides are generally prepared by reaction of the appropriate sulfonium or phosphonium salt with a strong base, generally an organolithium. However, organolithiums can also act as nucleophiles to form σ-sulfuranes or phosphoranes; since both nucleophilic and base attack are reversible, there are obviously many possible side reactions. Such is the case, in what is probably the earliest recorded observation of a σ-sulfurane, when *Franzen* and *Mertz* (FM-61) reacted triphenylsulfonium bromide with n-butyllithium and found, among other things, phenyllithium and 1-butene. This is best explained as a nucleophilic attack by butyllithium at the sulfonium center to form butyltriphenylsulfurane, which, in turn, decomposed to phenyllithium (phenyl carbanion) and butyldiphenylsulfonium ion, followed by base attack to form the corresponding ylide from which 1-butene is formed by Hofmann degradation.

Sulfuranes are not limited to decomposition by carbanion extrusion as illustrated above; coupling of an apical and a basal substituent may also occur, generally with aromatic ligands. Thus, reaction of triphenylsulfonium tetrafluoroborate with carbon-14 labelled phenyllithium (HW&-72) gave biphenyl and diphenyl sulfide, with the radioactivity equally distributed between the two products. This result is strongly suggestive of an intermediate tetraphenylsulfurane. The specific radioactivity of the products increased with reaction time, a fact which is best explained by reversible sulfurane formation. In the initial stages of the reaction this leads to unlabelled phenyllithium, and consequently to unlabelled products.

As an example of the complexity of this reaction, we may cite the reaction of butyllithium with allyldiphenylsulfonium ion (*TL-68*), which gives seven products. The major hydrocarbons are propylene and benzene, formed by simple extrusion of allyl or phenyl carbanion from the sulfurane. The only coupling product was diphenyl (5.3% yield); neither allylbenzene, propenylbenzene, butylbenzene, nor l-heptene was observed, which suggests that the coupling reaction is limited in scope.

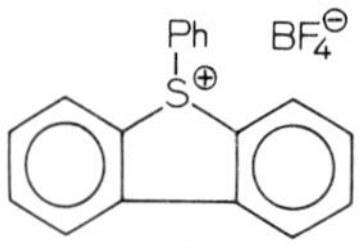

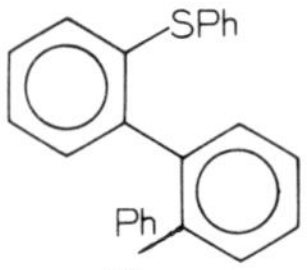

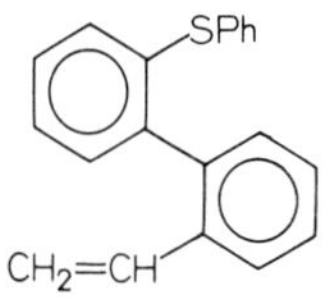

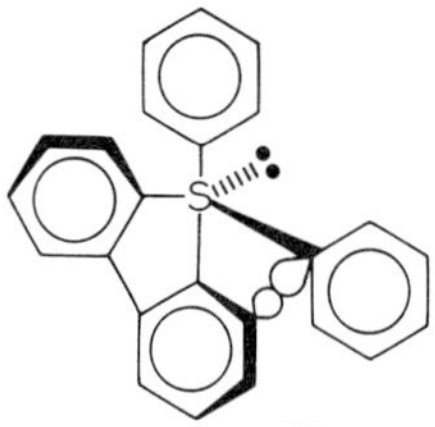

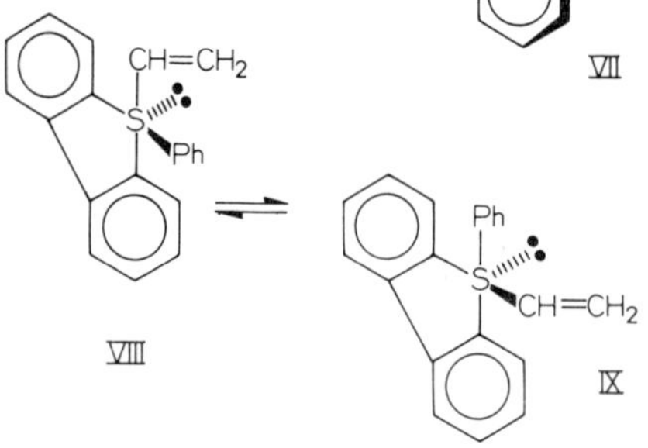

Trost (*LT-71*) has studied this coupling reaction in some of its details. He has indirectly proven, by a process of elimination, the occurrence of a sulfurane intermediate in the reaction of $Ar_3S^+ BF_4^-$ with RLi to give ArR and Ar_2S. Of the three possible mechanisms other than sulfurane formation, a benzyne type intermediate is eliminated by the stoichiometry, since only one mole of organolithium is required, and a free radical mechanism is eliminated by the stereospecificity, the *cis* or *trans* configuration of the R group being retained. A mechanism involving nucleophilic aromatic substitution is possible, but in that case reaction of IV with any organolithium should involve preferential attack on the biphenylyl group. In fact, reaction with phenyllithium gave 100% V, but reaction with vinyllithium gave only 8% VI, along with 60% styrene. These data are readily explained by sulfurane formation. Sulfurane VII must collapse to products by overlap of the π-electron systems of a basal and an apical group; since the electronic requirements for this occurrence are similar to those for nucleophilic aromatic substitution, the basal phenyl group should overlap preferentially with the biphenylyl group, as observed. On the other hand, vinyllithium would give VIII, which can pseudorotate to give IX. VIII can collapse only to styrene since overlap of the vinyl and biphenylyl group is here sterically impossible, whereas the vinyl group in IX would overlap preferentially with the biphenylyl group. Evidently, however, decomposition of VIII is faster than pseudorotation; in view (*B-60*) (Section V.A) of the high activation energy probably required for sulfurane formation, this is understandable when one considers that the reaction was carried out at −78 °C (*LT-71*).

Further work has shown (*TA-73*) that the coupling involves simple orbital overlap of an electron-rich and an electron-poor ligand; the biphenylyl group being considered as electron-poor since the o-phenyl group is electron withdrawing. For example, the sulfurane derived from X and Ar'Li gives 93.4% XI by coupling of Ar rather than Ar' with biphenylyl, when Ar = phenyl and Ar' = *p*-tolyl, but when Ar' = *p*-$CF_3C_6H_5$, 92.1% XI and only 3.2% XII is obtained. In the latter case, but not the former, biaryl Ar−Ar' is also formed since phenyl cannot couple with an electronrich ligand. The yield of biaryl is only 2.4% indicating that this coupling is slow in comparison with the pseudorotation process which places Ar' basal, as is necessary before it can couple with the biphenylyl group. On the other hand, vinyl is more electron rich than phenyl; thus, reaction of X, (Ar = m-trifluoromethylphenyl), with vinyllithium gives 22.3% XIII and 36.5% m-trifluoromethylstyrene, indicating that in this case coupling can compete with pseudorotation. However, only 3.8% *p*-methoxystyrene is formed with Ar = *p*-anisyl since coupling between two relatively electronrich ligands is slow.

Biaryls are also formed by reaction of aryllithiums with diaryl sulfoxides (*AA&-74*). In this case, however, the consumption of two equivalents of aryllithium and the formation of almost equal amounts of *pp'*-bitolyl and *mp'*-bitolyl from *p*-tolyllithium and bis(*p*-tolyl) sulfoxide indicate that an aryne intermediate is involved. The aryne cannot be derived from the triarylsulfonium salt, since the latter is converted by *p*-tolyllithium almost entirely to the *pp'*-bitolyl, in agreement with results observed earlier on IV (*LT-71*). A possible explanation is that the initial intermediate is sulfurane XIV which decomposes by two parallel paths: loss of OLi to form the sulfonium salt which reacts further to give *pp'*-bitolyl, and loss of aryne to give *mp'*- and *pp'*-bitolyl and XV, from which is derived the other main product bis(*p*-tolyl) sulfide. This is supported (*AA&-74*) by the formation of only 1−2% *mp'*-bitolyl in the reaction of $Ar_2S=NTos$

(Ar = p-tolyl) with p-tolyllithium. Since TosNH⁻ is a much better leaving group than OLi in XIV, the occurrence of the aryne decomposition path is much reduced.

Sulfuranes are apparently also involved in the reaction of sulfonium ylides with cyclopropenium salts (*TAH-73*). For R = Me, reaction of XVI and XVII gave only the thiophene XVIII as the major product, whereas for R = Ph, large amounts of XIX were also formed. This is best formulated as involving cyclization of ylide intermediate XX to sulfurane XXI; as expected, methane and ethane were formed in a combined yield equal to that of XVIII. Decomposition of XXI may be concerted, but methane formation would suggest that it is at least in part heterolytic involving methyl carbanion. However, an attempt to test this by reacting XVI with a 1:1 mixture of XVII and XVIIa gave all the possible deuterioethanes, indicating that ylide XX, the likely precursor, is more stable than sulfurane XXI and undergoes rapid exchange of its methyl groups (deuterium exchange). The stability of XX also explains formation of XIX by addition of XVI and XX followed by loss of ethane; this occurs for R = Ph but not R = Me because of the greater reactivity of the triphenylcyclopropenium salt. Although XXI is obviously very unstable, its decomposition by a heterolytic path would suggest that it is an intermediate rather than a transition state.

Coupling reactions have also been observed with phosphoranes. Thus, styrene was formed in 65% yield (*SFH-66*) by reaction of vinyllithium with tetraphenylphosphonium bromide, in all likelihood via an intermediate vinyltetraphenylphosphorane. Similarly tetra-p-tolylphosphonium bromide gave a 61% yield of p-methylstyrene. When cis- or trans-propenyllithium was substituted for vinyllithium, cis or trans-propenylbenzene were formed, respectively, i.e. retention of configuration, the same as observed by *Trost* (*LT-71*) with sulfuranes.

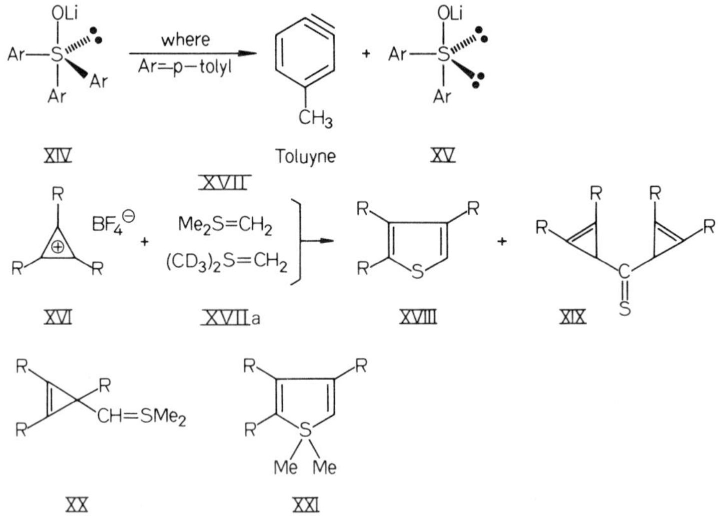

Probably more useful synthetically is sulfurane or phosphorane fragmentation, which involves simultaneous cleavage of two bonds to the central atom, generally to form an olefin. Since the reaction is concerted, its stereochemical restrictions can be determined by the Woodward-Hoffmann orbital symmetry conservation method (*WH-69*). This approach teaches that fragmentation of cyclic sulfuranes or phosphoranes may be either disrotatory or conrotatory depending on whether sulfur or phosphorus acquires its lone pair prior to departure as phosphane or sulfide from the symmetric (linear fragmentation) or antisymmetric (nonlinear fragmentation) S–C or P–C orbital. To illustrate, XXII (*TZ-71*) reacts with butyllithium to give 2,4-hexadiene, with *cis* XXII giving a 62% yield of the *cis, trans* diene and *trans* XXII giving 33% *trans,trans* and 58.5% *cis,cis*; eclipsing interactions apparently hinder formation of the *trans,trans* isomer. Thus, the intermediate sulfurane decomposes by concerted disrotatory fragmentation.

A similar intermediate is implicated in the desulfurization of episulfides by alkyllithium, which proceeds with almost complete retention of configuration (*TZ-73*). Since decomposition of the presumed intermediates, e.g. the threo and erythro carbanions XXIX and XXX from 2-butene episulfide, was found to proceed with much less stereospecificity than the overall desulfurization, the only reasonable assumption is that the actual intermediate or transition state XXXV formed by attack of butyllithium on sulfur, has a sulfurane structure and undergoes concerted disrotatory fragmentation to eliminate the olefin 2-butene.

According to the Woodward-Hoffmann rules, these results imply linear fragmentation of XXII but nonlinear fragmentation in episulfide desulfurization. Similar results occur with phosphoranes; the phosphiran XXIII (*DS-74*) reacts with XXIV at $-78\,^\circ\text{C}$ to give 95% *cis*-1,2-dideuteroethylene, apparently *via* disrotatory fragmentation of phosphorane XXV. Similarly, XXVI (*DD&-72*) formed by reaction of XXVII with diethyl peroxide, decomposes to isoprene and $PhP(OEt)_2$.

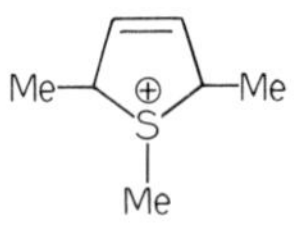

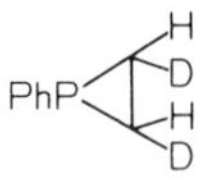

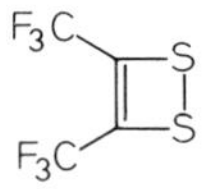

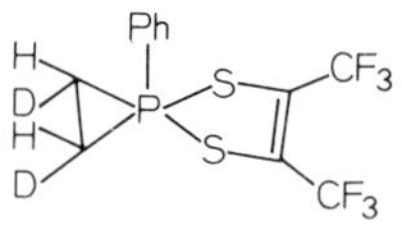

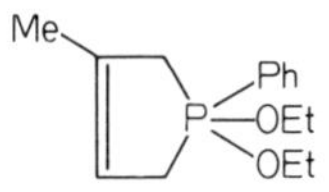

Unfortunately, *Hoffmann* (*HHM-72*) has shown that concerted disrotatory fragmentation of phosphoranes (and by extension sulfuranes) is symmetry forbidden unless the two bonds to be broken are both apical or both basal. Although these calculations were made for the elimination of H_2 from PH_5, the much longer time required for elimination of *trans,trans*-2,4-hexadiene from XXXI compared to XXXII (*HBL-72*) suggests that the same symmetry rules apply to formation of both π and σ

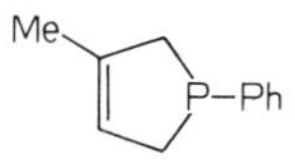

bonds in the elimination. In the reaction of XXXIV with diethyl peroxide, XXXII could not be isolated at room temperature, whereas the half-life of XXXI was 2 hours at 44 °C, corresponding to a free energy barrier of about 25 kcal/mole. Olefin formation from XXXI can only occur via the unfavorable isomer XXXIII, whereas in XXXII the strain induced by formation of the basal-basal ring formed by pseudorotation is compensated by the two apical ethoxy groups. Similarly olefin formation from XXXVa would require it to accept the structure of a highly-strained basal-basal ring and an apical lone pair. This possibly does occur since two-step fragmentation would only produce the carbanions XXIX and XXX, which have already been eliminated from consideration. Although the low basal-basal angle in sulfuranes (*PMP-71*) means that ring strain on pseudorotation of XXXV would be no greater than in episulfones, which are isolable at room temperature (*P-68*), and although fragmentation and pseudorotation could be concerted, one would still expect a high energy barrier for butene formation from XXXVa. The occurrence of ~ 90% conversion after 1 hour at 40 °C does correspond to a free energy barrier of over 20 kcal/mole.

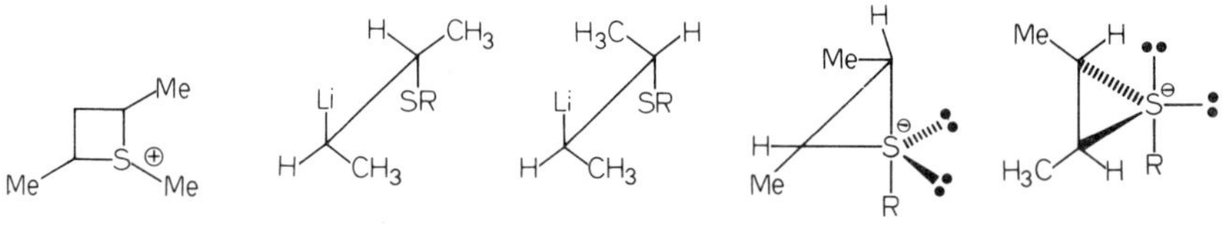

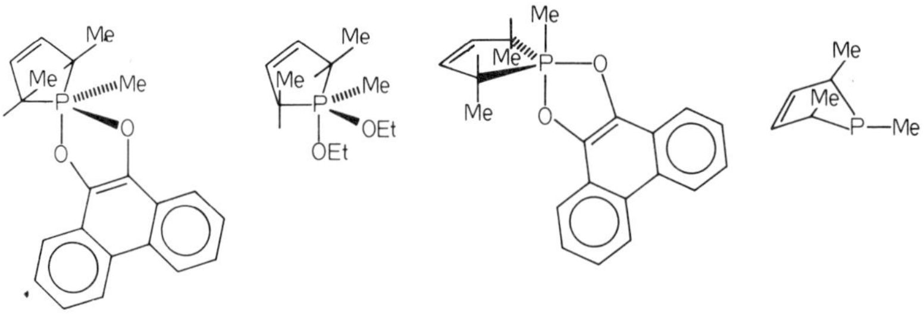

An interesting example of fragmentation occurs in the reaction of XXVIII (*TSM-69*) (*TS&-71*) with butyllithium to give 1,2-dimethylcyclopropane in 22% yield, but with almost complete retention of configuration, i.e. cis-XXVIII giving the cis cyclopropane and trans XXVIII the trans cyclopropane. This stereochemistry would require conrotatory fragmentation, a path of high strain energy in small rings. Consequently, this reaction may well involve a diradical intermediate.

Cyclic oxyphosphoranes decompose thermally to epoxides. Since this occurs with inversion of configuration, as shown by *Ramierez* (*RGS-68*) who obtained trans-dinitrostilbene epoxide XXXVI from cis XXXVII, it very likely involves P—O cleavage followed by $S_N 2$ attack of $-O^\ominus$ on carbon. Similarly triphenylphosphane reacts with XXXVIII (*BBL-73*) to give epoxide XXXIX.

Reaction of phosphanes with aldehydes or ketones is a common method of synthesis of oxyphosphoranes (*R-68, R-70*). Since two molecules of aldehyde add in separate steps, the stereochemistry is complex with the ratio of the two diastereomeric products depending on solvent, temperature, and phosphane-aldehyde ratio (*RGS-68*). Each diastereomer contains two isomers interconverting by pseudorotation or turnstile rotation, but the two diastereomers cannot interconvert without C–C cleavage. At 140°, 80°, and 150 °C, respectively, the three phosphorane species derived from benzoyl cyanide (*WBM-75*) and either triethyl phosphite, XL, or XLI decomposed entirely to *cis* XLII. Either the *trans* diastereomer of the phosphorane is formed in all cases, or the intermediate betaine XLIII survives long enough to undergo rotation around the C–C bond and assume the most stable conformation, as shown. This is supported by the reaction of benzoyl cyanide with tris(dimethylamino)phosphane, which gives the *cis* epoxide XLII directly at 0 °C. In this case the betaine is formed directly from the phosphane and benzoyl cyanide since steric hindrance prevents phosphorane formation.

A similar betaine is apparently involved in the conversion of diols to epoxides by reaction with sulfuranes containing the hexafluorocumyloxy group, –ORf. Sulfuranes of the type $Ar_2S(ORf)_2$ are stable enough to be isolated in crystalline form (*MA-71*) yet are reactive enough to serve as extremely powerful dehydrating agents, converting diols to epoxides and alcohols to olefins or ethers under very mild conditions. This apparently contradictory behaviour is possibly related to the generally low stabilization of sulfuranes compared to phosphoranes by $dp-\pi$ bonding as a result of the lower oxidation state of sulfur (see Section VA). This is supported by the long S–O bonds observed in X-ray diffraction of $Ph_2S(ORf)_2$ (*PMP-71*) and confirmed by correlation of the ^{19}F NMR spectra of the *p*- and *m*-fluoro isomers of XLIV compared to fluorobenzene with the Taft σ_I and σ_R values (*KM-73*). This particular sulfuranyl group was found to have negligible $dp-\pi$ bonding ability, about the same as the SF_5 group (*ST-72*) and, perhaps surprisingly, to be less electron withdrawing by induction than the phenylsulfinyl group. Electron withdrawal by induction and resonance was also much less than for the $-SF_3$ group. Moreover, the positive charge on sulfur was quite low, as befits the low electronegativity of the ligands in $Ar_2S(ORf)_2$ compared to fluorine.

A Hammett plot of tert.-butanol dehydration rate by these sulfuranes for various Ar substituents gave $\rho = -1.68$ compared to $\rho = -3.0$ for the rapid exchange of $Ar_2S(ORf)_2$ with RfOH, as determined by NMR. This proves that the transition state for the dehydration must have less positive charge on sulfur than that for the exchange, which involves the sulfonium ion $[Ar_2SORf]^+$. It also suggests that the rate-determining step for tert.-butanol dehydration is decomposition of the sulfonium ion $[Ar_2SOCMe_3]^+$ to sulfoxide Ar_2SO and tert.-butyl cation (*KM-73*). For secondary alcohols the dehydration follows an E–2 mechanism (*AM-72*) in which attack of RfO^- on a C–H bond assists sulfonium ion decomposition to sulfoxide, and there is no discrete carbocation. Similarly the transition state for epoxide formation from diols (*MFA-74*) can be written as XLV and that for sulfilimine formation from amides (*MF-75*) as XLVI. On the other hand, $Ar_2S(ORf)_2$ converts primary alcohols to ethers; this involves displacement at carbon by RfO⁻, for which transition state XLVIa can be written (*AM-72*).

Reaction of sulfonium ylides with carbonyl compounds has become a very common method of generating epoxides. This reaction proceeds through an intermediate betaine entirely analogous to that generated in the Wittig reaction of phosphonium ylides, e.g. XLIII. However, because of the much lower affinity of sulfur than phosphorus for oxygen, the oxyanion attacks the adjacent carbon in contradistinction to the Wittig reaction, where this occurs at phosphorus. Since this reaction does not involve a tetravalent sulfurane intermediate, it will not be discussed further here. For an extensive discussion of the multitude of applications of this reaction, and of many other reactions of sulfonium ylides, including the very interesting electrocyclic rearrangement of ylides containing allylic groups which proceeds with retention of configuration at chiral sulfur (*TH-73*), the reader is referred to the recent book by *Trost* and *Melvin* (*TM-75*).

The Hofmann elimination is a very well known synthetic method, and it is known to be more facile with sulfonium than with ammonium salts, possibly because of the greater polarizability of sulfur. This reaction is known to involve nucleophilic attack of the anion, usually hydroxide, on carbon; attack on sulfur would be meaningless since the polarity of the bond formed would cause the resulting sulfurane to decompose only to starting sulfonium salt. In this connection it is interesting that attack of chloride or bromide ion on $Me_2SCX(COOMe)_2$ (*MMK-75*) occurs at the carbon containing the carbomethoxy groups when X = H but at the methyl groups when X = Br, Cl, or PhS; in $Me_2CH(Me)SCX(COOMe)$ attack occurs at the isopropyl group when X = PhS. This has been attributed to poor overlap (*MMK-75*) of X with the sulfonium sulfur, which weakens the Me–S (or preferably Me_2CH–S) bond and makes the alkyl group more susceptible to nucleophilic attack. A similar sort of intramolecular sulfurane formation has been proposed by *Kwart* (*GK-68*) to explain the change in course of acetolytic chlorination of ArSCl to $ArSO_2Cl$, when Ar = o-nitrophenyl. Analysis of the observed kinetics indicates that the rate-determining step is generally chlorine addition to the sulfenyl chloride to form the sulfurane $ArSCl_3$ (or the corresponding tight ion pair) with subsequent nucleophilic attack by the solvent acetic acid. For o-nitrobenzenesulfenyl chloride the rate-determining step is nucleophilic attack by water (no reaction occurs in glacial acetic acid) to form ArSOH, followed by fast chlorine addition. This has been attributed to the existence of o-nitrobenzenesulfenyl chloride as an "intramolecular sulfurane", XLVII, formed by overlap with the nitro group, which makes the sulfur resistant to chlorine addition (*GK-68*). This structure is supported by X-ray diffraction of XLVIII (*HL-64*), which shows a trigonal bipyramidal configuration at sulfur, an S–O length of 2.44 Å, far shorter than the sum of the van der Waals radii, and unequal N–O bond lengths.

Mixtures of tertiary phosphanes and carbon tetrachloride act as versatile reagents for conversion of alcohols and carboxylic acids to the corresponding chlorides; for dehydration of carboxamides to nitriles, of formamides to isocyanides, of disubstituted ureas to carbodiimides, and of β amino alcohols to aziridines; for conversion of amino acids to peptides; and for a number of other reactions. The various applications of this reagent have been recently reviewed (*A-75*) and will not be discussed in detail here except to note that the rather complex mechanism involves phosphorane intermediates of the type R_3PCl_2, (a class of excellent phosphorylating agents), and IL, (or the corresponding phosphonium salts), as well as the ylides $R_3P=CCl_2$, and in the presence of alcohols or amines, IL a or IL b.

XLVII   XLVIII   IL   IL a   IL b

Although phosphoranes are probably involved in the preliminary steps here, the main step in, e.g., conversion of ROH to RCl is an Arbuzov type of reaction involving nucleophilic attack of Cl^- on R in the phosphonium cation $R_3'POR$. As expected for this mechanism, the chlorination occurs with inversion of configuration at R (*A-75*).

Pentacovalency

$$R-S\overset{O}{\underset{R'}{\diagup}} + (MeO)_3P \rightleftharpoons \left[(MeO)_3P\overset{OR}{\underset{SR}{\diagup}} \rightleftharpoons (MeO)_3\overset{+}{P}OR + \bar{S}R \atop (MeO)_3\overset{+}{P}SR + \bar{O}R \right]$$

L

LI LIa

The resistance of these phosphonium salts toward the Arbuzov reaction depends on the anion, increasing with decreasing anion nucleophilicity, and on the ligands, being highest with ligands containing nitrogen or oxygen and so capable of $dp-\pi$ bonding. Thus, phosphonium salts derived from tris(dimethylamino)phosphane are much more stable than those derived from triphenylphosphane, and this fact has been used by *Castro* to promote some unusual reactions. For example the $(Me_2N)_3P/CCl_4$ reagent appears to be selective for primary over secondary hydroxyl groups, a fact which has been of considerable utility in sugar chemistry (*CCG-73*); with primary 1,3-diols, only one —OH group is attacked. The chloride anion of the alkoxyphosphonium salts so obtained can be exchanged for less nucleophilic anions, e.g. perchlorate or tetrafluoroborate, and then subjected to an Arbuzov type reaction with any desired nucleophile, e.g. iodide, azide, or thiols (*CS-74*). Where the phosphonium salt is sufficiently stable, nucleophilic attack on phosphorus, resulting in ligand exchange, can also be observed. This fact forms the basis of Evans' use of trimethyl phosphite to trap *via* a Cope type of rearrangement the allylic sulfenate esters which exist in equilibrium with allylic sulfoxide. This involves as intermediate the equilibrating phosphorane/-phosphonium salt system L, where R is an allylic group. In tetrahydrofuran L decomposes in the course of an Arbuzov displacement of $R'S^\ominus$ on R, which occurs in preference to attack on Me (*EA-72*), but in excess methanol the RO ligand is completely replaced by MeO prior to occurrence of the Arbuzov mechanism of decomposition of L. In a typical case the sulfoxide LI is converted by this procedure to the allylic alcohol LIa in 91% yield. The many synthetic applications of this method of converting allylic sulfoxides to allylic alcohols have been reviewed by *Evans* (*E-74*).

Up to now we have considered only reactions in which phosphorus or sulfur is eliminated. However, phosphorylation of alcohols is of major importance in biochemistry and in the laboratory synthesis of biologically important compounds. Actually the reaction is a simple esterification and may occur by nucleophilic attack of phosphate oxygen on an appropriate carbon, e.g. the α carbon in epoxides or alkyl halides, (which will not be discussed further), or by nucleophilic attack of ROH or RO^- on phosphorus, a reaction which often involves a pseudorotating phosphorane intermediate, as discussed earlier; see Section V.C. Since the alcohols of interest are

Synthetic Applications of Pentacoordinate Intermediates

weak nucleophiles and/or contain groups unstable to the reaction conditions, the problem is to find a suitable, active phosphorylating agent. The more common phosphorylating agents and their uses have been discussed by *Khorana* (*K-61*); most are chloride or anhydride derivatives of phosphate esters with organic groups chosen to be easily removable in the second step, i.e. conversion of the phosphate diester formed by phosphorylation to a monoester. Ordinary phosphate esters are inactive by themselves, possibly because of strong P–O, dp–π bonding, which reduces the positive charge on phosphorus. They can nonetheless serve as phosphorylating agents in the presence of carbodiimides, which convert acids to anhydrides. Two more recently proposed phosphorylating agents are LII (*KRS-70*) and LIII (*TC-75*), which are exceptionally active, in line with the lability of cyclic phosphates to nucleophilic solvolysis arising from ready formation of pseudorotating phosphorane intermediates. Both organic groups are easily removed after the phosphorylation step, the phenylenedioxy group of LII by oxidation with lead tetraacetate and the ethylenedioxy group of LIII by reaction with sodium cyanide. The conversion of geraniol and cholesterol to the corresponding monophosphates in 70% yield are good examples of the use of LIII (*TC-75*). *Ramierez* (*RU-74*) has recently reviewed the necessary properties of a good phosphorylating agent and has recommended the use of alkylated amine (usually N-methylpyridinium) salts LIIIb. These are selective for primary alcohols and can be prepared by amine alkylation with LIIIa, in turn prepared from phosgene and the stable oxyphosphorane derived from biacetyl and trimethyl phosphite.

In view of the scarcity of stable pentacoordinate siliconium species, all of which must be ionic, (see Section VE), and in view of the general difficulty of cleaving Si–C bonds (*EB-69*), it is not surprising to find that relatively few synthetically useful reactions involving pentacoordinate silicon intermediates are known. However Si–C bonds are easily cleaved by strong acids, particularly aluminium halides in the presence of proton donors or alkyl or acyl halides. Thus, hexamethyldisilane is inert to anhydrous aluminium chloride but loses methane rapidly when traces of water are added (*EB-69*). Similarly aluminium chloride catalyzes the reaction of 2-chloropropane with tetraethylsilane (*VRD-63*) to give chlorotriethylsilane, isopentane, ethylene, and propane. Two cyclic, concerted mechanisms can be written for this reaction, LIV leading to isopentane and LV to ethylene and propane.

Aluminium chloride also catalyzes coupling with acyl halides; for instance, vinyltrimethylsilane is converted by acetyl chloride to vinyl methyl ketone, the low yield of 45% being due to polymerization of the product (*PDC-74*). Similarly, *trans* Me$_3$SiCH=CHSiMe$_3$ is converted to *trans* Me$_3$SiCH=CHCOR in 88% yield for R = methyl, 65% yield for R = tert.-butyl, and 80% yield for R = phenyl (*PDC-74*). By an extension of the same process, vinyltrimethylsilane and Me$_2$C=CHCOCl give 53% Me$_2$C=CHCOCH=CH$_2$; *trans* Me$_3$SiCH=CHSiMe$_3$ and MeOCHCl$_2$ give, after reaction with water, 20% Me$_3$SiCH=CHCHO (*PDC-75*). Pillot (*PDC-75*) proposes a stepwise mechanism involving attack of acylium ion on the carbon to form a β-silyl carbonium ion, which would probably be stabilized by overlap with silicon, followed by nucleophilic attack of chloride on silicon with regeneration of the double bond.

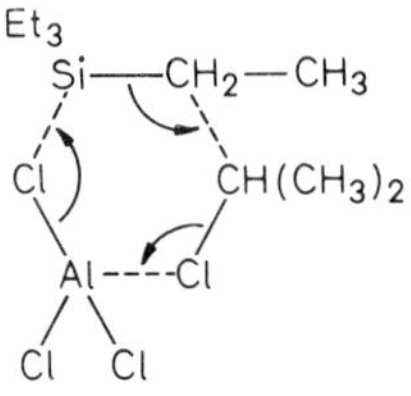

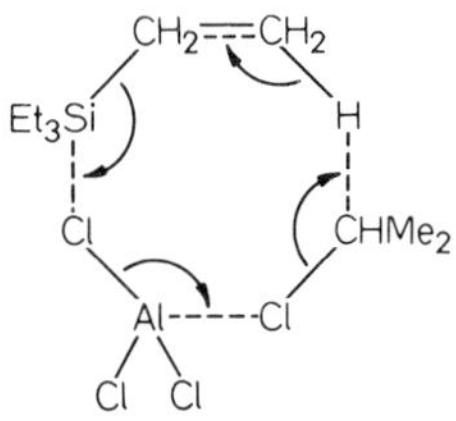

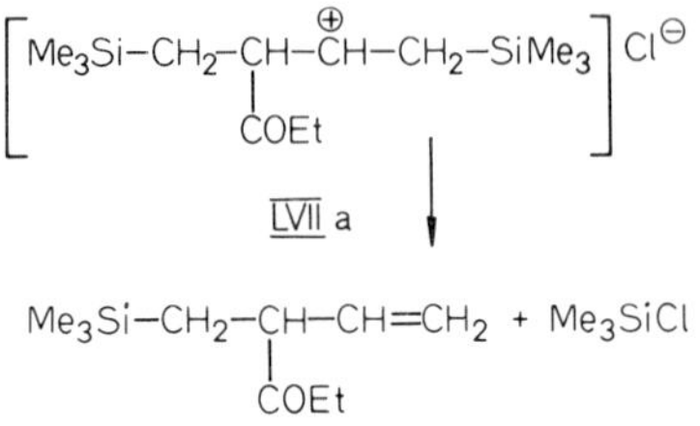

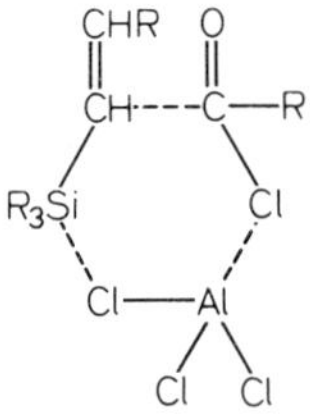

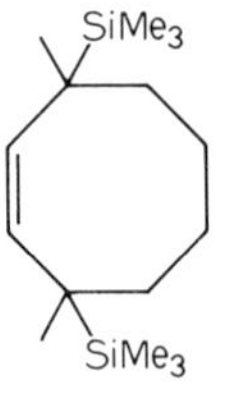

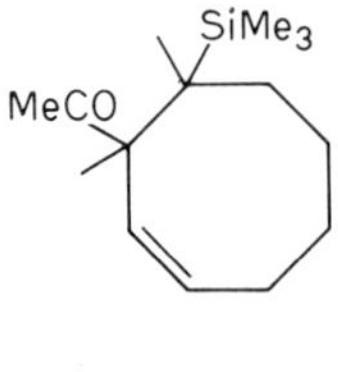

On the other hand it seems unlikely that the aluminium chloride catalyzed reactions with acyl and alkyl chlorides would proceed by different mechanisms; indeed, a cyclic mechanism LVI similar to LIV and LV can conceivably be written for attack of acyl chloride on an alkenylsilane. Such a mechanism would explain the formation of *trans* alkenyl ketone from *trans* silane. However, this is not a necessary mechanistic feature since *cis* alkenyl ketones are known (*CG-69*) to isomerize to the *trans* form in the presence of aluminium chloride by what is apparently a carbonium ion mechanism. In view of the ease of formation of carbonium ions by reaction with aluminium chloride and in view of the high stability of β-silyl carbonium ions, which are known to occur in the loss of ethylene by solvolysis of Me$_3$SiCH$_2$CH$_2$Cl in aqueous methanol (*CEW-70, VH&-73*, see Section V.E.), there is every reason to expect an intermediate carbonium ion in this reaction. Loss of ethylene from tetraethylsilane by reaction with 2-chloropropane (*VRD-63*) can also be described as a result of abstraction of a β-hydrogen by isopropyl carbonium ion to form a β-silyl carbonium ion. Final proof of the existence of a carbonium ion intermediate comes in the aluminium chloride catalyzed reaction of acyl chlorides with allylsilanes, which occurs with rearrangement. Thus, EtCOCl and $Me_3Si-CH_2CH=CHCH_2SiMe_3$ give LVII, presumably *via* the carbonium ion LVIIa, and acetyl chloride converts LVIII to LIX (*CD&-75*).

Aldehydes and ketones are readily reduced by hydridosilanes (i.e. compounds with Si–H bonds) in acidic media. An interesting feature of the reaction is the wide variety of products which may be obtained depending on conditions (*DD&-74*). Benzaldehyde is reduced primarily to benzyl alcohol by triethylsilane in acetonitrile in the presence of aqueous hydrochloric or sulfuric acid. When the acetonitrile was omitted, dibenzyl ether was the major product; however, cyclohexanone was reduced mainly to the alcohol even without the organic solvent. At long reaction times in acetonitrile, alcohol products underwent the Ritter reaction, giving N-substituted acetamides. Similarly in carboxylic acid solvents, the alcohol products were converted to the corresponding esters; however, in strong acids, e.g. trifluoroacetic, ether formation competed strongly with esterification.

The degree of stereospecificity in formation of the less stable, more hindered product, e.g. *cis*-4-tert.-butylcyclohexanol from 4-tert.-butylcyclohexanone (*DW-75a*), depends upon the structures of ketone and silane, increasing with increasing substitution at silicon and with branching of the alkyl groups attached to silicon, and being greater for 2- than for 4-methylcyclohexanone. Consequently, formation of the more hindered alcohol increases with increasing steric hindrance in the transition state. Approximately the same degree of stereospecificity was observed in ether formation, with *cis*, *trans* and *cis,cis*-ethers from 4-substituted cyclohexanones being favored over the *trans,trans*-ethers. In general, as stereospecificity increased with increasing steric hindrance, reactivity of the silanes decreased, but exceptions were observed with highly hindered silanes. To illustrate, the rate of hydride transfer is greater from tri- than from di-tert.-butyl-silane (*DW-75b*), and $(tBu)_2MeSiH$ is 20–40 times more reactive than $(tBu)_2SiH_2$. However, reduction of 4-alkylcyclohexanones in trifluoroacetic acid with these silanes gave only the corresponding 4-alkylcyclohexene and 4-alkylcyclohexyl trifluoroacetate (with, of course, the expected high preference for the *cis* ester), although at shorter reaction times some of the corresponding 4-alkylcyclohexyl silyl ether was also isolated. The increased reactivity of these hindered silanes can be attributed to the inductive and/or hyperconjugative effect of the alkyl groups attached to silicon, which implies positive charge on silicon in the transition state.

Careful analysis of the kinetics of reaction of cyclohexanone and 4-tert.-butyl-cyclohexanone with $(tBu)_2SiH_2$ (*DW-75c*), taking into account the equilibrium solvolysis of the silyl ether, showed that the silyl ether is definitely an intermediate in the formation of alcohol, trifluoroacetate ester, and olefin. The silyl ether can only be formed by a four-center reaction process, LX, in which the carbonyl oxygen acts as a nucleophile to attack silicon. Other work on related reactions suggests that Si–H cleavage is not extensive in the transition state (*OB&-72*) and that LX may be an intermediate (*EJ-74*). In fact, if we assume apical entry and departure as in analogous nucleophilic attack at phosphorus, then LX would have to be an intermediate capable of pseudorotation or turnstile rotation. For less hindered silanes the reaction is faster and the silyl ether is not isolable; in these cases the operative mechanism appears to employ simultaneous attack of hydride on the carbonyl carbon and of chloride or a nucleophile derived from the solvent on silicon. For formation of the symmetrical ether, which involves attack of the silane on LXI, formed by loss of water from LXII, this mechanism is, in fact, quite probable since silyl ether cannot be converted to symmetric ether under the reaction conditions used. Thus, the diminished extent of formation of symmetric ether from ketones compared to aldehydes can be attributed to steric hindrance in the formation of LXII by attack of solvent or product alcohol on the carbonyl carbon. The generally lower stereospecificity in the reductions by hydridosilanes compared to analogous reactions with hindered boron or aluminium hydrides argues in favor of the 4-center transition state like LX. In this case silicon would be less affected by steric hindrance arising from the substituents on the carbonyl, thereby reducing the energy difference between the transition states derived from hydride attack on the hindered and unhindered side of the carbonyl.

The reaction of hydridosilanes with carbonium ions (see p. 159) has been suggested (*DMW-76*) as a useful method of hydrocarbon synthesis. Thus, adamantane is prepared in 79% and 84% yields, respectively, from Et_3SiH and 1- or 2-bromoadamantane in the presence of catalytic quantities of aluminium chloride. Chlorides work just as well as bromides, and, as might be expected for a reaction involving carbonium ions, better results are obtained with tertiary than with secondary or primary halides. Formation of rearranged product from certain primary or secondary halides indicates the existence of the intermediate carbonium ion. For example, the conversion of bromocycloheptane to 65% methylcyclohexane with $n-BuSiH_3$ but only 26% methylcyclohexane with Et_3SiH can be attributed to slower hydrogen transfer with the trihydride, which thereby affords the carbonium ion a longer lifetime. Although the mechanism is not discussed by *Doyle* (*DMW-76*), we can propose with confidence in the analogy to other hydridosilane reactions a cyclic mechanism such as LXIII. Due to the $^{\delta+}Si-H^{\delta-}$ polarization of hydridosilanes, Et_3SiH would have more hydride character than $n-BuSiH_3$ and thus react faster with the carbonium ion in LXIII.

VI. References

A-60   Aksnes, G.: Acta. Chem. Scand., *14*, 1515 (1960).
A-68   Attridge, C.J.: J. Organomet. Chem., *13*, 259 (1968).
A-70   Allen, D.W.: J. Chem. Soc. *(B)*, 1490 (1970).
A-75   Appel, R.: Angew. Chem. Int. Ed. Engl., *14*, 801 (1975).
AAB-66   Abel, E.W., Armitage, D.A., Brady, D.B.: Trans. Faraday Soc., *62*, 3459 (1966).
AA&-74   Ackerman, B.K., Andersen, K.K., Karup-Nielsen, I., Peynircioglu, N.B., Yeager, S.A.: J. Org. Chem., *39*, 964 (1974).
AB-69   Alt, H., Bock, H.: Tetrahedron, *25*, 4825 (1969).
AB-70   Adamson, G.W., Bart, J.C.J.: J. Chem. Soc. *(A)*, 1452 (1970).
AB-73   D'Amore, M.B., Brauman, J.J.: Chem. Comm., 398 (1973).
AB-76   Arduengo, A.J., Burgess, E.M.: J. Am. Chem. Soc., *98*, 5020, 5021 (1976).
AB&-63   Almenningen, A., Bastiansen, C., Ewing, V., Hedberg, K., Traetteberg, M.: Acta. Chem. Scan., *17*, 2455 (1963).
AFB-69   Alt, H., Franke, E.R., Bock, H.: Angew. Chem., *81*, 538 (1969).
AFS-75   Albright, T.A., Freeman, W.J., Schweizer, E.E.: J. Am. Chem. Soc., *97*, 2946 (1975).
AFS-75a   Albright, T.A., Freeman, W.J., Schweizer, E.E.: J. Org. Chem., *40*, 3437 (1975).
AG&-70   Airey, W., Glidewell, C., Rankin, D.W.H., Robiette, A.G., Sheldrick, G.M.: Trans. Faraday Soc., *66*, 551 (1970).
AG&-71   Airey, W., Glidewell, C., Robiette, A.G., Sheldrick, G.M.: J. Mol. Struct. *8*, 413 (1971).
AH-72   Allen, D.W., Hutley, B.G.: J. Chem. Soc. Perkin II, *67* (1972).
AHM-72   Allen, D.W., Hutley, B.G., Mellor, M.T.J.: J. Chem. Soc. Perkin II, *63* (1972).
AK-67   Andrianov, K.A., Kotrelev, G.V.: J. Organomet. Chem., *7*, 217 (1967).
AK-70   Adcock, W., Kitching, W.: Chem. Comm., 1163 (1970).
AKK-63   Andrianov, K.A., Khaiduk, I., Khananashvili, L.M.: Dokl. Acad. Nauk. SSSR, *150*, 93 (1963); Dokl. Chem., *150*, 385 (1963).
AM-72   Arhart, R.J., Martin, J.C.: J. Am. Chem. Soc., *94*, 4997, 5003 (1972).
AM-76   Astrologes, G.W., Martin, J.C.: J. Am. Chem. Soc., *98*, 2895 (1976).
AP-65   Aylett, B.I., Peterson, L.K.: J. Chem. Soc., 4043 (1965).
AS-65   Aksnes, G., Songstad, J.: Acta. Chem. Scand., *14*, 893 (1965).
AS-76   Albright, T.A., Schweizer, E.E.: J. Org. Chem., *41*, 1168 (1976).
AT-70   Allen, D.W., Tebby, J.C.: J. Chem. Soc. *(B)*, 1527 (1970).
ATC-67   Anet, F.A.L., Trepka, R.D., Cram, D.J.: J. Am. Chem. Soc., *89*, 357 (1967).
ATM-68   Allen, C.W., Tsang, F.Y., Moeller, T.: Inorg. Chem. Z, 2183 (1968).
AV-72   Absar, I., Van Wazer, J.R.: J. Am. Chem. Soc., *94*, 2382 (1972).
AV-72a   Absar, I., Van Wazer, J.R.: J. Chem. Phys., *56*, 1284 (1972).
AW-72   Allcock, H.R., Walsh, E.J.: J. Am. Chem. Soc., *94*, 1195 (1972).
AW-72a   Allcock, H.R., Walsh, E.J.: J. Am. Chem. Soc., *94*, 4538 (1972).
AW-73   Archie, W.C., Westheimer, F.H.: J. Am. Chem. Soc., *95*, 5955 (1973).
AW&-70   Allegra, G., Wilson, G.E., Jr., Benedetti, E., Pedone, C., Albert, R.: J. Am. Chem. Soc., *92*, 4002 (1970).
AWC-75   Anastassiou, A.G., Wetzel, J.C., Chao, B.Y.H.: J. Am. Chem. Soc., *97*, 1124 (1975).
AWW-73   Ammom, H.L., Wheeler, G.L., Watts, P.H., Jr.: J. Am. Chem. Soc., *95*, 6158 (1973).
AYC-76   Altmann, J.A., Yates, K., Csizmadia, I.G.: J. Am. Chem. Soc., *98*, 1450 (1976).
B-58   Breslow, R.: J. Am. Chem. Soc., *80*, 3719 (1958).
B-60   Berry, R.S.: J. Chem. Phys., *32*, 933 (1960).
B-62   Brown, H.C.: "Hydroboration", (W.A. Benjamin, Inc., N.Y., N.Y., 1962).
B-64   Balli, H.: Angew. Chem., *76*, 995 (1964).
B-64a   Brune, H.A.: Chem. Ber., *97*, 2829 (1964).
B-64b   Brune, H.A.: Chem. Ber., *97*, 2848 (1964).
B-65   Brune, H.A.: Chem. Ber., *98*, 1998 (1965).
B-68   Brois, S.J.: J. Am. Chem. Soc., *90*, 506, 508 (1968).
B-68a   Broaddus, C.D.: J. Am. Chem. Soc., *90*, 5504 (1968).

References

B-68b   Broaddus, C.D.: Accts. Chem. Res., *1*, 231 (1968).
B-68c   Bürger, H.: Spectrochim. Acta., *24A*, 2015 (1968).
B-69   Bart, J.C.J.: J. Chem. Soc. *(B)*, 350 (1969).
B-69a   Budde, W.L.: J. Phys. Chem., *73*, 3718 (1969).
B-70a   Bunton, C.A.: Accts. Chem. Res., *3*, 257 (1970).
B-74   Brook, A.G.: Accts. Chem. Res., *7*, 77 (1974).
B-75   Batich, C.D.: private communication of unpublished results, Oct. 18, 1975.
B-75a   Brown, R.S.: Can. J. Chem., *53*, 2446 (1975).
B-76   Belkind, B.A.: Polytech. Inst. N.Y., Diss. (1976); Diss. Abstr. Int. B 37 #4:1685B (Oct. 1976).
BA-70   Bock, H., Alt, H.: J. Am. Chem. Soc., *92*, 1569 (1970).
BA-70a   Bock, H., Alt, H.: Chem. Ber., *103*, 1784 (1970).
BA-71   Biasotti, J.B., Andersen, K.K.: J. Am. Chem. Soc., *93*, 1178 (1971).
BAS-69   Bock, H., Alt, H., Seidl, H.: J. Am. Chem. Soc., *91*, 355 (1969).
BB-66   Bruice, T.C., Benkovic, S.J.: Bioorganic Mechanisms, Vol. 2, CH. 6, New York: W.A. Benjamin, Inc. 1966.
BB&-59   Boonstra, H.J., Brandsma, L., Wiegman, A.M., Arens, J.F.: Rec. Trav. Chim., *78*, 258 (1959).
BB&-69   Baenziger, N.C., Buckles, R.E., Maner, R.J., Simpson, T.D.: J. Am. Chem. Soc., *91*, 5749 (1969).
BB&-69a   Bürger, H., Burczyk, K., Höfler, F., Sawodny, W.: Spectrochim. Acta, 25A, 1891 (1969).
BB&-70   Belskii, V.E., Bezzubova, N.N., Elisenkov, V.N., Pudovic, A.N.: Zh. Obsch. Khim., *40*, 2557 (1970); J. Gen. Chem. USSR, *40*, 2548 (1970).
BB&-71   Bürger, H., Burczyk, K., Blaschette, A., Safari, H.: Spectrochim. Acta, *27A*, 1073 (1971).
BB&-74   Bartlett, P.D., Baumstark, A.L., Landis, N.E., Lerman, C.L.: J. Am. Chem. Soc., *96*, 5267 (1974).
BB&-75   Bassindale, A.R., Brook, A.G., Chen, P., Lennon, J.: J. Organometal. Chem., *94*, C21 (1975).
BB&-75a   Bassindale, A.R., Brook, A.G., Jones, P.F., Lennon, J.M.: Can. J. Chem., *53*, 332 (1975).
BB&-77   Bordwell, F.G., Bares, J.E., Bartmess, J.E., Drucker, G.E., Gerhold, J., McCollum, G.J., Van Der Puy, M., Vanier, N.R., Matthews, W.S.: J. Org. Chem., *42*, 326 (1977).
BB&-77a   Bordwell, F.G., Bares, J.E., Bartmess, J.E., McCollum, G.J., Van der Puy, M., Vanier, N.R., Matthews, W.S.: J. Org. Chem., *42*, 321 (1977).
BBL-73   Bartlett, P.D., Baumstark, A.L., Landis, M.E.: J. Am. Chem. Soc., *95*, 6486 (1973).
BC-69   Bank, K.C., Coffen, D.L.: Chem. Comm., 8 (1969).
BC&-71   Brown, M.D., Cook, M.J., Hutchinson, B.J, Katritzky, A.R.: Tetrahedron, *27*, 59B (1971).
BC&-73   Bragin, J., Chan, S., Mazzola, E., Goldwhite, H.: J. Phys. Chem., *77*, 1506 (1973).
BC&-75   Bernardi, F., Csizmadia, I.G., Mangini, A., Schlegel, H.B., Whangbo, M.H., Wolfe, S.: J. Am. Chem. Soc., *97*, 2209 (1975).
BCS-62   Bergson, G., Claeson, G., Schotte, L.: Acta Chem. Scand., *16*, 1159 (1962).
BD-73   Brook, A.G., Duff, J.M.: Can. J. Chem., *51*, 2024 (1973).
BD-74   Brook, A.G., Duff, D.M.: J. Am. Chem. Soc., *96*, 4692 (1974).
BD-74a   Bordwell, F.G., Doomes, E.: J. Org. Chem., *39*, 2526 (1974).
BD-74b   Bordwell, F.G., Doomes, E.: J. Org. Chem., *39*, 2531 (1974).
BDC-70   Bordwell, F.G., Doomes, E., Corfield, P.W.R.: J. Am. Chem. Soc., *92*, 2581 (1970).
BE-70   Belskii, V.E., Efremova, M.V.: Izv. Akad. Nauk., SSSR., 1542 (1970).
BE-71   Bock, H., Ensslin, W.: Angew. Chem. Intn'l. Ed. Engl., *10*, 404 (1971).
BE&-76   Bock, H., Ensslin, W., Feher, F., Freund, R.: J. Am. Chem. Soc., *98*, 668 (1976).
BEM-65   Bott, R.W., Eaborn, C., Rushton, B.M.: J. Organometal. Chem., *3*, 455 (1965).
BF-68   Brown, D.M., Frearson, M.J.: Chem. Comm., 1342 (1968).
BF-69   Boer, F.P., Flynn, J.J.: J. Am. Chem. Soc., *91*, 6604 (1969).
BF-70   Booth, M.R., Frankiss, S.G.: Spectrochim. Acta, *26A*, 859 (1970).
BG-63   Baliah, V., Ganapathy, K.: Trans. Faraday Soc., *59*, 1784 (1963).

BGM-75	Boudjebel, H., Gancalves, H., Mathis, F.: Bull. Soc. Chim. France, 628 (1975).
BGP-74	Bendazzoli, G.L., Gottarelli, G., Palmieri, P.: J. Am. Chem. Soc., 96, 11 (1974).
BGS-68	Bürger, H., Goetze, U., Sawodny, W.: Spectrochim. Acta., 24A, 2003 (1968).
BGS-70	Bürger, H., Goetze, U., Sawodny, W.: Spectrochim. Acta., 26A, 671, 685 (1970).
BH-65	Bartell, L.S., Hansen, K.W.: Inorg. Chem., 4, 1777 (1965).
BH-71	Boyd, D.B., Hoffman, R.: J. Am. Chem. Soc., 93, 1064 (1971).
BH&-76	Bentrude, W.G., Hargis, J.H., Johnson, N.A., Min, T.B., Rusek, P.E., Tan, H.W., Wielesek, R.A.: J. Am. Chem. Soc., 98, 5348 (1976).
BJB-74	Boekenstein, G., Jansen, E.H.J.M., Buck, H.M.: J.C.S. Chem. Comm., 118 (1974).
BJK-72	Bentrude, W.G., Johnson, W.D., Khan, W.A.: J. Am. Chem. Soc., 94, 923 (1972).
BK&-65	Bystrov, V.F., Kostyanovskii, R.G., Panshin, O.A., Stepanyants, A.V., Iuzhakova, O.A.: Opt. Spectrosc. (USSR) 19, 122 (1965).
BK&-74	Bentrude, W.G., Khan, W.A., Murakami, M., Tun, H.W.: J. Am. Chem. Soc., 96, 5566 (1974).
BLM-67	Brook, A.G., LeGrow, C.E., MacRae, D.M.: Can. J. Chem., 45, 239 (1967).
BM-62	Bentrude, W.G., Martin, J.C.: J. Am. Chem. Soc., 84, 1561 (1962).
BM-70	Baechler, R.D., Mislow, K.: J. Am. Chem. Soc., 92, 3090 (1970).
BM-70a	Baechler, R.D., Mislow, K.: J. Am. Chem. Soc., 92, 4758 (1970).
BM-70b	Bellama, J.M., MacDiarmid, A.G.: J. Organometal. Chem., 24, 91 (1970).
BM-71	Baechler, R.D., Mislow, K.: J. Am. Chem. Soc., 93, 773 (1971).
BM-71a	Biernbaum, M.S., Mosher, H.S.: J. Am. Chem. Soc., 93, 6221 (1971).
BM&-71	Bradamante, S., Mairoana, S., Mangia, A., Pagani, G.: J. Chem. Soc. (B), 74 (1971).
BM-72	Baechler, R.D., Mislow, K.: J. Chem. Soc., Chem. Comm., 185 (1972).
BM-76	Bentrude, W.G., Min, T.B.: J. Am. Chem. Soc., 98, 2918 (1976).
BMB-75	Brook, A.G., MacRae, D.M., Bassindale, A.R.: J. Organometal. Chem., 86, 185 (1975).
BMP-67	Banister, A.J., Moore, L.F., Padley, J.S.: Spectrochim. Acta, 23A, 2705 (1967).
BMP-70	Bradamante, S., Mangia, S.A., Pagani, G.: Tetrahedron Letters, 3381 (1970).
BO-74	Bordwell, F.G., O'Dwyer, J.B.: J. Org. Chem., 39, 2519 (1974).
BP-71	Brook, A.G., Pascoe, J.D.: J. Am. Chem. Soc., 93, 6224 (1971).
BP&-68	Belskii, V.E., Pudovik, A.N., Efremova, N.V., Elisenkov, V.N., Panteleeva, A.R.: Dokl. Adak. Nauk. SSSR, 180, 351 (1968) Dokl. Chem. 427 (1968).
BQ&-74	Bock, E., Queen, A., Brownlee, S., Nour, T.A., Paddon-Row, M.N.: Can. J. Chem., 52, 3113 (1974).
BRL-77	Bushweller, C.H., Ross, J.A., Lemal, D.M.: J. Am. Chem. Soc., 99, 629 (1977).
BS-68	Bock, H., Seidl, H.: J. Organometal. Chem., 13, 87 (1968).
BT-72	Bentrude, W.G., Tun, H.W.: J. Am. Chem. Soc., 94, 8222 (1972).
BT-75	Baccolini, G., Todesco, P.E.: J. Org. Chem., 40, 2318 (1975).
BTF-68	Boer, F.P., Turley, J.W., Flynn, J.J.: J. Am. Chem. Soc., 90, 5102 (1968).
BTW-68	Bristow, P.A., Tillett, J.G., Wiggins, D.E.: J. Chem. Soc. (B), 1360 (1968).
BU-63	Baliah, V., Una, M.: Tetrahedron, 19, 455 (1963).
BV&-75	Bordwell, F.G., Vanier, N.R., Matthews, W.S., Hendrickson, J.B., Skipper, P.L.: J. Am. Chem. Soc., 97, 7160 (1975).
BVV-76	Bordwell, F.G., Van Der Puy, M., Vanier, N.R.: J. Org. Chem., 41, 1883.
BW-71	Bordwell, F.G., Wolfinger, M.D.: J. Am. Chem. Soc., 93, 6303 (1971).
BW-71a	Boudjouk, P., West, R.: J. Am. Chem. Soc., 93, 5902 (1971).
BW-74	Bordwell, F.G., Wolfinger, M.D.: J. Org. Chem., 39, 2521 (1974).
BWH-68	Bordwell, F.G., Williams, J.M., Hoyt, E.B., Jr.: J. Am. Chem. Soc., 90, 429 (1968).
BWK-72	Bock, H., Wagner, G., Kroner, J.: Chem. Ber., 105, 3850 (1962).
BWL-67	Brook, A.G., Warner, C.M., Linburg, W.W.: Can. J. Chem., 45, 1231 (1967).
BW&-76	Bock, H., Wittel, K., Veith, M., Wiberg, N.: J. Am. Chem. Soc., 98, 109 (1976).
BZ-65	Bendazzoli, G.L., Zauli, C.: J. Chem. Soc., 6827 (1965).
C-60	Cilento, G.: Chem. Rev., 60, 147 (1960).
C-61	Cruickshank, D.W.J.: J. Chem. Soc., 5486 (1961).
C-66	Chvalovsky, V.: Pure & Applied Chem., 13, 231 (1966).

References

C-66a	Collin, R. L.: J. Am. Chem. Soc., *88*, 3281 (1966).
C-68	Clark, D. T.: Tetrahedron, *24*, 2663 (1968).
C-69	Coulson, C. A.: Nature, *221*, 1106 (1969).
CA&-71	Carpino, L. A., Adams, L. C., Rynbrandt, R. H., Spiewak, J. W.: J. Am. Chem. Soc., *93*, 476 (1971).
CB-54	Cason, L. F., Brooks, H. G.: J. Org. Chem., *19*, 1278 (1954).
CC-63	Chapman, A. C., Carroll, D. F.: J. Chem. Soc., 5505 (1963).
CC-71	Carpino, L. A., Chen, H. W.: J. Am. Chem. Soc., *93*, 785 (1971).
CC&-71	Chang, B. C., Conrad, W. E., Denney, D. B., Denney, D. Z., Edelman, R., Powell, R. L., White, D. W.: J. Am. Chem. Soc., *93*, 4004 (1971).
CC&-71a	Coffen, D. L., Chambers, J. Q., Williams, D. R., Garrett, P. E., Camfield, N. D.: J. Am. Chem. Soc., *93*, 2258 (1971).
CCG-73	Castro, B., Chapleur, Y., Gross, B.: Bull. Soc. Chim. France, 3034 (1973).
CCL-70	Cane, F., Corriu, R., Leard, M.: J. Organometal. Chem., *24*, 101 (1970).
CCT-69	Cremer, S. E., Chowat, R. J., Trivedi, B. C.: Chem. Comm., 769 (1969).
CCS-76	Christol, H., Cristau, H. J., Soleiman, M.: Bull. Soc. Chim. France, 161 (1976).
CD-68	Chioccola, G., Dalry, J. J.: J. Chem. Soc., (A), 568 (1968).
CDD-71	Chang, B. C., Denney, D. Z., Denney, D. B.: J. Org. Chem., *36*, 998 (1971).
CD&-70	Cowley, A. H., Dewar, M. J. S., Jackson, W. R., Jennings, W. B.: J. Am. Chem. Soc., *92*, 5206 (1970).
CD&-70a	Cram, D. J., Day, J., Rayner, D. R., Schriltz, D. M. von, Duchamp, D. J., Garwood, D. C.: J. Am. Chem. Soc., *92*, 7369 (1970).
CD&-73	Chan, S., Distefano, S., Fong, F., Goldwhite, H., Gysegem, P., Mazzola, E.: Inorg. Chem., *12*, 51 (1973).
CD&-73a	Cook, R. D., Diebert, C. E., Schwarz, W., Turley, P. C., Haake, P. C.: J. Am. Chem. Soc., *95*, 8088 (1973).
CD&-75	Calas, P., Dunogues, J., Pillot, J. P., Biran, C., Pisciotti, F., Arreguy, B.: J. Organometal. Chem. *85*, 149 (1975).
CDT-70	Corfield, J. R., De'Ath, N. J., Trippett, S.: Chem. Comm., *1502* (1970).
CE-71	Cradock, S., Ebsworth, E. A. V.: Chem. Comm., 57 (1971).
CEW-70	Cook, M. A., Eaborn, C., Walton, D. R. M.: J. Organometal. Chem., *24*, 301 (1970).
CF-68	Ciuffarin, E., Fava, A.: Prog. Phys. Org. Chem., *6*, 81 (1968).
CF-72	Chan, T. H., Finkenbine, J. R.: J. Am. Chem. Soc., *94*, 2880 (1972).
CFM-75	Cartledge, F. K., Fayssoux, J., McKinnie, B. G.: J. Organometal. Chem. *96*, 15 (1975).
CG-69	Combault, G., Giral, L.: Bull. Soc. Chim. France, 3258 (1969).
CG&-71	Cook, R. E., Glick, M. D., Rigau, J. J., Johnson, C. R.: J. Am. Chem. Soc., *93*, 924 (1971).
CH-69	Carey, F. S., Hsu, C. W.: J. Organometal. Chem., *19*, 29 (1969).
CH-76	Chen, M. M. L., Hoffmann, R.: J. Am. Chem. Soc., *98*, 1647 (1976).
CL-65	Corey, E. J., Lowry, T. H.: Tetrahedron Letters, 803 (1965).
CL-71	Corriu, R., Leard, M.: Chem. Comm., 1086 (1971).
CLM-68	Corriu, R., Leard, M., Masse, J.: Bull. Soc. Chim. (Fr.), 2555 (1968).
CM-56	Craig, D. P., Magnusson, E. A.: J. Chem. Soc., 4895 (1956).
CM-71	Cartledge, F. K., Mollere, P. D.: J. Organometal. Chem., *26*, 175 (1971).
CM&-54	Craig, D. P., Maccoll, A., Nyholm, R. S., Orgel, L. E., Sutton, L. E.: J. Chem. Soc., 332, 354 (1954).
CM&-66	Cumper, C. W. N., Melnikoff, A., Mooney, E. F., Vogel, A. L.: J. Chem. Soc. (B), 874 (1966).
CM&-71	Carpino, L. A., McAdams, L. V., Rynbrandt, R. H., Spiewak, J. W.: J. Am. Chem. Soc., *93*, 476 (1971).
CM&-76	Clardy, J. C., Milbrath, D. S., Springer, J. P., Verkade, J. G.: J. Am. Chem. Soc., *98*, 623 (1976).
CMW-70	Coffen, D. L., McEntee, T. E., Jr., Williams, D. R.: Chem. Comm., 913 (1970).
CP-64	Collin, R. L., Pullman, B.: Arch. Biochem. Biophys., *103*, 535 (1964).
CP-72	Chivers, T., Paddock, N. L.: Inorg. Chem., *11*, 848 (1972).
CPH-66	Caserio, M. C., Pratt, R. E., Holland, R. J.: J. Am. Chem. Soc., *88*, 5747 (1966).

CS-70	Cowley, A. H., Schweiger, J. R.: Chem. Comm., 1492 (1970).
CS-74	Castro, B., Selve, C.: Bull. Soc. Chim. France, 3004, 3009 (1974).
CS&-76	Collins, J. B., Schleyer, P. V. R., Binkley, J. S., Pople, J. A.: J. Chem. Phys. in press.
CSI-72	Ciuffarin, E., Senatore, L., Isola, M.: J.C.S. Perkin II, 468 (1972).
CST-69	Corfield, J. R., Schutt, J. R., Trippett, S.: J. C. S., Chem. Comm., 789 (1969).
CT-66	Craig, D. P., Thirunamachandran, T.: J. Chem. Phys., *45*, 3355 (1966).
CTS-66	Cram, D. J., Trepka, R. D., Janiak, P. St.: J. Am. Chem. Soc., *88*, 2749 (1966).
CTW-71	Cremer, S. E., Trivedi, B. C., Weitl, F. L.: J. Org. Chem., *36*, 3226 (1971).
CV-70	Claus, P., Vycudilik, M.: Monatshefte Chem., *101*, 405 (1970).
CW-63	Covitz, F., Westheimer, F. H.: J. Am. Chem. Soc., *85*, 1773 (1963).
CW-66	Cotton, F. A., Wilkinson, G.: "Advanced Inorganic Chemistry", 2nd ed., Interscience, New York, N.Y. (1966).
CW-72	Clipsham, R. M., Whitehead, M. A.: J.C.S. Faraday II, *55*, 72 (1972).
CWG-69	Carberry, E., West, R., Glass, G. E.: J. Am. Chem. Soc., *91*, 5446 (1969).
CWM-64	Cruickshank, D. W. J., Webster, B. C., Meyers, F. F.: J. Chem. Phys., *40*, 3733 (1964).
CY-57	Cornwall, C. D., Yamasaki, R. S.: J. Chem. Phys., *27*, 1060 (1957).
CZ-62	Craig, D. P., Zauli, C.: J. Chem. Phys., *37*, 601 (1962).
D-64	Duncan, J. L.: Spectrochim. Acta., *20*, 1807 (1964).
D-70	Daly, J. J.: J. Chem. Soc., *(A)*, 1832 (1970).
D-75	Durmaz, S.: J. Organometal. Chem., *96*, 331 (1975).
DA-76	Deakyne, C. A., Allen, L. C.: J. Amer. Chem. Soc., *98*, 4076 (1976).
DB-73	Duff, J. M., Brook, A. G.: Can. J. Chem., *51*, 2869 (1973).
DD&-69	Denney, D. B., Denney, D. Z., Chang, B. C., Marsi, K. L.: J. Am. Chem. Soc., *91*, 5243 (1969).
DD&-72	Denney, D. B., Denney, D. Z., Hall, C. D., Marsi, K. L.: J. Am. Chem. Soc., *94*, 245 (1972).
DD&-74	Doyle, M. P., DeBruyn, D. J., Donnelly, S. J., Kooistra, D. A., Odubela, A. A., West, C. T., Zonnebelt, S. M.: J. Org. Chem., *39*, 2740 (1974).
DD&-76	De'Ath, N. J., Denney, D. B., Denney, D. Z., Hsu, Y. F.: J. Am. Chem. Soc., *98*, 768 (1976).
DDH-73	Denney, D. B., Denney, D. Z., Hsu, Y. F.: J. Am. Chem. Soc., *95*, 4064 (1973).
DF&-70	Durst, T., Fraser, R. R., McClory, M. R., Swingle, R. B., Wigfield, Y. Y.: Can. J. Chem., *48*, 2148 (1970).
DG&-64	Driscoll, J. S., Grisley, D. W., Jr., Pustinger, J. V., Harris, J. E., Matthews, C. N.: J. Org. Chem., *29*, 2427 (1964).
DG&-73	Durig, J. R., George, W. O., Li, Y. S., Carter, R. O.: J. Mol. Struct. *16*, 47 (1973).
DH-69	Durig, J. R., Hallams, K. L.: Inorg. Chem., *8*, 944 (1969).
DK-76	Davis, F. A., Kluger, E. W.: J. Am. Chem. Soc., *98*, 302 (1976).
DKF-75	Durig, J. R., Kalasinsky, V. F., Flanagan, M. J.: Inorg. Chem., *14*, 2839 (1975).
DLW-60	Dewar, M. J. S., Lucken, E. A. C., Whitehead, M. A.: J. Chem. Soc., 2423 (1960).
DM-75	Durst, T., Molin, M.: Tetrahedron Letters, 63 (1975).
DMW-76	Doyle, M. P., McOsker, C. C., West, C. T.: J. Org. Chem., *41*, 1393 (1976).
DP-68	Daniel, H., Paetsch, J.: Ber., *101*, 1451 (1968).
DP-72	DeBruin, K. E., Petersen, J. R.: J. Org. Chem., *37*, 2772 (1972).
DPB-75	van Dijk, J. M. F., Pennings, J. F. M., Buck, H. M.: J. Am. Chem. Soc., *96*, 4836 (1975).
DR-64	Denney, D. B., Relles, H. M.: J. Am. Chem. Soc., *86*, 3897 (1964).
DR-70	Davison, A., Rakita, R. E.: Inorg. Chem., *9*, 289 (1970).
DS-62	Davies, W. O., Stanley, E.: Acta. Cryst., *15*, 1092 (1962).
DS-66	Denney, D. B., Saferstein, S. L.: J. Am. Chem. Soc., *88*, 1839 (1966).
DS-68	Dimroth, K., Stade, W.: Angew. Chem. Int. Ed. Engl., *7*, 881 (1968).
DS-74	Denney, D. B., Shih, L. S.: J. Am. Chem. Soc., *96*, 317 (1974).
DT-66	Darwish, D., Tourigny, G.: J. Am. Chem. Soc., *88*, 4303 (1966).
DTV-62	DeLaMare, P. D. B., Tillett, J. G., Van Woerden, H. F.: J. Chem. Soc., 4888 (1962).
DU-69	Darcy, R., Ullman, E. F.: J. Am. Chem. Soc., *91*, 1024 (1969).
DVM-71	Durst, T., Viau, R., McClory, M. R.: J. Am. Chem. Soc., *93*, 3077 (1971).

References

DW-66   Dennis, E. A., Westheimer, F. H.: J. Am. Chem. Soc., *88*, 3431 (1966).
DW-66a Dennis, E. A., Westheimer, F. H.: J. Am. Chem. Soc., *88*, 3432 (1966).
DW-75a Doyle, M. P., West, C. T.: J. Org. Chem., *40*, 3821 (1975).
DW-75b Doyle, M. P., West, C. T.: J. Org. Chem., *40*, 3829 (1975).
DW-75c Doyle, M. P., West, C. T.: J. Org. Chem., *40*, 3835 (1975).
DW&-76 Delker, G. L., Wang, Y., Stucky, G. D., Lambert, R. L., Haas, C. K., Seyferth, D.: J. Am. Chem. Soc., *98*, 1779 (1976).
DWD-71 Denney, D. Z., White, D. W., Denney, D. B.: J. Am. Chem. Soc., *93*, 2066 (1971).
DWJ-72 Drews, M. J., Wong, P. S., Jones, P. R.: J. Am. Chem. Soc., *94*, 9122 (1972).
DZ&-69 DeBruin, K. E., Zon, G., Naumann, K., Mislow, K.: J. Am. Chem. Soc., *91*, 7027 (1969).
E-60   Eaborn, C.: Organosilicon Compounds, p. 103. New York, N.Y.: Academic Press 1960.
E-63   Ebsworth, E. A. V.: Volatile Silicon Compounds. New York, N.Y.: MacMillan Co. 1963.
E-68   Eckstein, F.: FEBS Letters, *2*, 85 (1968).
E-69   Ebsworth, E. A. V.: The Bond to Carbon (ed. A. G. McDiarmid), Part I, New York: Marcel Dekker 1969.
E-70   Eckstein, F.: J. Am. Chem. Soc., *92*, 4718 (1970).
E-74   Evans, D. A.: Accounts Chem. Res., *7*, 147 (1974).
EA-72   Evans, D. A., Andrews, G. C.: J. Am. Chem. Soc., *94*, 3672 (1972).
EB-69   Eaborn, C., Bott, R. W.: The Bond to Carbon (ed. A. G. MacDiarmid), Part 1, New York: Marcel Dekker 1969.
EBB-74 Ensslin, W., Bock, H., Becker, G.: J. Am. Chem. Soc., *96*, 2757 (1974).
EC&-70 Egan, W., Chauviere, G., Mislow, K., Clark, R. T., Marsi, K. L.: J.C.S. Chem. Comm., 733 (1970).
EE&-75 Eaborn, C., Eidenschink, R., Jackson, P. M., Walton, D. R. M.: J. Organometal. Chem., *101*, C40 (1975).
EF-63   Ebsworth, E. A. V., Frankiss, S. G.: J. Am. Chem. Soc., *85*, 3516 (1963).
EG-76   Eisch, J. J., Galle, J. E.: J. Org. Chem., *41*, 2615 (1976).
EHA-74 Eliel, E. L., Hartmann, A. A., Abatjoglou, A. G.: J. Am. Chem. Soc., *96*, 1807 (1974).
EJ-74   Eaborn, C., Jenkins, I. D.: J. Organometal. Chem., *69*, 185 (1974).
EK&-76 Eliel, E. L., Koskimies, J., McPhail, A. T., Swern, D.: J. Org. Chem., *41*, 2137 (1976).
ELU-75 ElGomati, T., Lenoir, D., Ugi, I.: Angew. Chem. Int. Ed. Engl., *14*, 59.
EM-67   Eaton, D. R., McClellan, W. R.: Inorg. Chem., *6*, 2134 (1967).
EM&-74 Eisenhut, M., Mitchell, H. L., Traficante, D. D., Kaufman, R. J., Deutch, J. M., Whitesides, G. M.: J. Am. Chem. Soc., *96*, 5385 (1974).
EPN-67 Engelhard, E., Pichal, K., Nenner, M.: Angew. Chem. Int. Ed. Engl., *6*, 615 (1967).
EP&-76 Effenberger, F., Podszun, W., Schoeller, W. W., Seufert, W. G., Stohrer, W. D.: Chem. Ber., *109*, 306 (1976).
ER-73   Eargle, D. H., Jr., Ramos de Carvalho, M. C.: J. Phys. Chem., *77*, 1716 (1973).
ER-75   Eisch, J. J., Rhee, S. G.: J. Am. Chem. Soc., *97*, 4673 (1975).
ER&-73 Engelhardt, G., Radeglia, R., Jancke, H., Lippmaa, E., Maegi, M.: Org. Magn. Res., *5*, 561 (1973).
ES-63   Ewing, V. G., Sutton, L. E.: Trans. Faraday Soc., *59*, 1241 (1963).
ESS-73 Eisenhut, M., Schmutzler, R., Sheldrick, W. S.: Chem. Comm., 144 (1973).
ES&-74 Ernst, C. R., Spialter, L., Buell, G. R., Wilhite, D. L.: J. Am. Chem. Soc., *96*, 5375 (1974).
ET-73   Eisch, J. J., Tsai, M. R.: J. Am. Chem. Soc., *95*, 4065 (1973).
EV&-70 Egorochkin, A. N., Vyazankin, N. S., Burov, A. I., Khorshev, S. Ya.: Izv. Akad. Nauk. SSSR, Ser, Khim., 1279 (1970).
EVK-72 Egorochkin, A. N., Vyazankin, N. S., Khorshev, S. Ya.: Uspekhii Khim, *41*, 828 (1972).
EY&-76 Epiotis, N. D., Yates, R. L., Bernardi, F., Wolfe, S.: J. Am. Chem. Soc., *98*, 5435 (1976).
F-72   Froeyen, P.: Acta Chem. Scand. *26*, 2163 (1972).
F-64   Frye, C. L.: J. Am. Chem. Soc., *86*, 3170 (1964).
FCI-61 Fukui, K., Chikayoshi, M., Imamura, A.: Bull. Chem. Soc. Japan, *34*, 1224 (1961).
FDC-70 Furtsch, T. A., Dierdorf, D. S., Cowley, A. H.: J. Am. Chem. Soc., *92*, 5759 (1970).

FG-70	Feugas, C., Galy, J.P.: C.R. Acad. Sci. Paris, 270C (1970).
FH&-71	Frost, D.C., Herring, F.G., Katrib, A., McLean, R.A.N., Drake, J.E., Westwood, N.P.C.: Can. J. Chem., 49, 4033 (1971).
FH&-72	Frost, D.C., Herring, F.G., Katrib, A., McDowell, C.A., McLean, R.A.N.: J. Phys. Chem., 76, 1030 (1972).
FK&-73	Fueno, T., Kajimoto, O., Izawa, K., Masago, M.: Bull. Chem. Soc. Japan, 46, 1418, (1973).
FM-61	Franzen, V., Mertz, C.: Ann., 643, 24 (1961).
FN-76	Fraser, R.R., Ng, L.K.: J. Am. Chem. Soc., 98, 4335 (1976).
FS-68	Frey, H.M., Solly, R.K.: Trans. Faraday Soc., 1858 (1968).
FS-69	Fraser, R.R., Schuber, F.J.: Chem. Comm., 1474 (1969).
FS-69a	Fraser, R.R., Schuber, F.J.: Chem. Comm., 397 (1969).
FS-70	Fraser, R.R., Schuber, F.J.: Can. J. Chem., 48, 633 (1970).
FS-71	Freeburger, M.E., Spialter, L.: J. Am. Chem. Soc., 93, 1894 (1971).
FSW-72	Fraser, R.R., Schuber, F.J., Wigfield, Y.Y.: J. Am. Chem. Soc., 94, 8795 (1972).
FU-67	Frank, D.S., Usher, D.S.: J. Am. Chem. Soc., 89, 6360 (1967).
FVF-71	Frye, C.L., Vincent, G.A., Finzel, W.A.: J. Am. Chem. Soc., 93, 6805 (1971).
G-59	Gould, E.S.: Mechanism & Structure in Organic Chemistry, Holt, Rinehart & Winston, New York, 1959.
G-60	Gillespie, R.J.: Can. J. Chem., 38, 818 (1960).
G-63	Gillespie, R.J.: J. Chem. Soc., 4672 (1963).
G-67	Gervais, H.P.: Compt. Rend. Acad. Sci., Paris, 264C, 1027 (1967).
G-69	Gresser, M.: Mechanisms of Reactions of Sulfur Compounds, 4, 29 (1969).
G-70	Gervais, H.P.: J. Chim. Phys., 67, 1402 (1970).
G-71	Gray, G.A.: J. Am. Chem. Soc., 93, 2132 (1971).
GB-66	Ginjaar, L., Blasse-Vel, S.: Rec. Trav. Chim. 85, 694 (1966).
GBC-74	Gilje, J.W., Braun, R.W., Cowley, A.H.: J. Chem. Soc.; Chem. Commun., 15 (1974).
GCA-64	Gilman, H., Cottis, S.G., Atwell, W.H.: J. Am. Chem. Soc., 86, 1596 (1964).
GCF-70	Garbesi, A., Corsi, N., Fava, A.: Helv. Chim. Acta., 53, 1499 (1970).
GC&-76	Gray, R.W., Chapleo, C.B., Vergnani, T., Dreiding, A.S., Liesner, M., Seebach, D.: Helv. Chim. Acta, 59, 1547 (1976).
GDV-71	Gilow, H.M., DeShazo, M., Van Cleave, W.C.: J. Org. Chem., 36, 1745 (1971).
GD-76	Glass, R.S., Duchek, J.R.: J. Am. Chem. Soc., 98, 965 (1976).
GH-68	Gleiter, R., Hoffman, R.: Tetrahedron, 24, 5899 (1968).
GH&-68	Gerson, F., Heinzer, J., Bock, H., Alt, H.: Helv. Chim. Acta, 51, 707 (1968).
GH-70	Gay, D.C., Hamer, N.K.: Chem. Comm., 1564 (1970).
GH&-71	Gillespie, P., Hoffman, P., Klusacek, H., Marquarding, D., Pfohl, S., Ramierez, F., Tsolis, E.A., Ugi, I.: Angew. Chem. Int. Ed. Engl., 10, 687 (1971).
GHB-70	Gerson, F., Heinzer, J., Bock, H.: Mol. Physics, 18, 461 (1970).
GI-73	Griller, D., Ingold, K.U.: J. Am. Chem. Soc., 95, 6459 (1973).
GJC-73	Garwood, D.C., Jones, M.R., Cram, D.J.: J. Am. Chem. Soc., 95, 1925 (1973).
GK-68	Givens, E.N., Kwart, H.: J. Am. Chem. Soc., 90, 378, 386 (1968).
GK-69	Goetz, H., Klabuhn, B.: Ann., 724, 1 (1969).
GKB-69	Gerson, F., Krynitz, U., Bock, H.: Angew. Chem. Int. Ed. Engl., 8, 767 (1969).
GKB-69a	Gerson, F., Krynitz, U., Bock, H.: Helv. Chim. Acta, 52, 2512 (1969).
GKJ-70	Goetz, H., Klabuhn, B., Juds, H.: Ann., 735, 88 (1970).
GKS-65	Goodman, L., Konstan, A.H., Sommer, L.H.: J. Am. Chem. Soc., 87, 1012 (1965).
GL-66	Gerdil, R., Lucken, E.A.C.: J. Am. Chem. Soc., 88, 733 (1966).
GL-71	Goubeau, J., Lentz, A.: Spectrochim. Acta., 27A, 1703 (1971).
GLS-73	Guillerm, G., Leguan, M., Simonnin, M.P.: Bull. Soc. Chim. France, 1649 (1973).
GM&-68	Gastaminza, A., Modro, T.A., Ridd, J.H., Utley, J.H.P.: J. Chem. Soc., B, 534 (1968).
GM&-68a	Glemser, O., Müller, A., Bohler, D., Krebs, B.: Z. Anorg. Allg. Chem., 357, 184 (1968).
GP-69	Grant, M.W., Prince, R.H.: Nature, 222, 1163 (1969).
GPB-70	Gerson, F., Plattner, G., Bock, H.: Helv. Chim. Acta, 53, 1629 (1970).

References

GR-62	Goldstein, J.H., Reddy, G.S.: J. Chem. Phys., *36*, 2644 (1962).
GR-63	Gillespie, R.J., Robinson, E.A.: Can. J. Chem., *41*, 2074 (1963).
GR&-70	Glidewell, C., Rankin, D.W.H., Robiette, A.G., Sheldrick, G.M.: J. Chem. Soc., *(A)*, 318 (1970).
GR&-73	Gillespie, P.D., Ramirez, F., Ugi, I., Marquarding, D.: Angew. Chem. Int. Ed. Engl., *12*, 91 (1973).
GT-65	Goodman, L., Taft, R.W.: J. Am. Chem. Soc., *87*, 4385 (1965).
GU-71	Gillespie, P.D., Ugi, I.: Angew. Chem. Int. Ed., *10*, 503 (1971).
GW-67	Greenhalgh, R., Weinberger, M.A.: Can. J. Chem., *45*, 495 (1967).
GW-74	Gilbro, T., Williams, F.: J. Am. Chem. Soc., *96*, 5032 (1974).
H-55	Hedberg, K.: J. Am. Chem. Soc., *77*, 6491 (1955).
H-56	Heath, D.F.: J. Chem. Soc., 3796 (1956).
H-60	Handbook of Chemistry & Physics, 41st Ed. (1960).
H-64	Hamer, F.M.: "The Cyanine Dyes & Related Compounds" Interscience, New York, 1964.
H-65	Hellwinkel, D.: Ber., *98*, 576 (1965).
H-65a	Hudson, R.F.: "Structure & Mechanism in Organophosphorus Chemistry", Academic Press, New York, N.Y., 1965, Chapter 3.
H-66	Hellwinkel, D.: Chem. Ber., *99*, 3628 (1966).
H-66a	Hellwinkel, D.: Chem. Ber., *99*, 3642 (1966).
H-66b	Hellwinkel, D.: Chem. Ber., *99*, 3668 (1966).
H-71	Hays, H.R.: J. Org. Chem., *36*, 98 (1971).
H-74	Holmes, R.R.: J. Am. Chem. Soc., *96*, 4143 (1974).
H-75	Howell, J.M.: J. Am. Chem. Soc., *97*, 3930 (1975).
HA-70	Haake, P.C., Allen, G.W.: Tetrahedron Lett., 3113 (1970).
HAK-69	Hutchinson, B.J., Andersen, K.K., Katritsky, A.R.: J. Am. Chem. Soc., *91*, 3839 (1969).
HB-72	Hudson, R.F., Brown, C.: Accts. Chem. Res., *5*, 204 (1972).
HBG-70	Hoffmann, R., Boyd, D.B., Goldberg, D.Z.: J. Am. Chem. Soc., *92*, 3929 (1970).
HBL-72	Hall, C.D., Bramblett, J.D., Lin, F.F.S.: J. Am. Chem. Soc., *94*, 9264 (1972).
HBM-69	Haake, P.C., Bausher, L.P., Miller, W.B.: J. Am. Chem. Soc., *91*, 1113 (1969).
HBM-71	Haake, P.C., Bausher, L.P., McNeal, J.P.: J. Am. Chem. Soc., *93*, 7045 (1971).
HC&-71	Haque, M.U., Caughlin, C.N., Ramirez, F., Pilot, J.F., Smith, C.P.: J. Am. Chem. Soc., *93*, 5229 (1971).
HD-68	Holmes, R.R., Deiters, R.M.: Inorg. Chem., *7*, 2229 (1968).
HF-68	Hellwinkel, D., Fahrbach, G.: Liebigs Ann. Chem., *712*, 1 (1968).
HH&-69	Heilbronner, E., Hornung, V., Bock, H., Alt, H.: Angew. Chem., *85*, 537 (1969).
HHH-74	Hedberg, E., Hedberg, L., Hedberg, K.: J. Am. Chem. Soc., *96*, 4417 (1974).
HHM-72	Hoffmann, R., Howell, J.M., Muetterties, E.L.: J. Am. Chem. Soc., *94*, 3047 (1972).
HHM-74	Hortmann, A.G., Harris, R.L., Miles, J.A.: J. Am. Chem. Soc., *96*, 6119 (1974).
HJ&-70	Hendra, P.J., Johns, R.A., Miles, T.S., Vear, C.J.: Spectrochim. Acta, *26A*, 2169 (1970).
HK-56	Hudson, R.F., Keay, L.: J. Chem. Soc., 2463 (1956).
HK-60	Hudson, R.F., Keay, L.: J. Chem. Soc., 1859 (1960).
HK-70	Hildbrand, J., Kaufmann, G.: Spectrochim. Acta, *26A*, 1407 (1970).
HL-64	Hamilton, W.C., LaPlaca, S.J.: J. Am. Chem. Soc., *86*, 2290 (1964).
HL-72	Hovanec, S.W., Lieske, C.N.: Biochemistry, *11*, 1051 (1972).
HLI-63	Hafferl, W.H., Lundin, R.E., Ingraham, L.L.: Biochemistry, *2*, 1293 (1963).
HLI-64	Hafferl, W.H., Lundin, R.E., Ingraham, L.L.: Biochemistry, *3*, 1072 (1964).
HMT-64	Haake, P.C., Miller, W.B., Tyssee, D.A.: J. Am. Chem. Soc., *86*, 3577 (1964).
HO-59	Horner, L., Oediger, H.: Liebigs Ann. Chem., *627*, 142 (1959).
HO-71	Hairston, T.J., O'Brien, D.H.: J. Organometal. Chem., *29*, 79 (1971).
HO-76	Howell, J.M. & Olsen, J.F.: J. Am. Chem. Soc., *98*, 7119 (1976).
HP-67	Hartzell, G.E., Paige, J.N.: J. Org. Chem., *32*, 459 (1967).
HPH-70	Henge, E., Pletka, H., Höfler, F.: Monatshefte Chem., *101*, 325 (1970).
HPM-70	Hine, J., Phillips, J.C., Maxwell, J.I.: J. Org. Chem., *35*, 3943 (1970).

HR-51	Hach, R.J., Rundle, R.E.: J. Am. Chem. Soc., 73, 4321 (1951).
HRT-73	Howard, J.A., Russell, D.R., Trippett, S.: Chem. Commun. 856 (1973).
HS-69	Hillier, I.H., Saunders, V.R.: Chem. Comm., 1181 (1969).
HS-70	Hillier, I.H., Saunders, V.R.: Trans. Faraday Soc., 66, 2401 (1970).
HS-70a	Hillier, I.H., Saunders, V.R.: Chem. Comm., 1183 (1970).
HS-73	Homer, G.D., Sommer, L.H.: J. Am. Chem. Soc., 95, 7700 (1973).
HSH-70	Höfler, F., Sawodny, W., Hengge, E.: Spectrochim. Acta, 26A, 819 (1970).
HSS-66	Hordvik, A., Sletten, E., Sletten, J.: Acta. Chem. Scand., 20, 2001 (1966).
HU&-69	Harris, M.R., Usher, D.A., Albrecht, H.P., Jones, G.H., Moffatt, J.G.: Proc. Nat. Acad. Sci., 63, 246 (1969).
HUZ-71	Harrison, B.G., Ulrich, S.E., Zuckerman, J.: J. Am. Chem. Soc., 93, 2307 (1971).
HV-74	Howell, J.M., Van Wazer, J.R.: J. Am. Chem. Soc., 96, 3064 (1974).
HVR-74	Howell, J.M., Van Wazer, J.R., Rossi, A.R.: Inorg. Chem., 13, 1747 (1974).
HW-59	Havinga, E.E., Wiebinga, E.H.: Rec. Trav. Chim., 78, 724 (1959).
HW-61	Haake, P.C., Westheimer, F.H.: J. Am. Chem. Soc., 83, 1102 (1961).
HW&-69	Houalla, D., Wolf, R., Gagnaire, D., Robert, J.B.: Chem. Comm., 443 (1969).
HW&-72	Harrington, D., Weston, J., Jacobus, J., Mislow, K.: Chem. Comm., 1079 (1972).
IYY-71	Iwata, K., Yoneda, S., Yoshida, Z.: J. Am. Chem. Soc., 93, 6745 (1971).
J-54	Jaffe, H.H.: J. Phys. Chem., 58, 185 (1954).
J-66	Johnson, A.W.: "Ylide Chemistry" (Academic Press, New York, 1966).
JA-68	Johnson, A.W., Amel, R.T.; Can. J. Chem., 46, 461 (1968).
JA-69	Johnson, W.A.W., Amel, R.T.: J. Org. Chem., 34, 1240 (1969).
JC-74	Jones, M.R., Cram, D.J.: J. Am. Chem. Soc., 96, 2183 (1974).
JD&-72	Jones, P.R., Drews, M.J., Johnson, J.K., Wong, P.S.: J. Am. Chem. Soc., 94, 4595 (1972).
JGN-63	Johnson, F., Gohlke, R.S. & Nasutavicus, W.H.: J. Organometal. Chem., 3, 233 (1963).
JHT-70	Jarvie, A.W.P., Holt, A., Thompson, J.: J. Chem. Soc., (B), 746 (1970).
JJ-71	Jonsson, E.U., Johnson, C.R.: J. Am. Chem. Soc., 93, 5308 (1971).
JJ-75	Jung, I.N., Jones, P.R.: J. Am. Chem. Soc., 97, 6102 (1975).
JJ-75a	Jung, I.N., Jones, P.R.: J. Organometal. Chem., 101, 27, 35 (1975).
JKS-62	Jenkins, D.R., Kewley, R., Snyder, T.M.: Trans. Faraday Soc., 58, 1284 (1962).
JM-72	Jones, P.R., Meyers, J.K.: J. Organometal. Chem., 34, C9 (1972).
JO-62	Jaffe, H.H., Orchin, M.: Theory and Applications of Ultraviolet Spectroscopy, New York: Wiley, 1962.
JS-67	Jacobsen, N.J., Senning, A.: Chem. Comm. 617 (1967).
JS-70	Jennings, W.B., Spratt, R.: Chem. Comm., 1418 (1970).
JS&-63	Jordan, T., Smith, H.W., Lohr, K.L., Jr., Lipscomb, W.N.: J. Am. Chem. Soc., 85, 846 (1963).
K-57	Kriegsmann, H.: Z. Electrochem., 61, 1088 (1957).
K-59	Kriegsmann, H.: Z. Anorg. Allg. Chem., 299, 138 (1959).
K-61	Khorana, H.G.: Some Recent Developments in the Chemistry of Phosphate Esters of Biological Interest (New York: Wiley, 1961).
K-68	Kice, J.L.: Accts. Chem. Res., 1, 58 (1968).
K-69	Klingsberg, E.: Quart. Rev., 23, 537 (1969).
KA-69	Kresze, G., Amann, W.: Spectrochim. Acta., 25A, 393 (1969).
KB-66	Karle, I.L., Britts, K.: Acta. Cryst., 20, 118 (1966).
KB-77	Kwart, H., Barnette, W.E.: J. Am. Chem. Soc., 99, 614 (1977).
KC&-69	Kluger, R., Covitz, F., Dennis, E.A., Williams, L.D., Westheimer, F.H.: J. Am. Chem. Soc., 91, 6066 (1969).
KD&-69	Kannengiesser, G., Damm, F., Deluzarche, A., Maillard, A.: Bull. Soc. Chim. France, 894 (1969).
KD-73	King, J.F., Dumanoir, J.R.: Can. J. Chem., 51, 4082 (1973).
KH-73	Koizumi, T., Haake, P.: J. Am. Chem. Soc., 95, 8073 (1973).
KJ-70	Kwart, H., Johnson, N.A.: J. Am. Chem. Soc., 92, 6064 (1970).

References

KJ-77   Kwart, H., Johnson, N.A.: J. Am. Chem. Soc., *99*, 3441, (1977).
KK-71   Krusic, P.J., Kochi, J.K.: J. Am. Chem. Soc., *93*, 846 (1971).
KK-72   Kawamura, T., Kochi, J.K.: J. Am. Chem. Soc., *94*, 648 (1972).
KK-75   Keil, F., Kutzelnigg, W.: J. Am. Chem. Soc., *97*, 3623 (1975).
KK&-75   Klemperer, W.G., Krieger, J.K., McCreary, M.D., Muetterties, E.L., Traficante, D.D., Whitesides, G.M.: J. Am. Chem. Soc., *97*, 7023 (1975).
KKF-73   Kajimoto, O., Kobayashi, M., Fueno, T.: Bull. Chem. Soc. Japan, *46*, 2316 (1973).
KKK-67   Kucsman, A., Kalaman, A., Kapovits, I.: Acta. Chim. Acad. Sci. Hung. *53*, 97 (1967).
KKW-65   Kaiser, E.T., Katz, I.R., Wulfers, T.F.: J. Am. Chem. Soc., *87*, 3781 (1965).
KM-51   Koch, J.H.F., Moffitt, W.E.: Trans. Faraday Soc., *47*, 7 (1951).
KM-68   Klanberg, F., Muertterties, E.L.: Inorg. Chem., *7*, 155 (1968).
KM-73   Kaplan, L.J., Martin, J.C.: J. Am. Chem. Soc., *95*, 793 (1973).
KMK-72   Krusic, P.J., Mahler, W., Kochi, J.K.: J. Am. Chem. Soc., *94*, 6033 (1972).
KO-70   Kemp, D.A., O'Brien, J.T.: J. Am. Chem. Soc., *92*, 2554 (1970).
KO-71   Kwart, H., Omura, H.: J. Am. Chem. Soc., *93*, 7250 (1971).
KOC-66   Kolodyazhnyi, Yu.V., Osipov. O.A., Chenskaya, T.B.: J. Gen. Chem. USSR, *36*, 2183 (1966).
KPB-72   Kuznesod, P.M., Pessini, F.B.T., Bruns, R.E.: J. Am. Chem. Soc., *94*, 9087 (1972).
KPW-63   Kaiser, E.T., Panar, M., Westheimer, F.H.: J. Am. Chem. Soc., *85*, 602 (1963).
KRS-70   Khwaja, T.A., Reese, C.B., Stewart, J.C.M.: J. Chem. Soc., (C) 2092 (1970).
KS-72   Kwart, H., Slutsky, J.J.: J. Am. Chem. Soc., *94*, 2515 (1972).
KS-76   Kwart, H., Stanulonis, J.J.: J. Am. Chem. Soc., *98*, 1009 (1976).
KSB-76   Kwart, H., Stanulonis, J., Benko, D.: To be published.
KT-75   Kokot, Z., Tamres, M.: Inorg. Chem., *14*, 2441 (1975).
KW-69   Kluger, R., Westheimer, F.H.: J. Am. Chem. Soc., *91*, 4143 (1969).
KW-72   Kice, J.L., Walters, C.A.: J. Am. Chem. Soc., *94*, 590 (1972).
KY-72   Komoriya, A., Yoder, C.H.: J. Am. Chem. Soc., *94*, 5285 (1972).
KZ-75   Kost, U., Zeichner, A.: Tetrahedron Letters, 3239 (1975).
L-49   Longuet-Higgins, H.C.: Trans. Faraday Soc., *4*, 173 (1949).
L-57   Larsson, L.: Acta, Chem. Scand., *11*, 1137 (1957).
L-69   Lautenschlaeger, F.: J. Org. Chem., *34*, 3998 (1969).
L-70   Lehn, J.M.: Fortschr. Chem. Forsch., *15*, 311 (1970), see p. 368.
L-72   Lipowitz, J.: J. Am. Chem. Soc., *94*, 1582 (1972).
LC-69   Lepeska, B., Chvalowsky, V.: Coll. Chem. Commun., *34*, 3553 (1969).
LCM-76   Lett, R., Chassaing, G., Marquet, A.: J. Organometal. Chem., *111*, C17 (1976).
LD-61   Larive, H., Dennilauler, R.: Chimia, *15*, 115 (1961).
LF-75   Larson, G.L., Fernandez, Y.V.: J. Organometal. Chem., *86*, 193 (1975).
LG-76   Lindner, H.J., Gross, B.K.: Chem. Ber., *109*, 314 (1976).
LH-71   Levin, I.W., Harris, W.C.: J. Chem. Phys., *55*, 3048 (1971).
LM-67   Luken, E.A.C., Mazeline, C.: J. Chem. Soc., *(A)*, 439 (1967).
LM-70   Lehn, J.M., Munsch, B.: Chem. Comm., 994 (1970).
LM-72   Lehn, J.M., Munsch, B.: Mol. Phys., *23*, 91 (1972).
LM-75   Limouzin, Y., Maire, J.C.: J. Organometallic Chem., *92*, 169 (1975).
LM-76   Livant, P., Martin, J.C.: J. Am. Chem. Soc., *98*, 7851 (1976).
LM&-66   Loshadkin, N.A., Markov, S.M., Polekhin, A.M., Neimysheva, A.A., Makaeva, F.F., Knunyants, I.L.: Zh. Obshch. Chim., *36*, 1105 (1966). J. Gen. Chem. USSR, *36*, 1118 (1966).
LM&-70   Landini, D., Modena, G., Montanari, F., Scorrano, G.: J. Am. Chem. Soc., *92*, 7168 (1970).
LN-71   Leung, F., Nyburg, S.C.: Can. J. Chem., *49*, 167 (1971).
LPO-71   Lambert, J.B., Packard, B.S., Oliver, W.L.: J. Org. Chem., *36*, 1309 (1971).
LS-72   Lyons, A.R., Symons, M.C.R.: J. Chem. Soc. Faraday Trans; *II*, 622 (1972).
LSD-70   Landau, M.A., Sheluchenko. V.V., Dubov, S.S.: Zh. Strukt, Khim., *11*, 513 (1970).

LT-71	LaRochelle, R.W., Trost, B.M.: J. Am. Chem. Soc., *93*, 6077 (1971).
LW-65	Lüttke, W., Wilhelm, K.: Angew. Chem., *77*, 867 (1965).
LW-75	Limouzin, Y., Maire, J.C.: J. Organometallic Chem., *92*, 169 (1975).
LW-76	Lerman, C.L., Westheimer, F.H.: J. Am. Chem. Soc., *98*, 179 (1976).
LW-76a	Lehn, J.M., Wipff, G.: J. Am. Chem. Soc., *98*, 7498 (1976).
LZ-69	Lonardo, G. di., Zauli, C.: J. Chem. Soc. (A) 1305 (1969).
M-55	Mulliken, R.S.: J. Am. Chem. Soc., *77*, 884 (1955).
M-57	Mortimer, F.S.: Spectrochim. Acta, *9*, 270 (1957).
M-63	Markl, G.: Angew. Chem., *75*, 168 (1963).
M-63a	Morton, J.R.: Can. J. Phys., *41*, 706 (1963).
M-64	Moriarty, R.M.: Tet. Letters, 509 (1964).
M-68	Mitchell, K.A.R.: Can. J. Chem., *46*, 3499 (1968).
M-69	Mitchell, K.A.R.: Chem. Rev., *69*, 157 (1969).
M-69a	Musher, J.I.: Angew. Chem. Int. Ed. Engl., *8*, 54 (1969).
M-70	Muetterties, E.L.: Accts. Chem. Res., *3*, 266 (1970).
M-70a	Mislow, K.: Accts. Chem. Res., *3*, 321 (1970).
M-71	Marsi, K.L.: J. Am. Chem. Soc., *93*, 6341 (1971).
M-72	Musher, J.I.: J. Am. Chem. Soc., *94*, 1370 (1972).
M-73	McLean, R.A.N.: Can. J. Chem., *51*, 2089 (1973).
M-74	Marsi, K.L.: J. Org. Chem., *39*, 265 (1974).
M-75	Marsi, K.L.: J. Org. Chem., *40*, 1779 (1975).
M-75a	MacLagen, R.G.A.R.: Unpublished, reported in WH&-75.
MA-71	Martin, J.C., Arhart, R.J.: J. Am. Chem. Soc., *93*, 2341 (1971).
MA&-65	McEwèn, W.E., Axelrad, G., Zanger, M., Van der Werf, C.A.: J. Am. Chem. Soc., *87*, 3948 (1965).
MB&-72	Mollere, P., Bock, H., Becker, G., Fritz, G.: J. Organometal. Chem., *46*, 89 (1972).
MB&-75	Matthews, W.S., Bartmess, J.E., Bordwell, F.G., Cornforth, F.J., Drucker, G.E., Margolin, Z., McCollum, R.J. & Vanier, N.R.: J. Am. Chem. Soc., *97*, 7006 (1975).
MBC-72	Marsi, K.L., Burns, F.B., Clark, R.T.: J. Org. Chem., *37*, 238 (1972).
MBK-70	Meppelder, F.H., Benschap, H.P., Kraay, G.W.: Chem. Comm., 431 (1970).
MC-67	McDonald, W.S., Cruickshank, D.W.J.: Acta. Cryst., *22*, 37 (1967).
MC-70	Marsi, K.L., Clark, R.T.: J. Am. Chem. Soc., *92*, 3791 (1970).
MC-74	Martin, J.C., Chau, M.M.: J. Am. Chem. Soc., *96*, 3319 (1974).
MD-66	McDaniel, D.H., Deiters, R.M.: J. Am. Chem. Soc., *88*, 2607 (1966).
MD&-76	Moreland, C.G., Doak, G.O., Littlefield, L.B., Walker, N.S., Gilje, J.W., Braun, R.W., Cowley, A.H.: J. Am. Chem. Soc., *98*, 2161 (1976).
MDL-73	Moreland, C.G., Doak, G.O., Littlefield, L.B.: J. Am. Chem. Soc., *95*, 255 (1973).
ME&-71	Maria, P.C., Élégant, L., Azzaro, M., Majoral, J.P., Navech, J.: Bull. Soc. Chim. Fr., 3750 (1971).
MF-75	Martin, J.C., Franz, J.A.: J. Am. Chem. Soc., *97*, 6137 (1975).
MFA-74	Martin, J.C., Franz, J.A., Arhart, R.J.: J. Am. Chem. Soc., *96*, 4604 (1974).
MG&-72	Meyer, B., Gouterman, M., Jensen, D., Oommen, T.V., Spitzer, K., Stroyer-Hansen, T.: Adv. Chem. Ser., *110*, 53 (1972).
MH-68	Murdock, L.L., Hopkins, T.L.: J. Agr. Food Chem., *16*, 954 (1968).
MH-75	Mollere, P.D., Hoffmann, R.: J. Am. Chem. Soc., *97*, 3680 (1975).
MK&-67	Müller, A., Krebs, B., Niecke, E., Ruoff, A.: Ber. Bunsenges Phys. Chem., *71*, 571 (1967).
MM-65	Mahler, W., Muetterties, E.L.: Inorg. Chem., *4*, 1520 (1965).
MM-68	Markl, G., Merz, A.: Tetrahedron Letters, 3611 (1968).
MM-72	Mathey, F., Mueller, G.: Tetrahedron, *28*, 5645 (1972).
MMK-75	Matsuyama, H., Minato, H., Kobayashi, M.: Bull. Chem. Soc. Japan, *48*, 3287 (1975).
MM&-68	Mavel, G., Mankowski-Favelier, R., Lavielle, G., Sturtz, G.: J. Chim. Phys., *64*, 1698 (1968).

References

MMS-66	McEachan, C.E., MacPhail, A.T., Simm, G.A.: J. Chem. Soc., *(B)*, 579 (1966).
MMS-68	Mavel, G., Mankowski-Favelier, R., Sturtz, G.: J. Chim. Phys., *64*, 1686 (1968).
MMS-69	Mackle, H., McNally, D.V., Steele, W.V.: Trans. Faraday Soc., *65*, 2060 (1969).
MNG-66	Müller, A., Niecke, E., Glesmer, O.: Z. Anorg. Allg. Chem., *347*, 275 (1966).
MNG-67	Müller, A., Niecke, E., Glesmer, O Z. Anorg. Allg. Chem., *350*, 246 (1967).
MO-73	Marsi, K.L., Oberlander, J.E.: J. Am. Chem. Soc., *95*, 200 (1973).
MOM-75	Maryanoff, C.A., Ogura, F., Mislow, K.: Tet. Letters, 4095 (1975).
MP-59	Muetterties, E.L., Phillips, W.D.: J. Am. Chem. Soc., *81*, 1084 (1959).
MP-59a	Muller, N., Pritchard, D.E.: J. Chem. Phys., *31*, 1471 (1959).
MP-72	McGrath, L., Prebble, K.A.: J. Appl. Chem. Biotechnol., *22*, 823 (1972).
MP-76	Martin, J.C., Perozzi, E.F.: Science, *191*, 154 (1976).
MPM-75	Malinski, E., Piekos, A., Modro, T.A.: Can. J. Chem., *53*, 1468 (1975).
MS-72	Meyer, B., Spitzer, K.: J. Phys. Chem., *76*, 2274 (1972).
MS-74	Mishra, S.P., Symons, M.C.R.: J. Chem. Soc., Chem. Commun., 279 (1974).
MS&-64	Mislow, K., Simmons, T., Melillo, J.T., Ternay, A.L., Jr.: J. Am. Chem. Soc., *86*, 1452 (1964).
MS&-71	Morgan, W.E., Stec, W.J., Albridge, R.G., Van Wazer, J.R.: Inorg. Chem., *10*, 926 (1971).
MS&-74	Maryanoff, B.E., Senkler, G.H., Stackhouse, J., Mislow, K.: J. Am. Chem. Soc., *96*, 5651 (1974).
MS&-75	Maryanoff, B.E., Stackhouse, J., Senkler, G.H., Mislow, K.: J. Am. Chem. Soc., *97*, 2718 (1975).
MS&-76	Milbrath, D.S., Springer, J.P., Clardy, J.C., Verkade, J.G.: J. Am. Chem. Soc., *98*, 5493 (1976).
MSE-63	Mironov, V.A., Sobolev, E.V., Elizarova, A.N.: Izv. Akad. Nauk. SSSR, 1607 (1963); Engl. Transl., p. 1467.
MSV-75	Mueller, C., Schweig, A., Vermeer, H.: J. Am. Chem. Soc., *97*, 982 (1975).
MU&-69	Minamido, I., Uneyama, K., Tagaki, W., Oae, S.: Bull. Chem. Soc., Japan, *42*, 3609 (1969).
N-66	Neureiter, N.P.: J. Am. Chem. Soc., *88*, 558 (1966).
NCB-66	Newton, M.G., Cox, J.R., Jr., Bertrand, J.A.: J. Am. Chem. Soc., *88*, 1503 (1966).
NEK-70	Neimysheva, A.A., Ermolaeva, M.V., Knunyants, I.L.: Zh. Obshch. Khim., *40*, 798 (1970); J. Gen. Chem. USSR, *40*, 774 (1970).
NFV-68	Nimrod, D.M., Fritzwater, D.R., Verkade, J.G.: J. Am. Chem. Soc., *90*, 2780 (1968).
NH-70	Nagy, J., Hencsei, P.: J. Organometal. Chem., *24*, 285 (1970).
NH-72	Nagy, J., Hencsei, P.: J. Organometal. Chem., *38*, 261 (1972).
NH-76	Nelson, R.M., Haake, P.C.: private communication (12/27/76) to be published
NP&-67	Neimysheva, A.A., Palm, U.A., Semin, G.K., Loshadkin, N.A., Knunyants, I.L.: Zh. Obshch. Khim., *37*, 2255 (1967), J. Gen. Chem., USSR, *37*, 2140 (1967).
NS&-68	Neimysheva, A.A., Sauchik, Y.I., Ermolaeva, M.V., Knunyants, I.L.: Izv. Akad. Nauk. SSSR, 2222 (1968), Bull. Acad. Sci. USSR, 2104 (1968).
NV-71	Nagy, J., Vandorffy, T.: J. Organometal. Chem., *31*, 205, 217 (1971).
NW-76	Nowakowski, P., West, R.: J. Am. Chem. Soc., *98*, 5616 (1976).
O-72	Olah, G.A.: J. Am. Chem. Soc., *94*, 808 (1972).
OB&-72	O'Donnell, K., Bacon, R., Chellappa, K.L., Schowen, R.L., Lee, K.J.K.: J. Am Chem. Soc., *94*, 2500 (1972).
OD-67	Oakland, R.L., Duffey, G.H.: J. Chem. Phys., *46*, 19 (1967).
OK-71	Ohkubo, K., Kanezawa, H.: Bull. Japan Petrol. Inst., *13*, 177 (1971).
OK-72	Ohkubo, K., Kaneda, H.: JCS-Faraday Trans., *I*, 1164 (1972).
OK-72a	Ohkubo, K., Kaneda, H.: Bull. Chem. Soc., Japan, *45*, 322 (1972).
OK-73	Ojima, I., Kondo, K.: Bull. Chem. Soc. Japan, *46*, 2571 (1973).
OKS-69	Olah, G.A., Klopman, G., Schlosberg, R.H.: J. Am. Chem. Soc., *91*, 3261 (1969).
OL-66	Olofson, R.A., Landesberg, J.M.: J. Am. Chem. Soc., *88*, 4263 (1966).
OL&-66	Olofson, R.A., Landesberg, J.M., Houk, K.M., Michelman, J.S.: J. Am. Chem. Soc., *88*, 4265 (1966).

OM-67	O'Connell, A.M., Maslen, E.N.: Acta. Cryst., *22*, 134 (1967).
OM-71	Olah, G.A., McFarland, C.W.: J. Org. Chem., *36*, 1374 (1971).
OOK-73	Osaki, T., Otera, J., Kawasaki, Y.: Bull. Chem. Soc., Japan, *46*, 1803 (1973).
OS-70	Oehling, H., Schweig, A.: Tetrahedron Letters, 4941 (1970).
OSN-76	Osamura, Y., Sayanagi, O., Nishimoto, K.: Bull. Chem. Soc. Jap., *49*, 845 (1976).
OTO-64	Oae, S., Tagaki, T.W., Ohno, A.: Tetrahedron, *20*, 437 (1964).
OTO-64a	Oae, S., Tagaki, T.W., Ohno, A.: Tetrahedron, *20*, 427 (1964).
OY&-68	Oae, S., Yokayama, M., Kise, M., Furukawa, N.: Tetrahedron Letters, 4131 (1968).
OY-70	Ohkubo, K., Yamabe, T.: Bull. Japan Petrol. Inst., *12*, 130 (1970).
OY-71	Ohkubo, K., Yamabe, T.: J. Org. Chem., *36*, 3149 (1971).
OYF-69	Ohkubo, K., Yamabe, T., Fukui, K.: Bull. Chem. Soc., Japan, *42*, 1800 (1969).
P-50	Pauling, L.: Nature of the Chemical Bond, New York: Cornell Press, Ithaca, 1950, p. 178.
P-64	Paddock, N.L.: Quart, Rev., *18*, 168 (1964).
P-66	Peterson, D.J.: J. Org. Chem., *31*, 950 (1966).
P-67	Perkins, P.G.: Chem. Commun., 268 (1967).
P-68	Paquette, L.A.: Accts. Chem. Res., *1*, 209 (1968).
P-70	Pitt, C.G.: J. Organometal. Chem., *23*, C35 (1970).
P-71	Pitt, C.G.: Chem. Commun., 816 (1971).
PC-68	Pant, A.K., Cruickshank, D.W.J.: Acta Cryst., *24B*, 13 (1968).
PCH-70	Pinnavaia, T.J., Collins, W.T., Howe, J.J.: J. Am. Chem. Soc., *92*, 4544 (1970).
PCT-72	Pitt, C.G., Carey, R.N., Toren, E.C., Jr.: J. Am. Chem. Soc., *94*, 3806 (1972).
PDC-74	Pillot, J.P., Dunogues, J., Calas, R.: C.R. Acad. Sci., *278*, 789 (1974).
PDC-75	Pillot, J.P., Dunogues, J., Calas, R.: Bull. Soc. Chim. France, 2143 (1975).
PFW-71	Paquette, L.A., Freeman, J.P., Wyvratt, M.J.: J. Am. Chem. Soc., *93*, 3216 (1971).
PGB-60	Pearson, R.G., Gray, H.B., Basolo, F.: J. Am. Chem. Soc., *82*, 787 (1960).
PH&-63	Price, C.C., Hori, M., Parasaran, T., Polk, M.: J. Am. Chem. Soc., *85*, 2278 (1963).
PH-69	Paquette, L.A., Houser, R.W.: J. Am. Chem. Soc., *91*, 3870 (1969).
PH-71	Paquette, L.A., Houser, R.W.: J. Am. Chem. Soc., *93*, 4522 (1971).
PI-74	Porter, N.A., Iloff, J.M.: J. Am. Chem. Soc., *96*, 6200 (1974).
PM-72	Perozzi, E.F., Martin, J.C.: J. Am. Chem. Soc., *94*, 5519 (1972).
PMP-71	Paul, I.C., Martin, J.C., Perozzi, E.F.: J. Am. Chem. Soc., *93*, 6674 (1971).
PMP-74	Perozzi, E.F., Martin, J.C., Paul, I.C.: J. Am. Chem. Soc., *96*, 578 (1974).
PO-62	Price, C.C., Oae, S.: Sulfur Bonding N.Y.: Ronald Press, 1962.
PP-68	Pillai, M.G.K., Pillai, P.P.: Can. J. Chem., *46*, 2393 (1968).
PPW-71	Paquette, L.A., Philips, J.C., Wingard, R.E., Jr.: J. Am. Chem. Soc., *93*, 4516 (1971).
PR-70	Phillips, W.G., Ratts, K.W.: J. Org. Chem., *35*, 3144 (1970).
PZ-67	Palmieri, P., Zauli, C.: J. Chem. Soc. *(A)*, 813 (1967).
QBM-71	Quin, L.D., Breen, J.J., Myers, D.K.: J. Org. Chem., *36*, 1297 (1971).
QW-73	Quinn, C.B., Wiseman, J.R.: J. Am. Chem. Soc., *95*, 6120 (1973).
R-59	Rebane, T.K.: Optika Spektrosk.: *6*, 11 (1959), C.A., 53, 7759.
R-63	Robinson, E.A.: Can. J. Chem., *41*, 3021 (1963).
R-63a	Rundle, R.E., J. Am. Chem. Soc., *85*, 112 (1963).
R-68	Ramirez, F.: Accts. Chem. Res., *1*, 168 (1968).
R-68a	Roseky, H.W.: Z. Naturforsch, *23B*, 103 (1968).
R-69	Ramsey, B.G.: Electronic Transitions of Organometaloids, New York: Academic Press, 1969.
R-69a	Rankin, D.W.H.: J. Chem. Soc. (A) 1926 (1969).
R-70	Ramirez, F.: Bull. Soc. Chim. France, 3491 (1970).
R-73	Reetz, M.T.: Tetrahedron, *29*, 2189 (1973).
R-75	Reffy, J.: J. Organometal. Chem., *96*, 187 (1975).
RAM-72	Rauk, A., Allen, L.C., Mislow, K.: J. Am. Chem. Soc., *94*, 3035 (1972).
RB&-65	Rauk, A., Buncel, E., Moir, R.Y., Wolfe, S.: J. Am. Chem. Soc., *87*, 5498 (1965).
RB&-74	Ramsey, B.G., Brook, A., Bassendale, A.R., Bock, H.: J. Organometal. Chem., *74*, C41 (1974).

References

RBS-68	Ramierez, F., Bigler, A.J., Smith, C.P.: Tetrahedron, *24*, 5041 (1968). and J. Am. Chem. Soc., *90*, 3507 (1968).
RC-51	Roberts, J.D., Chambers, V.C.: J. Am. Chem. Soc., *73*, 5034 (1951).
RC&-72	Raban, M., Carlson, E.H., Lauderback, S.K., Moldowan, J.M., Jones, F.B., Jr.: J. Am. Chem. Soc., *94*, 2738 (1972).
RFB-73	Rothuis, R., Font Freide, J.J.H.M., Buck, H.M.: Rec. Trav. Chim., *92*, 1308 (1973).
RF&-74	Rothuis, R., Font Freide, J.J.H.M., Dijk, J.M.F. van, Buck, H.M.: ibid. *93*, 128 (1974).
RGM-68	Rayner, D.R., Gordon, A.J., Mislow, K.: J. Am. Chem. Soc., *90*, 4854 (1968).
RGS-68	Ramierez, F., Gulati, A.S., Smith, C.P.: J. Org. Chem., *33*, 13 (1968).
RH&-67	Rudman, R., Hamilton, W.C., Novick, S., Goldfarb, T.D.: J. Am. Chem. Soc., *89*, 5157 (1967).
RJ-71	Raban, M., Jones, F.B., Jr.: J. Am. Chem. Soc., *93*, 2692 (1971).
RKJ-69	Raban, M., Kenney, G.W.J., Jr., Jones, F.B., Jr.: J. Am. Chem. Soc., *91*, 6677 (1969).
RKP-74	Reetz, M.T., Kliment, M., Plachky, M.: Angew. Chem. Int. Ed. Engl., *13*, 813 (1974).
RLB-72	Rothuis, R., Luderer, T.K.J., Buck, H.M.: Rec. Irav. Chim., *91*, 836 (1972).
RM-51	Roberts, J.D., Mazur, R.H.: J. Am. Chem. Soc., *73*, 3542 (1951).
RMA-70	Rauk, A., Mislow, K., Allen, L.C.: Angew. Chem. Int. Ed. Engl., *9*, 400 (1970).
RNW-75	Rojhantalab, H., Nibler, J.W., Wilkins, C.J.: Spectrochim. Acta, in press.
RPE-67	Rüchardt, C., Panse, P., Eichler, S.: Chem. Ber., *100*, 1144 (1967).
RP&-68	Ramirez, R., Patwardhan, A.V., Kugler, H.J., Smith, C.P.: Tetrahedron, *24*, 2275 (1968).
RP&-71	Ramirez, F., Pfohl, S., Tsolis, G.A., Pilot, J.F., Smith, C.P., Ugi, I., Marquarding, D., Gillespie, P., Hoffman, P.: Phosphorus, *1*, 1 (1971).
RP&-71a	Rünchle, R., Pohl, W., Blaich, B., Goubeau, J.: Ber. Bunsen Ges. Phys. Chem., *75*, 66 (1971).
RR&-67	Rerat, B., Rerat, C., Dauphin, G., Kergomard, A.: Compt. Rend. Acad. Sci., Paris, *264C*, 500 (1967).
RR&-69	Rankin, D.W.H., Robiette, A.G., Sheldrick, G.M., Sheldrick, W.S., Aylett, B.J., Ellis, I.A., Monaghan, J.J.: J. Chem. Soc., (A), 1224 (1969).
RS&-68	Robiette, A.G., Sheldrick, G.M., Sheldrick, W.S., Beagley, B., Cruickshank, D.W.J., Monaghan, J.J., Aylett, B.J., Ellis, I.A.: Chem. Commun., 909 (1968).
RSL-76	Ross, J.A., Seiders, R.P., Lemal, D.M.: J. Am. Chem. Soc., *98*, 4325 (1976).
RTS-68	Rakshys, J.W., Taft, R.W., Sheppard, W.A.: J. Am. Chem. Soc., *90*, 5236 (1968).
RU-74	Ramirez, F., Ugi, I.: Bull. Soc. Chim. France, 453 (1974).
RWC-69	Rauk, A., Wolfe, S., Csizmadia, I.G.: Can. J. Chem., *47*, 113 (1969).
RWM-76	Rieke, R.D., White, C.K., Milliren, O.M.: J. Am. Chem. Soc., *98*, 6872 (1976).
RYZ-67	Randall, E.W., Yoder, C.H., Zuckerman, J.J.: Inorg. Chem., *6*, 744 (1967).
RZ-66	Randall, E.W., Zuckerman, J.J.: Chem. Comm., 732 (1966).
RZ-68	Randall, E.W., Zuckerman, J.J.: J. Am. Chem. Soc., *90*, 3167 (1968).
S-30	Slater, J.C.: Phys. Rev., *36*, 57 (1930).
S-53	Siebert, H.: Z. Anorg. Allg. Chem., *273*, 170 (1953).
S-62	Sheppard, W.A.: J. Am. Chem. Soc., *84*, 3058 (1962).
S-62a	Streitwieser, A.: Molecular Orbital Theory for Organic Chemistry, New York: Wiley, 1962.
S-63	Sands, D.E.: Z. Krist., *119*, 245 (1963).
S-63a	Schmidbauer, H.: J. Am. Chem. Soc., *85*, 2336 (1963).
S-65	Sommer, L.H.: Stereochemistry and Mechanisms in Silicon, New York: McGraw-Hill, 1965.
S-65a	Schmidt, M.: Elemental Sulfur, Chemistry and Physics, (B. Meyer, ed.) New York: Wiley Interscience, 1965, ch. 15.
S-65b	Stephens, F.S.: J. Chem. Soc., 5658 (1965).
S-68	Santry, D.P.: J. Am. Chem. Soc., *90*, 3309 (1968).
S-68a	Schmidt, A.: Z. Anorg. Allg. Chem., *362*, 129 (1968).
S-69	Sawodny, C.W.: Z. Anorg. Allg. Chem., *368*, 284 (1969).
S-71	Schiemenz, G.P.: Tetrahedron, *27*, 5723 (1971).
S-71a	Sheppard, W.A.: J. Am. Chem. Soc., *93*, 5597 (1971).

S-72	Symonds, M.C.R.: J. Am. Chem. Soc., *94*, 8589 (1972).
S-74	Stohrer, W.D.: Chem. Ber., *107*, 1795 (1974).
S-75	Seyferth, D.: J. Organometal. Chem., *100*, 237 (1975).
S-75a	Salomon, M.: J. Phys. Chem., *79*, 429, 2000 (1975).
SA-59	Smith, A.L., Angelotti, N.C.: Spectrochím. Acta, *15*, 412 (1959).
SA-60	Simes, J.G., Abrahams, S.C.: Acta. Cryst., *13*, 1 (1960).
SAV-76	Seyferth, D., Annarelli, D.C., Vick, S.C.: J. Am. Chem. Soc., *98*, 6382 (1976).
SB-69	Sommer, L.H., Baumann, D.L.: J. Am. Chem. Soc., *91*, 7045 (1969).
SB-71	Schary, K.J., Benkovic, S.J.: J. Am. Chem. Soc., *93*, 2522 (1971).
SBG-67	Schmidpeter, A., Brecht, R.H., Groeger, H.: Chem. Ber., *100*, 3063 (1967).
SC&-67	Swank, D., Caughlan, C.N., Ramierez, F., Madan, O.P., Smith, C.P.: J. Am. Chem. Soc., *89*, 6503 (1967).
SC&-71	Swank, D.D., Caughlan, C.N., Ramierez, F., Pilot, J.F.: J. Am. Chem. Soc., *93*, 5236 (1971).
SDJ-71	Steward, O.W., Dziedzic, J.E., Johnson, J.S.: J. Org. Chem., *36*, 3475 (1971).
SE-70	Saenger, W., Eckstein, F.: J. Am. Chem. Soc., *92*, 4712 (1970).
SF-68	Sommer, L.H., Fujimoto, H.: J. Am. Chem. Soc., *90*, 982 (1968).
SFH-66	Seyferth, D., Fogel, J., Heeren, J.K.: J. Am. Chem. Soc., *88*, 2207 (1966).
SG-66	Saunders, M., Gold, E.H.: J. Am. Chem. Soc., *88*, 3376 (1966).
SG-69	Shatenstein, A.I., Grozdeva, H.A.: Tetrahedron, *25*, 2749 (1969).
SH&-71	Sommer, L.H., Homer, G.D., Messing, A.W., Kutschinski, J.L., Stark, F.O., Michael, K.W.: J. Am. Chem. Soc., *93*, 2093 (1971).
SH&-71a	Streitwieser, A., Hollyhead, W.B., Pudjaatmaka, H.H., Owens, P.H., Kruger, T.L., Rubenstein, P.A., MacQuarrie, R.H., Browska, M.L., Chu, W.K.C., Niemeyer, H.M.: J. Am. Chem. Soc., *93*, 5088 (1971).
SHL-67	Spratley, R.D., Hamilton, W.C., Ladell, J.: J. Am. Chem. Soc., *87*, 2272 (1967).
SI-69	Sorenson, H.C., Ingraham, L.L.: Arch. Biochem. Biophys., *134*, 214 (1969).
SIK-72	Smith, J.H., Inove, T., Kaiser, E.T.: J. Am. Chem. Soc., *94*, 3098 (1972).
SJB-75	Schipper, P., Jansen, E.H.J.M., Buck, H.M.: unpublished review, Eindhoven Tech. University, The Netherlands.
SK-73	Slutsky, J., Kwart, H.: J. Am. Chem. Soc., *95*, 8678 (1973).
SK-73a	Slutsky, J., Kwart, H.: J. Org. Chem., *38*, 3659 (1973).
SKF-72	Sommer, L.H., Korte, W.D., Frye, C.L.: J. Am. Chem. Soc., *94*, 3463 (1972).
SKW-70	Stewart, H.F., Koepsell, D.G., West, R.: J. Am. Chem. Soc., *92*, 846 (1970).
SKW-76	Szele, I., Kubisen, S.J., Westheimer, F.H.: J. Am. Chem. Soc., *98*, 3533 (1976).
SL-65	Steitz, T.N., Lipscomb, W.N.: J. Am. Chem. Soc., *87*, 2488 (1965).
SM-70	Schmidbauer, H., Malisch, W.: Chem. Ber., *103*, 3007 (1970).
SM-72	Saito, S., Makimo, F.: Bull. Chem. Soc., Japan, *45*, 92 (1972).
SM&-74	Stackhouse, J., Maryanoff, B.E., Senkler, G.H., Mislow, K.: J. Am. Chem. Soc., *96*, 5650 (1974).
SMG-72	Sommer, L.H., McLick, J., Golino, C.M.: J. Am. Chem. Soc., *94*, 669 (1972).
SO-72	Sheppard, W.A., Ovenall, D.W.: Org. Magn. Resonance, *4*, 695 (1972).
SOM-76	Stec, W.J., Okruszek, A., Michalski, J.: J. Org. Chem., *41*, 233 (1976).
SP&-71	Spialter, L.A., Pazdernik, L., Bernstein, S., Swansinger, W.A., Buell, G.R., Freeburger, M.E.: J. Am. Chem. Soc., *93*, 5682 (1971).
SP&-74	Swain, C.G., Porschke, K.R., Ahmed, W., Schowen, R.L.: J. Am. Chem. Soc., *96*, 4700 (1974).
SR-63	Schneider, F.W., Rabinovitch, B.S.: J. Am. Chem. Soc., *85*, 2365 (1963).
SR-64	Schmidbaur, H., Ruidisch, I.: Inorg. Chem., *3*, 599 (1964).
SR-65	Speziale, A.J., Ratts, K.W.: J. Am. Chem. Soc., *87*, 5603 (1965).
SR&-76	Sarma, R., Ramirez, F., McKeever, B., Maracek, J.F., Lee, S.: J. Am. Chem. Soc., *98*, 581 (1976).
SRM-76	Sarma, R., Ramirez, F., Maracek, J.F.: J. Org. Chem., *41*, 473 (1976).
SS-41	Schomaker, V., Stevenson, D.P.: J. Am. Chem. Soc., *63*, 37 (1941).
SS-67	Santry, D.P., Segal, G.A.: J. Chem. Phys., *47*, 158 (1967).

References

SS-72   Schafer, W., Schweig, A.: Angew. Chem. Int. Ed. Engl., *11*, 836 (1972).
SS-73   Silhite, D.L., Spialter, L.: J. Am. Chem. Soc., *95*, 2100 (1973).
SS-75   Schwenzer, G.M., Schaefer, H.F.: J. Am. Chem. Soc., *97*, 13983 (1975).
SS-76   Stohrer, W.D., Schmieder, K.R.: Chem. Ber., *109*, 285 (1976).
SS&-74 Senkler, G.H., Stackhouse, J., Maryanoff, B.E., Mislow, K.: J. Am. Chem. Soc., *96*, 5649 (1974).
SS&-76 Schaefer, W., Schweig, A., Dimroth, K., Kanter, H.: J. Am. Chem. Soc., *98*, 4410 (1976).
SSH-75 Sheldrick, W.S., Schmidpeter, A., Helmut, J.: Angew. Chem. Int. Ed. Eng., *14*, 490 (1975).
SSK-68 Sato, T., Shiro, M., Koyama, H.: J. Chem. Soc., *(B)*, 935 (1968).
ST-68   Schmidbaur, H., Tronich, W.: Tetrahedron Lett., 5335 (1968).
ST-68a Schmidbaur, H., Tronich, W.: Chem. Ber., *101*, 595 (1968).
ST-72   Sheppard, W.A., Taft, R.W.: J. Am. Chem. Soc., *94*, 1919 (1972).
ST-76   Seckar, J.A., Thayer, J.S.: Inorg. Chem., *15*, 501 (1976).
ST&-75 Seno, M., Tsuchiya, S., Kise, H., Asahara, T.: Bull. Chem. Soc. Japan, *48*, 2001 (1975).
SUP-72 Sommer, L.H., Ulland, L.A., Parker, G.A.: J. Am. Chem. Soc., *94*, 3469 (1972).
SV-73   Strich, A., Veillard, A.: J. Am. Chem. Soc., *95*, 5574 (1973).
SW-69   Sheppard, W.A., Wudl, F.: Private communication.
T-60   Thompson, H.W.: Spectrochim. Acta., *16*, 238 (1960).
T-69   Thyagarajan, B.S.: Mechanisms of Reactions of Sulfur Compounds, *4*, 115 (1969).
T-70   Tebby, J.C.: Organophosphorus Chemistry Vol. 1 (ed. S. Trippett) London: Chem. Society, 1970, Ch. 11.
T-76   Trost, B.M.: Private communication, February 1976.
T-76a   Tillett, J.G.: Chem. Rev., *76*, 747 (1976).
TA-73   Trost, B.M., Arndt, H.C.: J. Am. Chem. Soc., *95*, 5288 (1973).
TAH-73 Trost, B.M., Atkins, R.C., Hoffman, L.: J. Am. Chem. Soc., *95*, 1285 (1973).
TB-68   Turley, J.W., Boer, F.P.: J. Am. Chem. Soc., *90*, 4026 (1968).
TB-73   Trost, B.M., Bogdanowicz, M.J.: J. Am. Chem. Soc., *95*, 5298 (1973).
TBM-63 Tuleen, D.L., Bentrude, W.G., Martin, J.C.: J. Am. Chem. Soc., *85*, 1938 (1963).
TC-75   Thuong, N.T., Chabrier, P.: Bull. Soc. Chim. France, 2033 (1975).
TCM-64 Tullock, C.W., Coffman, D.D., Muetterties, E.L.: J. Am. Chem. Soc., *86*, 357 (1964).
TEB-70 Tavares, D.F., Estep, R.E., Blezard, M.: Tetrahedron Letters, 2373 (1970).
TFO-70 Tsujihara, K., Furukawa, N., Oae, S.: Bull. Chem. Soc., Japan, *43*, 2153 (1970).
TF&-70 Tenud, L., Farooq, S., Seibl, J., Eschenmoser, A.: Helv. Chim. Acta., *53*, 2059 (1970).
TH-73   Trost, B.M., Hammen, R.F.: J. Am. Chem. Soc., *95*, 962 (1973).
TH&-67 Takamizawa, A., Hirai, K., Hamashima, Y., Matsumoto, S.: Tetrahedron Lett., *5071* (1967).
TK-71   Tamao, K., Kumada, M.: J. Organometal. Chem., *30*, 339 (1971).
TK-71a Turnblom, E.W., Katz, T.J.: J. Am. Chem. Soc., *93*, 4065 (1971).
TL-68   Trost, B.M., LaRochelle, R.: Tetrahedron Letters, 3327 (1968).
TL&-69 Tsvetkov, E.N., Lobanov, D.I., Makhmatkhanov, M.M., Kabachnik, M.I.: Tetrahedron, *25*, 5623 (1969).
TLA-69 Trost, B.M., LaRochelle, R., Atkins, R.C.: J. Am. Chem. Soc., *91*, 2175 (1969).
TM-69   Tang, R., Mislow, K.: J. Am. Chem. Soc., *91*, 5644 (1969).
TM-75   Trost, B.M., Melvin, L.S.: Sulfur Ylides, New York: Academic Press, 1975.
TN-60   Thorson, W.R., Nakagawa, I.: J. Chem. Phys. *33*, 996 (1960).
TP&-70 Torok, F., Paldi, E., Dobos, S., Fogarasi, G.: Acta Chim. Sci. Hung., *63*, 417 (1970).
TS-68   Tamura, C., Sim, G.A.: J. Chem. Soc., *(B)*, 8 (1968).
TS-69   Terauchi, K., Sakurai, H.: Bull. Chem. Soc., Japan, *42*, 821 (1969).
TS&-71 Trost, B.M., Shinski, W.L., Mantz, I.B., Chen, F.: J. Am. Chem. Soc., *93*, 676 (1971).
TSK-68 Tamura, C., Sato, S., Kishida, Y.: Tetrahedron Lett. 2739 (1968).
TSM-69 Trost, B.M., Schinski, W.L., Mantz, I.B.: J. Am. Chem. Soc., *91*, 4320 (1969).
TT&-68 Tagaki, W., Tada, T., Nomura, R., Oae, S.: Bull. Chem. Soc. Japan, *41*, 1696 (1968).
TZ-69   Trost, B.M., Ziman, S.: Chem. Comm., 181 (1969).

TZ-71	Trost, B.M., Ziman, S.D.: J. Am. Chem. Soc., *93*, 3825 (1971).
TZ-73	Trost, B.M., Ziman, S.: J. Org. Chem., *38*, 932 (1973).
U-69	Usher, D.A.: Proc. U.S. Nat. Acad. Sci., *62*, 661 (1969).
UEE-72	Usher, D.A., Erenrich, E.S., Eckstein, F.: Proc. U.S. Nat. Acad. Sci., *69*, 115 (1972).
UJC-68	Ulbricht, K., Jakoubkova, J., Chvalovsky, V.: Coll. Czech. Comm., *33*, 1691 (1968).
UM&-70	Ugi, I., Marquarding, D., Klusacek, H., Gokel, G., Gillespie, P.: Angew. Chem. Int. Ed. Engl., *9*, 703 (1970).
UR-70	Usher, D.A., Richardson, D.I., Jr.: Nature, *228*, 663 (1970).
UR-72	Ugi, I., Ramierez, F.: Chem. Brit., *8*, 198 (1972).
URO-70	Usher, D.A., Richardson, D.I., Jr., Oakenfull, D.G.: J. Am. Chem. Soc., *92*, 4699 (1970).
V-66	Voronkov, M.G.; Pure & Appl. Chem., *13*, 35 (1966).
V-68	Vorsanger, J.J.: Bull. Soc. Chim. France, 964 (1968).
V-68a	Vorsanger, J.J.: Bull. Soc. Chim. France, 971 (1968).
VB-74	Voncken, W.G., Buck, H.M.: Rec. Trav. Chim., *93*, 210 (1974).
VBS-68	Vilceanu, R., Balint, A., Simon, Z.: Nature, *217*, 61 (1968).
VH&-73	Vencl, J., Hetflejs, J., Cermak, J., Chvalovsky, V.: Coll. Czech. Chem. Comm., *38*, 1256 (1973).
VK&-62	Vol'pin, M.E., Koreshkov, H.Yu.D., Dulova, V.G., Kursanov, D.N.: Tetrahedron, *18*, 107 (1962).
VN-68	Vahrenkamp, H., Noth, H.: J. Organomet. Chem., *12*, 281 (1968).
VRD-63	Vyazankin, N.S., Razubaev, G.A., D'yachkovskaya, O.S.: Zh. Obsch. Khim., *33*, 613 (1963).
VS-73	Vedejs, E., Snoble, K.A.J.: J. Am. Chem. Soc., *95*, 5579 (1973).
VT-69	Vilkov, L.V., Tarasenko, N.A.: Chem. Comm., 1176 (1969).
W-62	Wanzlick, H.W.: Angew. Chem. Int. Ed. Engl., *1*, 75 (1962).
W-62a	Wells, P.R.: Chem. Rev. *62*, 171 (1962).
W-68	Westheimer, F.H.: Accts. Chem. Res., *1*, 70 (1968).
W-69	West, R.: Pure & Appl. Chem., *19*, 291 (1969).
W-72	Wolfe, S.: Accts. Chem. Res., *5*, 105 (1972).
W-73	Wadsworth, W.S.: J. Org. Chem., *38*, 2921 (1973).
WA-69	White, J.A., Anderson, R.C.: J. Heterocyclic Chem., *6*, 199 (1969).
WB-59	West, R., Baney, R.H.: J. Am. Chem. Soc., *81*, 6145 (1959).
WB-63	West, R., Bailey, R.E.: J. Am. Chem. Soc., *85*, 2871 (1963).
WB-67	Whitesides, G.M., Bunting, M.: J. Am. Chem. Soc., *89*, 6801 (1967).
WB&-70	Whitefield, G.F., Beilan, H.S., Saika, D., Swern, D.: Tetrahedron Letters, 3543 (1970).
WB-72	West, R., Bichlmeir, B.: J. Am. Chem. Soc., *94*, 1649 (1972).
WB-73	West, R., Boudjouk, P.: J. Am. Chem. Soc., *95*, 3983 (1973).
WB-73a	West, R., Boudjouk, P.: J. Am. Chem. Soc., *95*, 3987 (1973).
WBM-75	Willson, M., Burgada, R., Mathis, F.: Bull. Soc. Chim. France, 2241 (1975).
WC-63	West, R., Corey, J.Y.: J. Am. Chem. Soc., *85*, 4030 (1963).
WEB-74	Whitesides, G.M., Eisenhut, M., Bunting, W.M.: J. Am. Chem. Soc., *96*, 5398 (1974).
WF-52	Wittig, G., Fritz, H.: Ann., *577*, 39 (1952).
WH-69	Woodward, R.B., Hoffmann, R.: Angew. Chem. Int. Ed., Engl., *8*, 781 (1969).
WH-70	Wilson, E.G.E., Jr., Huang, M.G.: J. Org. Chem., *35*, 3002 (1970).
WH&-75	Wilkins, C.J., Hagen, K., Hedberg, L., Shen, Q., Hedberg, K.: J. Am. Chem. Soc., *97*, 6352 (1975).
WK-64	Wittig, G., Kochendoerfer, E.: Ber., *97*, 741 (1964).
WK-71	Williams, D.R., Kontnik, L.T.: J. Chem. Soc., *(B)*, 312 (1971).
WK-74	Wetzel, R.B., Kenyon, G.L.: J. Am. Chem. Soc., *96*, 5199 (1974).
WL&-71	West, R., Lowe, R., Stewart, H.F., Wright, A.: J. Am. Chem. Soc., *93*, 282 (1971).
WNB-76	West, R., Nowakowski, P., Boudjouk, P.: J. Am. Chem. Soc., *98*, 5620 (1976).
WM&-74	Wunderlich, H., Mootz, D., Schmutzler, R., Wieber, M.: Z. Naturforsch., *B29*, 32 (1974).

References

WP-67   Weiner, M.A., Pasternack, G.: J. Org. Chem., *32*, 3707 (1967).
WP-69   Weiner, M.A., Pasternack, G.: ibid., *34*, 1130 (1969).
WR-49   Wittig, G., Rieber, M.: Liebigs Ann. Chem., *562*, 187 (1949).
WRC-69   Wolfe, W., Rauk, A., Csizmadia, I.G.: J. Am. Chem. Soc., *91*, 1567 (1969).
WS&-70   Whitesides, G.M., Selgestad, J.M., Thomas, S.P., Andrew, D.W., Morrison, B.A., Panek, E.J., Filippo, J.S., Jr.: J. Organometal. Chem., *22*, 365 (1970).
WS-72   Weidner, U., Schweig, A.: J. Organometal. Chem., *39*, 261 (1972).
WS-72a   Weidner, U., Schweig, A.: Angew. Chem. Int. Ed. Engl., *11*, 538 (1972).
WS-72b   Weidner, U., Schweig, A.: Angew. Chem. Int. Ed. Engl., *11*, 146 (1972).
WSG-47   Whitmore, F.C., Sommer, L.H., Gold, J.: J. Am. Chem. Soc., *69*, 1976 (1947).
WT-74   Wadsworth, W.S., Tsay, Y.G.: J. Org. Chem., *39*, 984 (1974).
WW-76   Wadsworth, W.S., Wilde, R.L.: J. Org. Chem., *41*, 1264 (1976).
WWB-75   Whangbo, M.H., Wolfe, S., Bernardi, F.: Can. J. Chem., *53*, 3040 (1975).
WWL-61   West, R., Whatley, L.S., Lake, K.J.: J. Am. Chem. Soc., *83*, 761 (1961).
WWS-61   Wittig, G., Weigmann, H.D., Schlosser, M.: Chem. Ber., *94*, 76 676 (1961).
Y-57   Yokoi, M.: Bull. Chem. Soc., Japan, *30*, 100 (1957).
YK&-76   Yamaguchi, H., Kawada, K., Okamoto, T., Egert, E., Lindner, H.J., Braun, M., Dammann, R., Liesner, M., Neumann, H., Seebach, D.: Chem. Ber., *109*, 1589 (1976).
YO-69   Yano, Y., Oae, S.: Mechanisms of Reactions of Sulfur Compounds, *4*, 167 (1969).
YO-70   Yano, Y., Oae, S.: Tetrahedron, *26*, 67 (1970).
YR&-73   Yamagishi, I.G., Rayner, D.R., Zwicker, E.T., Cram, D.J.: J. Am. Chem. Soc., *95*, 1916 (1973).
Z-65   Zahradnik, R.: Adv. Heterocyclic Chem., *5*, 1 (1965).
ZP-65   Zahradnik, R., Parkanyi, C.: Coll. Czech. Chem. Somm., *30*, 195 (1965).

VII. Appendix of Recent Results, Interpretations and References

Many of the subjects discussed in this monograph, particularly the pseudorotation or turnstile rotation of phorphorane or sulfurane reaction intermediates, represent areas in which research is continuing actively. New papers are constantly appearing, and many of them may negate conclusions reached earlier. The following works, which appeared too late for inclusion in the text, represent examples of this, shedding new light on subjects already discussed.

Based on Westheimer's (GWS-75) measurement of the heats of hydrolysis of dimethyl and ethylene phosphates, Gorenstein (GL&-77) has offered a new explanation for the rapid hydrolysis of cyclic phosphates (text p.121 ff). Since hydrolysis of the diester ethylene phosphate involves ring cleavage, the difference in heats of hydrolysis between dimethyl and ethylene phosphate should represent the ring strain in the latter; this difference (4.4 kcal) is much less than the hydrolysis activation free energy difference (about 10.0 kcal). Gorenstein's MO calculations (GL&-77) on a form of $(MeO)_2P(OH)_3$ in which one methoxy group is basal and the other apical show that the stability depends strongly on conformation and the apical P–O overlap is reduced by antiperiplanar *(app)* overlap between oxygen lone pairs and a P–O* antibonding orbital (LW-75, text p. 19). For the phosphorane derived from ethylene phosphate, the lowest energy conformation is one with maximum *app* overlap; this conformation is sterically inaccessible to acyclic phosphoranes. This hypothesis also explains the much less negative entropies of activation for hydrolysis of cyclic than acyclic phosphate esters, since the phosphoranes from the former are rigidly held in a conformation which is ideal for reaction.

Guthrie (G-77) calculated the free energies of the phosphorane intermediates in hydrolysis of mono-, di-, and trialkyl phosphates based on the measured thermodynamics of hydrolysis of $(EtO)_5P$ and various assumptions as to the energy involved in forming $(EtO)_nP(OH)_{5-n}$ from $(EtO)_5P$. The results ranged from 4 kcal for formation of $(EtO)_3P(OH)O^-$ by hydroxide attack on triethyl phosphate to 33 kcal for formation of $(EtO)P(OH)_2(O^-)_2$ by attack of water on monoethyl phosphate dianion; 14–17 kcal was required for formation of neutral $(EtO)_nP(OH)_{5-n}$ depending on n. Although rate data for the overall hydrolyses corresponding to the above phosphoranes are unavailable, free energies of activation at 25 °C ranged from 22.8 kcal for hydroxide attack on trimethyl phosphate to 41.4 kcal for H_2O attack on monomethyl phosphate dianion. These data clearly imply that P–O formation and cleavage require high energy and that these phosphoranes are true intermediates. This is surprising in view of the fact that attack of $^{18}OH^-$ on trimethyl phosphate occurs without incorporation of ^{18}O into the unreacted ester (BB&-61). However, in this case isotope exchange would involve pseudorotation of $(MeO)_3P(OH)O^-$ to form a phosphorane in which the $-O^-$ is apical. Due to the strong basal preference of the $-O^-$ group, such pseudorotation is impossible without prior protonation; however $(EtO)_3P(OH)_2$ is 10 kcal less stable than $(EtO)_3P(OH)O^-$ (G-77). Since the overall free energy of activation for the hydrolysis is 22.8 kcal, the lack of exchange implies that the free energy of activation for pseudorotation of $(MeO)_3P(OH)_2$ is greater than 8.8 kcal. In view of pseudorotation the observed energy barriers for R_2PF_3 (see text p.100), this is surely not unreasonable.

Several recent papers illustrate differences of opinion in the interpretation of the structures of non-trigonal bipyramidal phosphoranes; (see text pp.102–104). Thus, Ramirez (NB&-77) has interpreted the rather low symmetry structure of I as an x^0TR, i.e. one in which the direction but not the magnitude of the deviation from the ideal trigonal bipyramidal or square pyramidal geometry can be rationalized in terms of partial turnstile rotation. The value of x depends on the steric and electronic demands of the ligands, High x, with resultant loss of distinction between apical and basal bonds, is favored for spirobicyclic oxyphosphoranes in which the ring structure weakens dp–π bonding and lengthens the four endocyclic P–O bonds. On the other hand Holmes has correlated the structure of II (BH-77) and of many other spirobicyclic phosphoranes (HD-77) in terms of Berry Pseudorotation. Thus, the deviations of the sum of the OPO angles from those for an ideal trigonal bipyramid or square pyramid (the pseudorotation transition state) correlated linearly with the deviations expected along the Berry pseudorotation reaction coordinate. No such correlation was found with the turnstile rotation reaction coordinate. For II the structure

Appendix of Recent Results, Interpretations and References

was 72% along the pseudorotation reaction coordinate towards a square pyramid. A similar correlation was derived based on the sum of the dihedral angles.

Tricoordinate hypervalent sulfur compounds similar to those proposed by Kwart as intermediates (see text pp. 141 ff) have recently (AB-77) been isolated and found to be quite stable. Thus, reaction of Br_2 with III gave IV, mp 173°. X-ray diffraction showed the Br–S–Br group in IV to be linear and the C–S–Br angle to be 87°. The apical S–Br bonds were 2.47 and 2.52 Å in length, and the basal C–S bond was 1.73 Å long compared to 1.69 Å in III. Presumably IV is stabilized by apical electronegative ligands and by a π-electron acceptor in the basal position.

It is well known that incorporation of phosphorus or sulfur into a small ring greatly stabilizes phosphoranes or sulfuranes. An interesting illustration of this is V (AM-77), which behaves quite differently from $Ph_2S(ORf)_2$ (Rf = hexafluorocumyl), converting pinacol and ethylene glycol to VI (R = CH_3 or H, respectively) and methanol to VII. Note that an analogous reaction of $Ph_2S(ORf)_2$ with diols is less likely since it would involve formation of a sulfurane with phenyl apical. The NMR spectrum of V was temperature-invariant and showed the ring to be basal-basal. Such structures are much more likely (although still not common) with sulfuranes than with phosphoranes (See p. 138), and are here favored by the high electronegativity (and thus apical preference) of the $-ORf$ group. On the other hand the ring in VII is apical-basal, and the methoxy groups appear in the NMR as a singlet above $-68°$, splitting below this temperature. This corresponds to a permutation in the M_4 mode of Musher (AM-77, M-72) with a free energy of activation of 11 kcal/mole. Martin (AM-77) proposes a mechanism involving ring S–O bond cleavage and reformation for this permutation, and this is supported by the NMR spectrum of VI, which shows a similar permutation but with a free energy of activation of 23 kcal/mole, the high barrier being due to sulfurane stabilization by the spirobicyclic ring system. It should be noted that a similar M_4 mode permutation has also been proposed to explain the temperature-dependent ESR spectrum of the phosphoranyl radical VIII (DR-75), and may also explain the rapid fluorine exchange in $ROPF_3$ radical (EPR-75). The work of Bentrude (BH&-76, BA&-77) on the stereochemistry of the free radical Arbuzov reaction, which involves an intermediate phosphoranyl radical formed from a cyclic phosphite, permits, even if it does not require, this permutational mode but it does exclude ordinary pseudorotation or turnstile rotation, which are both of the M_1 mode (see text p. 8).

Appendix of Recent Results, Interpretations and References

A hydroxyphosphorane has recently been observed for the first time (RNM-77), but IX exists in solution in equilibrium with phosphate ester X. This equilibrium is solvent dependent, the phosphorane/phosphate ratio being 1.5 at $-48°$ in acetonitrile but 3.0 in acetone, a more basic and less polar solvent, at the same temperature. This suggests that phosphate formation is acid catalyzed, with proton attack on a ring oxygen, and that the P–OH group has very low acidity, lower than that of phenol.

References

AB-77   Arduengo, A.J., Burgess, E.M.: J. Am. Chem. Soc., *99*, 2376 (1977).
AM-77   Astrologes, G.W., Martin, J.C.: J. Am. Chem. Soc., *99*, 4390 (1977).
BB&-61 Barnard, P.W.C., Bunton, C.A., Llwewllyn, D.R., Vernon, C.A., and Welch, V.A.: J. Chem. Soc., 2670 (1961).
BA&-77 Bentrude, W.G., Del Alley, W., Johnson, N.A., Murakami, M., Nishikida, K., and Tan, H.W.: J. Am. Chem. Soc., *99*, 4383 (1977).
BH-77   Brown, R.K., and Holmes, R.R.: J. Am. Chem. Soc., *99*, 3326 (1977).
DR-75   Dennis, R.W., and Roberts, B.P.: JCS, Perkins Trans. II, 2140 (1975).
EPR-75 Elson, I.H., Parrott, M.J., and Roberts, B.P.: JCS, Chem. Commun., 586 (1975).
G-77   Guthrie, J.P.: J. Am. Chem. Soc., *99*, 3991 (1977).
GL&-77 Gorenstein, D.G., Luxon, B.A., Findlay, J.B., and Momii, R.: J. Am. Chem. Soc., *99*, 4170 (1977).
GWS-75 Gerlt, J.A., Westheimer, F.H., and Sturtevant, J.M.: J. Biol. Chem., *250*, 5059 (1975).
HD-77   Holmes, R.R., and Deiters, J.A.: J. Am. Chem. Soc., *99*, 3318 (1977).
NB&-77 Narayanan, P., Berman, H.M., Ramirez, F., Marecek, J.F., Chaw, Y.F., and Prasad, V.A.V.: J. Am. Chem. Soc., *99*, 3336 (1977).
RNM-77 Ramirez, F., Nowakowski, M., and Marecek, J.F.: J. Am. Chem. Soc., *99*, 4514 (1977).

VIII. Index of Authors and References

Abatjoglou, A.G., (EHA-74)
Abel, E.W., (AAB-65)
Abrahams, S.C., (SA-60)
Absar, I., (AV-72), (AV-72a)
Ackerman, B.K., (AA&-74)
Adams, L.C., (CA&-71)
Adamson, G.W., (AB-70)
Adcock, W., (AK-70)
Ahmed, W., (SP&-74)
Airey, W., (AG&-70), (AG&-71)
Aksnes, G., (A-60), (AS-65)
Albert, R., (AW&-70)
Albrecht, H.P., (HU&-69)
Aldridge, R.G., (MS&-71)
Albright, T.A., (AFS-75), (AFS-75a), (AS-76)
Allcock, H.R., (AW-72), (AW-72a)
Allegra, G., (AW&-70)
Allen, C.W., (ATM-68)
Allen, D.W., (A-70), (AH-72), (AHM-72), (AT-70)
Allen, G.W., (HA-70)
Allen, L.C., (DA-76), (RAM-72), (RMA-70)
Almenningen, A., (AB&-63)
Alt, H., (AB-69), (AFB-69), (BA-70), (BA-70a), BAS-69), GH&-68), HH&-69)
Altmann, J.A., (AYC-76)
Amann, W., (KA-69)
Amel, R.T., (JA-68), (JA-69)
Ammom, H.L., (AWW-73)
D'Amore, M.B., (AB-73)
Anastassiou, A.G., (AWC-75)
Andersen, K.K., (AA&-74), (HAK-69)
Andrews, G.C., (EA-72)
Andrianov, K.A., (AK-67), (AKK-63)
Anet, F.A.L., (ATC-67)
Angelotti, N.C., (SA-59)
Annarelli, D.C., (SAV-76)
Appel, R., (A-75)
Archie, W.C., (AW-73)
Arduengo, A.J., (AB-76)
Arens, J.F., (BB&-59)
Arhart, R.J., (AM-72), (MA-71), (MFA-74)
Armitage, D.A., (AAB-66)
Arndt, H.C., (TA-73)
Arreguy, B., (CD&-75)
Asahara, T., (ST&-75)
Astrologes, G.W., (AM-76)
Atkins, R.C., (TAH-73), (TLA-69)
Attridge, C.J., (A-68)
Atwell, W.H., (GCA-64)
Axelrad, G., (MA&-65)

Aylett, B.J., (AP-65), (RR&-69), (RS&-68)
Azzaro, M., (ME&-71)

Baccolini, G., (BT-75)
Bacon, R., (OB&-72)
Baechler, R.D., (BM-70), (BM-70a), (BM-71), (BM-72)
Baenziger, N.C., (BB&-69)
Baliah, V., (BG-63), (BU-63)
Balint, A., (VBS-68)
Balli, H., (B-64)
Banister, A.J., (BMP-67)
Bank, K.C., (BC-69)
Barnette, W.E., (KB-76)
Bart, J.C.J., (AB-70), (B-69)
Bartell, L.S., (BH-65)
Bartlett, P.D., (BB&-74), (BBL-73)
Basolo, F., (PGB-60)
Bassindale, A.R. (BB&-75), (BB&-75a), (BMB-75), (RB&-74)
Bastiansen, C., (AB&-63)
Batich, C.D., (B-75)
Baumann, D.L., (SB-69)
Baumstark, A.L., (BB&-74), (BBL-73)
Bausher, L.P., (HBM-69), (HBM-71)
Beagley, B., (RS&-68)
Becker, G., (EBB-74), (MB&-72)
Belkind, B.A., (B-76)
Bellama, J.M., (BM-70b)
Belskii, V.E., (BB&-70), (BE-70), (BP&-68)
Bendazzoli, G.L., (BGP-74), (BZ-65)
Benedetti, E., (AW&-70)
Benko, D., (KSB-76)
Benkovic, S.J., (BB-66), (SB-71)
Benschap, H.P., (MBK-70)
Bentrude, W.G., (BJK-72), (BK&-74), (BM-76), (BT-72), (TBM-63)
Bergson, G., (BCS-62)
Bernardi, F., (BC&-75), (EY&-76), (WWB-75)
Bernstein, S., (SP&-71)
Berry, R.S., (B-60)
Bertrand, J.A., (NCB-66)
Bezzubova, N.N., (BB&-70)
Biasotti, J.B., (BA-71)
Bichlmeir, B., (WB-72)
Biernbaum, M.S., (BM-71a)
Bigler, A.J., (RBS-68)
Binkley, J.S., (CS&-76)
Biran, C., (CD&-75)
Blaich, B., (RP&-71a)
Blaschette, A., (BB&-71)
Blasse, V.E.L., (GB-66)

Blezard, M., (TEB-70)
Bock, E., (BQ&-74)
Bock, H., (AB-69), (AFB-69), (BA-70),
  (BA-70a), (BAS-69), (BE-71), (BE&-76),
  (BS-68), (BWK-72), (BW&-76), (EBB-74),
  (GH&-68), (GHB-70), (GKB-69),
  (GKB-69a), (GPB-70), (HH&-69),
  (MB&-72), (RB&-74)
Boekenstein, G., (BJB-74)
Boer, F.P., (BJ-69), (BTF-68), (TB-68)
Bogdanowicz, M.J., (TB-73)
Bohler, D., (GM&-68a)
Boonstra, H.J., (BB&-59)
Booth, M.R., (BF-70)
Bordwell, F.G., (BD-74a), (BD-74b),
  (BDC-70), (BO-74), (BW-71), (BW-74),
  (BWH-68)
Bott, R.W., (BEM-65), (EB-69)
Boudjebel, H., (BGM-75)
Boudjouk, P., (WB-73), (WB-73a), (WNB-76)
Boyd, D.B., (BH-71), (HBG-70)
Bradamante, S., (BMP-70), (BM&-71)
Brady, D.B., (AAB-66)
Bragin, J., (BC&-73)
Bramblett, J.D., (HBL-72)
Brandsma, L., (BB&-59)
Brauman, J.J., (AB-73)
Braun, M., (YK&-76)
Braun, R.W., (GBC-74), (MD&-76)
Brecht, R.H., (SBG-67)
Breen, J.J., (QBM-71)
Breslow, R., (B-58)
Bristow, P.A., (BTW-68)
Britts, K., (KB-66)
Broaddus, C.D., (B-68a), (B-69b)
Broise, S.J., (B-68)
Brook, A.G., (B-74), (BB&-75), (BB&-75a),
  (BD-73), (BD-74), (BLM-67), (BMB-75),
  (BP-71), (BWL-67), (DB-73), (RB&-74)
Brooks, H.G., (CB-54)
Brown, C., (HB-72)
Brown, D.M. (BF-68)
Brown, H.C., (B-62)
Brown, M.D., (BC&-71)
Brown, R.S., (B-75a)
Brownlee, S., (BQ&-74)
Browska, M.L., (SH&-71a)
Bruice, T.C., (BB-66)
Brune, H.A., (B-64a), (B-64b), (B-65)
Bruns, R.E., (KPB-72)
Buck, H.M., (BJB-74), (DPB-75), (RF&-74),
  (RFB-73), (RLB-72), (SJB-75), (VB-74)
Buckles, R.E., (BB&-69)
Budde, W.L., (B-69a)
Buell, G.R., (ES&-74), (SP&-71)
Buncel, E., (RB&-65)

Bunting, W.M., (WEB-74)
Bunton, C.A., (B-70)
Burczyk, K., (BB&-69a), (BB&-71)
Burgada, R., (WBM-75)
Burgess, E.M., (AB-76)
Bürger, H., (B-68a), (BB&-69a), (BB&-71),
  (BGS-68), (BGS-70)
Burns, F.B., (MBC-72)
Burov, A.I., (EV&-70)
Bushweller, C.H., (BRL-77)
Bystrov, V.F., (BK&-65)

Calas, R., (CD&-75), (PDC-74), (PDC-75)
Camfield, N.D., (CC&-71a)
Cane, F., (CCL-70)
Carberry, E., (CWG-69)
Carey, F.S., (CH-69)
Carey, R.N., (PCT-72)
Carlson, E.H., (RC&-72)
Carpino, L.A., (CA&-71), (CC-71), (CM&-71)
Carroll, D.F., (CC-63)
Carter, R.O., (DG&-73)
Cartledge, F.K., (CMF-75), (CM-71)
Caserio, M.C., (CPH-66)
Cason, L.F., (CB-54)
Castro, B., (CCG-73), (CS-74)
Caughlin, C.N., (HC&-71), (SC&-67),
  (SC&-71)
Cermak, J., (VH&-73)
Chabrier, P., (TC-75)
Chambers, J.Q., (CC&-71a)
Chambers, V.C., (RC-51)
Chan, S., (BC&-73), (CD&-73)
Chan, T.H., (CF-72)
Chang, B.C., (CC&-71), (CDD-71), (DD&-69)
Chao, B.Y.H., (AWC-75)
Chapleur, Y., (CCG-73)
Chapleo, C.B., (GC&-76)
Chapman, A.C., (CC-63)
Chau, M.M., (MC-74)
Chauviere, G., (EC&-70)
Chellappa, K.L., (OB&-72)
Chen, H.W., (CC-71)
Chen, M.M.L., (CH-76)
Chen, P., (BB&-75)
Chenskaya, T.B., (KOC-66)
Chikayoshi, M., (FCI-61)
Chioccola, G., (CD-68)
Chivers, T., (CP-72)
Chowat, R.J., (CCT-69)
Christol, H., (CCS-76)
Chu, W.K.C., (SH&-71a)
Chvalovsky, V., (C-66), (LC-69),
  (UJC-68), (VH&-73)
Cilento, G., (C-60)
Ciuffarin, E., (CF-68), (CSI-72)

Index of Authors and References

Claeson, G., (BCS-62)
Clardy, J.C., (CM&-76), (MS&-76)
Clark, D.T., (C-68)
Clark, R.T., (EC&-70), (MBC-72), (MC-70)
Claus, P., (CV-70)
Clipsham, R.M., (CW-72)
Coffen, D.L., (BC-69), (CC&-71a), (CMW-70)
Coffman, D.D., (TCM-64)
Collin, R.L., (C-66a), (CP-64)
Collins, J.B., (CS&-76)
Collins, W.T., (PCH-70)
Combault, G., (CC-69)
Conrad, W.E., (CC&-71)
Cook, M.A., (CEW-70)
Cook, M.J., (BC&-71)
Cook, R.D., (CD&-73a)
Cook, R.E., (CG&-71)
Corey, E.J., (CL-35)
Corey, J.Y., (WC-63)
Corfield, J.R., (CDT-70), (CST-69)
Corfield, P.W.R., (BDC-70)
Cornwall, C.D., (CY-57)
Corriu, R., (CCL-70), (CL-71), (CLM-68)
Corsi, N., (GCF-70)
Cottis, S.G., (GCA-64)
Cotton, F.A., (CW-66)
Coulson, C.A., (C-69)
Covitz, F., (CW-63), (KC&-69)
Cowley, A.H., (CD&-70), (CS-70), (FDC-70), (GBC-74), (MD&-76),
Cox, Jr., J.R., (NCB-66)
Cradock, S., (CE-71)
Craig, D.P., (CM-56), (CM&-54), (CT-66), (CZ-62)
Cram, D.J., (ATC-67), (CD&-70a), (CTS-66), (GJC-73), (JC-74), (YR&-73)
Cremer, S.E., (CCT-69), (CTW-71)
Cristau, H.J., (CCS-76)
Cruickshank, D.W.J., (C-61), (CWM-64), (MC-67), (PC-68), (RS&-68)
Csizmadia, I.G., (AYC-76), (BC&-75), (RWC-69), (WRC-69)
Cumper, C.W.N., (CM&-66)

Daly, J.J., (D-70), (CD-68)
D'Amore, M.B., (AB-73)
Damm, F., (KD&-69)
Dammann, R., (YK&-76)
Daniel, H., (DP-68)
Darwish, D., (DT-66)
Davies, W.O., (DS-62)
Davis, F.A., (DK-76)
Davison, A., (DR-70)
Dauphin, G., (RR&-67)
Day, J., (CD&-70a)

Deakyne, C.A., (DA-76)
De'Ath, N.J., (CDT-70), (DD&-76)
DeBruin, K.E., (DP-72), DZ&-69)
DeBruyn, D.J., (DD&-74)
Deiters, R.M., (HD-68), (MD-66)
De La Mare, P.D.B., (DTV-62)
Delker, G.L., (DW&-76)
Deluzarche, A. (KD&-69)
Denney, D.B., (CC&-71), (CDD-71), (DD&-69), (DD&-72), (DD&-76), (DDH-73), (DR-64), (DS-66), (DS-74), (DWD-71)
Denney, D.Z., (CC&-71), (CDD-71), (DD&-69), (DD&-72), (DD&-76), (DDH-73), (DWD-71)
Dennilauler, R., (LD-61)
Dennis, E.A., (DW-66), (DW-66a), (KC&-69)
DeShazu, M., (GDV-71)
Deutch, J.M., (EM&-74)
Dewar, M.J.S., (DC&-70), (DLW-60)
Diebert, C.E., (CD&-73a)
Dierdorf, D.S., (FDC-70)
Dijk, J.M.F., van, (DPB-75), (RF&-74)
Dimroth, K. (DS-68), (SS&-76)
Distefano, S., (CD&-73)
Doak, G.O. (MD&-76), (MDL-73)
Dobos, S., (TP&-70)
Donnelly, S.J., (DD&-74)
Doomes, E., (BD-74a), (BD-74b), (BDC-70)
Doyle, M.P., (DD&-74), (DMW-76), (DW-75a), (DW-75b), (DW-75c)
Drake, J.E., (FH&-71)
Dreiding, A.S., (GC&-76)
Drews, M.J., (DWJ-72), (JD&-72)
Driscoll, J.S., (DG&-64)
Dubov, S.S., (LSD-70)
Duchamp, D.J., (CD&-70a)
Duchek, J.R., (GD-76)
Duff, J.M., (BD-73), (BD-74), (DB-73)
Duffey, G.H., (OD-67)
Dulova, V.G., (VK&-62)
Dumanoir, J.R., (KD-73)
Duncan, J.L., (D-64)
Dunoques, J., (CD&-75), (PDC-74), (PDC-75)
Durig, J.R., (DH-69), (DG&-73), (DKF-75)
Durmaz, S., (D-75)
Durst, T., (DF&-70), (DVM-71)
D'yachkovskaya, O.S., (VRD-63)
Dziedzic, J.E., (SDJ-71)

Eaborn, C., (BEM-65), (CEW-70), (E-60), (EB-69), (EE&-75), (EJ-74)
Eargle, D.H., (ER-73)
Ebsworth, E.A.V., (CE-71), (E-63), (E-69), (EF-63)

Eckstein, F., (E-68), (E-70), (SE-70)
Edelman, R., (CC&-71)
Effenberger, F., (EP&-76)
Efremova, M.V., (BE-70), (BP&-68)
Egan, W., (EC&-70)
Egert, E., (YK&-76)
Egorochkin, A.N., (EVK-72)
Eichler, S., (RPE-67)
Eidenschink, R., (EE&-75)
Eisch, J.J., (EG-76), (ER-75), (ET-73)
Eisenhut, M., (EM&-74), (ESS-73), (WEB-74)
Élégant, L., (ME&-71)
El Gomati, T., (ELU-75)
Eliel, E.L., (EHA-74), (EK&-76)
Elisenkov, V.N., (BB&-70), (BP&-68)
Elizarova, A.N., (MSE-63)
Ellis, I.A., (RR&-69), (RS&-68)
Englehard, E., (EPN-67)
Engelhardt, G., (ER&-73)
Ennsslin, W., (BE-71), (BE&-76), (EBB-74)
Epiotis, N.D., (EY&-76)
Erenrich, E.S., (UEE-72)
Ermolaeva, M.V., (NEK-70), (NS&-68)
Ernst, C.R. (ES&-74)
Eschenmoser, A., (TF&-70)
Estep, R.E., (TEB-70)
Evans, D.A., (E-74), (EA-72)
Ewing, V., (AB&-63), (ES-63)

Fahrbach, G., (HF-68)
Farooq, S., (TF&-70)
Fava, A., (CF-68), (GCF-70)
Fayssoux, J., (CFM-75)
Fernandez, Y.V., (LF-75)
Feugas, C., (FG-70)
Finkenbine, J.R., (CF-72)
Finzel, W.A., (FVF-71)
Flanagan, M.J., (DKF-75)
Flynn, J.J., (BF-69), (BTF-68)
Fogarasi, G., (TP&-70)
Fogel, J., (SFH-66)
Fong, F., (CD&-73)
Fontfreide, J.J.H.M., (RF&-74), (RFB-73)
Frank, D.S., (FU-67)
Franke, E.R., (AFB-69)
Frankiss, S.G., (BF-70), (EF-63)
Fraser, R.R., (DF&-70), (FN-76), (FS-69), (FS-69a), (FS-70), (FSW-72)
Franz, J.A., (MF-75), (MFA-74)
Franzen, V., (FM-61)
Frearson, M.J., (BF-68)
Freeburger, M.E., (FS-71), (SP&-71)
Freeman, J.P., (PFW-71)
Freeman, W.J., (AFS-75), (AFS-75a)
Frey, H.M., (FS-68)
Fritz, G., (MB&-72)

Fritz, H., (WF-52)
Fritzwater, D.R., (NFV-68)
Froeyen, P., (F-72)
Frost, D.C., (FH&-71)
Frye, C.L., (F-64), (FUF-71), (SKF-72)
Fueno, T., (FK&-73), (KKF-73)
Fujimoto, H., (SF-68)
Fukui, K., (FCI-61), (OYF-69)
Furtsch, T.A., (FDC-70)
Furokawa, N., (OY&-68), (TFO-70)

Gagnaire, D., (HW&-69)
Galle, J.E., (EG-76)
Galy, J.P., (FG-70)
Ganapathy, K., (BG-63)
Gancalves, H., (BGM-75)
Garbesi, A., (GCF-70)
Garrett, P.E., (CC&-71a)
Garwood, D.C., (CD&-70a), (GJC-73)
Gastamina, A., (GM&-68)
Gay, D.C., (GH-70)
George, W.O., (DG&-73)
Gerdil, R., (GL-66)
Gerson, F., (GH&-68), (GHB-70), (GKB-69) (GKB-69a), (GPB-70)
Gervais, H.P., (G-67), (G-70)
Gilbro, T., (GW-74)
Gilje, J.W., (GBC-74), (MD&-76)
Gillespie, P.D., (GH&-71), (GR&-73), (RP&-71), (UM&-70)
Gillespie, R.J., (G-60), (G-63), (GR-63)
Gilman, H., (GCA-64)
Gilow, H.M., (GDV-71)
Ginjaar, L., (GB-66)
Giral, L., (CG-69)
Givens, E.N., (GK-68)
Glass, G.E., (CWG-69)
Glass, R.S., (GD-76)
Gleiter, R., (GH-68)
Glesmer, O., (GM&-68a), (MNG-66), (MNG-67)
Glick, M.D., (CG&-71)
Glidewell, C., (AG&-70), (AG&-71), (GR&-70)
Goetz, H., (GK-69), (GKJ-70)
Goetze, U., (BGS-68), (BGS-70)
Gohlke, R.S., (JGN-65)
Gokel, G., (UM&-70)
Gold, E.H., (SG-66)
Gold, J., (WSG-47)
Goldberg, D.Z., (HBG-70)
Goldfrab, T.D., (RH&-67)
Goldstein, J.H., (GR-62)
Goldwhite, H., (BC&-73), (CD&-73)
Golini, C.M., (SMG-72)
Goodman, L., (GKS-65), (GT-65)

Index of Authors and References

Gordon, A.J., (RGM-68)
Gottarelli, G., (BGP-74)
Goubeau, J., (GL-71), (RP&-71a)
Gould, E.S., (G-59)
Gouterman, M., (MG&-72)
Grant, M.W., (GP-69)
Gray, G.A., (G-71)
Gray, H.B., (PGB-60)
Gray, R.W., (GC&-76)
Greenhalgh, R., (GW-67)
Gresser, M., (G-69)
Griller, D., (GI-73)
Grisley, Jr., D.W., (DG&-64)
Groeger, H., (SBG-67)
Gross, B., (CCG-73)
Gross, B.K. v., (LG-76)
Grozdeva, H.A., (SG-69)
Guillerm, G., (GLS-73)
Gulati, A.S., (RGS-69)
Gysegem, P., (CD&-73)

Haake, P.C., (CD&-73a), (HA-70), (HBM-69), (HBM-71), (HMT-64), (HW-61), (KH-73), (NH-76)
Haas, C.K., (DW&-76)
Hach, R.J., (HR-51)
Hafferl, W.H., (HLI-63), (HLI-64)
Hairston, T.J., (HO-71)
Hall, C.D., (DD&-72), (HBL-72)
Hallams, K.L., (DH-69)
Hamashima, Y., (TH&-67)
Hamer, F.M., (H-64)
Hamer, N.K., (GH-70)
Hamilton, W.C., (HL-64), (RH&-67), (SHL-67)
Hammen, R.F., (TH-73)
Handbook of Chemistry & Physics, (H-60)
Hansen, K.W., (BH-65)
Haque, M.U., (HC&-71)
Hargis, J.H., (BH&-76)
Harrington, D., (HW&-72)
Harris, J.E., (DG&-64)
Harris, M.R., (HU&-69)
Harris, R.L., (HHM-74)
Harris, W.C., (LH-71)
Harrison, B.G., (HUZ-71)
Hartmann, A.A., (EHA-74)
Hartzell, G.E., (HP-67)
Havinga, E.E., (HW-59)
Hays, H.R., (H-71)
Heath, D.F., (H-56)
Hedberg, E., (HHH-74)
Hedberg, K., (AB&-63), (H-55), (HHH-74)
Hedberg, L., (HHH-74)
Heeren, J.K., (SFH-66)

Heilbrunner, E., (HH&-69)
Heinzer, J., (GH&-68), (GHB-70)
Hellwinkel, D., (H-65), (H-66), (H-66a), (H-66b), (HF-68)
Helmut, J., (SSH-75)
Hencsei, P., (NH-70), (NH-72)
Hendra, P.J., (HJ&-70)
Hengge, E., (HPH-70), (HSH-70)
Herring, F.G., (FH&-71), (FH&-72)
Hetflejs, J., (VH&-73)
Hildbrand, J., (HK-70)
Hillier, I.H., (HS-69), (HS-70), (HS-70a)
Hine, J., (HPM-70)
Hirai, K., (TH&-67)
Hoffman, L., (TAH-73)
Hoffman, P., (GH&-71), (RP&-71)
Hoffmann, R., (BH-71), (CH-76), (GH-68), (HBG-70), (HHM-72), (MH-75)
Höfler, F., (BB&-69a), (HPH-70), (HSH-70)
Holland, R.J., (CPH-66)
Hollyhead, W.B., (SH&-71a)
Holmes, R.R., (H-74), (HD-68)
Holt, A., (JHT-70)
Homer, G.D., (HS-73), (SH&-71)
Hopkins, T.L., (MH-68)
Hordvik, A., (HSS-66)
Hori, M., (PH&-63)
Horner, L., (HO-59)
Hornung, V., (HH&-69)
Hortmann, A.G., (HHM-74)
Houalla, D., (HW&-69)
Houk, K.M., (OL&-66)
Houser, R.W., (PH-69), (PH-71)
Hovanec, S.W., (HL-72)
Howard, J.A., (HRT-73)
Howe, J.J., (PCH-70)
Howell, J.M., (H-75), (HHM-72), (HO-76), (HV-74), (HUR-74)
Hoyt, E.B., (BWH-68)
Hsu, C.W., (CH-69)
Hsu, Y.F., (DD&-76), (DDH-73)
Hudson, R.F., (H-65a), (HB-72), (HK-56), (HK-60)
Hutchinson, B.J., (BC&-71), (HAK-69)
Hutley, B.G., (AH-72), (AHM-72)

Iloff, J.M., (PI-74)
Imamura, A., (FCI-61)
Ingold, K.U., (GI-73)
Ingraham, L.L., (HLI-63), (HLI-64), (SI-69)
Inove, T., (SIK-72)
Isola, M., (CSI-72)
Iuzhakova, O.A., (BK&-65)
Iwata, K., (IYY-71)
Izawa, K., (FK&-73)

Jackson, P.M., (EE&-75)
Jackson, W.R., (CD&-70)
Jacobsen, N.J., (JS-67)
Jacobus, J., (HW&-72)
Jaffe, H.H., (H-54), (JO-62)
Jakoubkova, J., (UJC-68)
Jancke, H., (ER&-73)
Jansen, E.H.J.M., (BJB-74), (SJB-75)
Jarvie, A.W.P., (JHT-70)
Jenkins, D.R., (JKS-62)
Jenkins, I.D., (EJ-74)
Jennings, W.B., (CD&-70), (JS-70)
Jensen, D., (MG&-72)
Johns, R.A., (HJ&-70)
Johnson, A.W., (J-66), (JA-68), (JA-69)
Johnson, C.R., (CG&-71), (JJ-71)
Johnson, F., (JGN-65)
Johnson, J.K., (JD&-72)
Johnson, J.S., (SDJ-71)
Johnson, N.A., (BH&-76), (KJ-70), (KJ-76)
Johnson, W.D., (BJK-72)
Jones, Jr., F.B., (RC&-72), (RJ-71), (RKJ-69)
Jones, G.H., (HU&-69)
Jones, M.R., (GJC-73), (JC-74)
Jones, P.F., (BB&-75a)
Jones, P.R., (DWJ-72), (JD&-72), (JJ-75), (JJ-75a), (JM-72)
Jonsson, E.U., (JJ-71)
Jordan, T., (JS&-63)
Jung, I.N., (JJ-75), (JJ-75a)

Kabachnik, M.I., (TL&-69)
Kaiser, E.T., (KKW-65), (KPW-63), (SIK-72)
Kajimoto, O., (FK&-73), (KKF-73)
Kalaman, A., (KKK-67)
Kalasinsky, V.F., (DKF-75)
Kaneda, H., (OK-72), (OK-72a)
Kanezawa, H., (OK-71)
Kannengiesser, G., (KD&-69)
Kanter, H., (SS&-76)
Kaplan, L.J., (KM-73)
Karle, I.L., (KB-66)
Karup-Nielsen, I., (AA&-74)
Katrib, A., (FH&-71), (FH&-72)
Katritzky, A.R., (BC&-71), (HAK-69)
Katz, I.R., (KKW-65)
Katz, J.T., (TK-71a)
Kaufman, G., (HK-70)
Kaufman, R.J., (EM&-74)
Kawada, K., (YK&-76)
Kawamura, T., (KK-72)
Kawasaki, Y., (OOK-73)
Keay, L., (HK-56), (HK-60)
Keil, F., (KK-75)
Kemp, D.A., (KO-70)
Kenney, Jr., G.W.T., (RKJ-69)

Kergomard, A., (RR&-67)
Kewley, R., (JKS-62)
Khaiduk, I., (AKK-63)
Khan, W.A., (BJK-72), (BK&-74)
Khananashrili, L.M., (AKK-63)
Khorana, H.G., (K-61)
Khorsher, S.Y., (EV&-70), (EVK-72)
Khwaia, T.A., (KRS-70)
Kice, J.L., (K-68), (KW-72)
King, J.F., (KD-73)
Kise, H., (ST&-75)
Kise, M., (OY&-68)
Kishida, Y., (TSK-68)
Kitching, W., (AK-70)
Klabuhn, B., (GK-69), (GKJ-70)
Klanberg, F., (KM-68)
Klemperer, W.G., (KK&-75)
Kliment, M., (RKP-74)
Klingsberg, E., (K-69)
Klopman, G., (OKS-69)
Kluger, E.W., (DK-76)
Kluger, R., (KC&-69), (KW-69)
Klusacek, H., (GH&-71), (UM&-70)
Knunyants, I.L., (LM&-66), (NEK-70), (NP&-67), (NS&-68)
Kobayashi, M., (KKF-73), (MML-75)
Koch, J.H.F., (KM-51)
Kochendoerfer, E., (WK-64)
Kochi, J.K., (KK-71), (KK-72), (KMK-72)
Koepsell, D.G., (SKW-70)
Koizumi, T., (KH-73)
Kokot, Z., (KT-75)
Kolodyazhnyi, Yu., V., (KOC-66)
Komoriya, A., (KY-72)
Kondo, K., (OK-73)
Konstan, A.H., (GKS-65)
Kooistra, D.A., (DD&-74)
Koreshkov, H. Yu. D., (VK&-62)
Korte, W.D., (SKF-72)
Koskimies, J., (EK&-76)
Kost, U., (KZ-75)
Kostyanovskii, R.G., (BK&-65)
Kotrelev, G.V., (AK-67)
Koyama, H., (SSK-68)
Kraay, G.W., (MBK-70)
Krebs, B., (GM&-68a), (MK&-67)
Kresze, G., (KA-69)
Krieger, J.K., (KK&-75)
Kriegsmann, H., (K-57), (K-59)
Kroner, J., (BWK-72)
Kruger, T.L., (SH&-71a)
Krusic, P.J., (KK-71), (KMK-72)
Krynitz, U., (GKB-69), (GKB-69a)
Kubisen, S.J., (SKW-76)
Kucsman, A., (KKK-67)
Kugler, H.J., (RP&-68)

Index of Authors and References

Kumada, M., (TK-71)
Kursanov, D.N., (VK&-62)
Kutschinski, J.L., (SH&-71)
Kutzelnigg, W., (KK-75)
Kuznesod, P.M., (KPB-72)
Kwart, H., (GK-68), KB-76), (KJ-70),
 (KJ-76), (KO-71), (KS-72), (KS-76),
 (KSB-76), (SK-73), (SK-73a)

Ladell, J., (SHL-67)
Lake, K.J., (WWL-61)
Lambert, J.B., (LPO-71)
Lambert, R.L., (DW&-76), (ML-73)
Landau, M.A., (LSD-70)
Landesberg, J.M., (OL-66), (OL&-66)
Landini, D., (LM&-70)
Landis, M.E., (BB&-74), (BBL-73)
LaPlaca, S.J., (HL-64)
Larive, H., (LD-61)
LaRochelle, R.W., (LT-71), (TL-68),
 (TLA-69)
Larson, G.L., (LF-75)
Larsson, L., (L-57)
Lauderback, S.K., (RC&-72)
Lautenschlaeger, F., (L-69)
Lavielle, G., (MM&-68)
Leard, M., (CCL-70), (CL-71), (CLM-68)
Lee, J.K., (OB&-72)
Lee, S., (SR&-76)
LeGrow, C.E., (BLM-67)
Leguan, M., (GLS-73)
Lehn, J.M., (L-70), (LM-70), (LM-72),
 (LW-76a)
Lemal, D.M., (BRL-77), (RSL-76)
Lennon, J., (BB&-75), (BB&-75a)
Lenoir, D., (ELU-75)
Lentz, A., (GL-71)
Lepeska, B., (LC-69)
Lerman, C.L., (BB&-74), (LW-76)
Leung, F., (LN-71)
Lerin, I.W., (LH-71)
Li, Y.S., (DG&-73)
Libscomb, W.N., (JS&-63)
Lieske, C.N., (HL-72)
Liesner, M., (GC&-76), (YK&-76)
Limouzin, Y., (LM-75), (LW-75)
Lin, F.F.S., (HBL-72)
Linburg, W.W., (BWL-67)
Lindner, H.J., (LG-76), (YK&-76)
Lipowitz, J., (L-72)
Lippmaa, E., (ER&-73)
Lipscomb, W.N., (SL-65)
Littlefield, L.B., (MD&-76), (MDL-73)
Livant, P., (LM-76)
Lobanov, D.I., (TL&-69)
Lohr, K.L., (JS&-63)

Lonardo, G. di, (LZ-69)
Longuet-Higgins, H.C., (L-49)
Loshadkin, N.A., (LM&-66), (NP&-67)
Lowry, T.H., (CL-65)
Lucken, E.A.C., (DLW-60), (GL-66), (LM-67)
Luderer, T.K.J., (RLB-72)
Lundin, R.E., (HLI-63), (HLI-64)
Lüttke, W., (LW-65)
Lyons, A.R., (LS-72)

Maccoll, A., (CM&-54)
MacDiarmid, A.G., (BM-70b)
Mackle, H., (MMS-69)
MacLagen, R.G.A.R., (M-75a)
MacQuarrie, R.H., (SH&-71a)
MacRae, D.M., (BMB-75), (BLM-67)
Madan, O.P., (SC&-67)
Maegi, M., (ER&-73)
Magnusson, E.A., (CM-56)
Mahler, W., (KMK-72), (MM-65),
Maillard, A., (KD&-69)
Maire, J.C., (LM-75), (LW-75)
Maiorana, S., (BM&-71)
Majoral, J.P., (ME&-71)
Makaeva, F., (LM&-66)
Makhmatkhanov, M.M., (TL&-69)
Makimo, F., (SM-72)
Malinski, E., (MPM-75)
Malisch, W., (SM-70)
Maner, R.J., (BB&-69)
Mangia, A., (BM&-71), (BMP-70)
Mangini, A., (BC&-75)
Mankowski-Favelier, R., (MM&-68), (MMS-68)
Mantz, I.B., (TSM-69)
Maracek, J.F., (SR&-76), (SRM-76)
Maria, P.C., (ME&-71)
Markl, G., (M-63), (MM-68)
Markov, S.M., (LM&-66)
Marsi, K.L., (DD&-69), (DD&-72), (EC&-70),
 (M-71), (M-74), (M-75), (MBC-72), (MC-70),
 (MO-73)
Martin, J.C., (AM-72), (AM-76), (KM-73),
 (LM-76), (MA-71), (MC-74), (MF-75),
 (MFA-74), (MP-76), (PM-72), (PMP-71),
 (PMP-74), (TBM-63)
Marquarding, D., (GH&-71), (GR&-73),
 (RP&-71), (UM&-70)
Maryanoff, B.E., (MS&-74), (MS&-75),
 (SM-74), (SS&-74)
Maryanoff, C.A., (MOM-75)
Masago, M., (FK&-73)
Maslen, E.N., (OM-67)
Masse, J., (CLM-68)
Mathey, F., (MM-72)
Mathis, F., (BGM-75), (WBM-75)
Matsumoto, S., (TH&-67)

Matsuyama, H., (MMK-75)
Matthews, N., (DG&-64)
Mavel, G., (MM&-65), (MM&-68)
Maxwell, J.I., (HPM-70)
Mazeline, C., (LM-67)
Mazur, R.H., (RM-51)
Mazzola, E., (BC&-73), (CD&-73)
McAdams, L.V., (CM&-71)
McClory, M.R., (DF&-70), (DVM-71)
McCreary, M.D., (KK&-75)
McDaniel, D.H., (MD-66)
McDonald, W.S., (MC-67)
McDowell, C.A., (FH&-72)
McEachan, C.E., (MMS-66)
McEntee, Jr., T.E., (CMW-70)
McEwen, W.E., (MA&-65)
McFarland, C.W., (OM-71)
McGrath, L., (MP-72)
McKeever, B., (SR&-76)
KcKinnie, B.G., (CFM-75)
McLean, R.A.N., (FH&-71), (FH&-72), (M-73)
McLick, J., (SMG-72)
McNally, D.V., (MMS-69)
McNeal, J.P., (HBM-71)
McOsker, C.C., (DMW-76)
McPhail, A.T., (EK&-76), (MMS-65)
Mellor, M.T.J., (AHM-72)
Melnikoff, A., (CM&-66)
Melvin, L.S., (TM-75)
Meppelder, F.H., (MBK-70)
Mertz, C., (FM-61)
Merz, A., (MM-68)
Messing, A.W., (SH&-71)
Meyer, B., (MG&-72), (MS-72)
Meyers, F.F., (CWM-64)
Meyers, J.K., (JM-72)
Michael, K.W., (SH&-71)
Michalski, J., (SOM-76)
Michelman, J.S., (OL&-66)
Miellillo, J.T., (MS&-64)
Milbrath, D.S., (CM&-76), (MS&-76)
Miles, J.A., (HHM-74)
Miles, T.S., (HJ&-70)
Miller, W.B., (HBM-69), (HMT-64)
Milliren, C.M., (RWM-76)
Min, T.B., (BH&-76), (BM-76)
Minamido, I., (MU&-69)
Minato, H., (MMK-75)
Mironov, V.A., (MSE-63)
Mishra, S.P., (MS-74)
Mislow, K., (BM-70), (BM-70a), (BM-71), (BM-72), (DZ&-69), (EC&-70), (HW&-72), (M-70a), (MOM-75), (MS&-64), (MS&-74), (MS-75), (RAM-72), (RGM-68), (RMA-70), (SM&-74), (SS&-74), (TM-69)

Mitchell, H.L., (EM&-74)
Mitchell, K.A.R., (M-68), (M-69)
Modena, G., (LM&-70)
Modro, T.A., (GM&-68), (MPM-75)
Moeller, T., (ATM-68)
Moffatt, J.G., (HU&-69)
Moffitt, W.E., (KM-51)
Moir, R.Y., (RB&-65)
Moldowan, J.M., (RC&-72)
Molléere, P., (MB&-72)
Molléere, P.D., (CM-71), (MH-75)
Monaghan, J.J., (RR&-69), (RS&-68)
Montanari, F., (LM&-70)
Mooney, E.F., (CM&-66)
Moore, L.F., (BMP-67)
Mootz, D., (WM&-74)
Moreland, C.G., (MD&-76), (MDL-73)
Morgan, W.E., (MS&-71)
Moriarty, R.M., (M-64)
Mortimer, F.S., (M-57)
Morton, J.R., (M-63a)
Mosher, H.S., (BM-71a)
Mueller, C., (MSV-75)
Mueller, G., (MM-72)
Muetterties, E.L., (HHM-72), (KK&-75), (KM-68), (M-70), (MM-65), (MP-59), (TCM-64)
Müller, A., (GM&-68a), (MK&-67), (MNG-66), (MNG-67)
Muller, N., (MP-59a)
Mulliken, R.S., (M-55)
Munsch, B., (LM-70), (LM-72)
Murakami, M., (BK&-74)
Murdock, L.L., (MH-68)
Musher, J.I., (M-69a), (M-72)
Myers, D.K., (QBM-71)

Nagy, J., (NH-70), (NH-72), (NV-71)
Nasutavicus, W.H., (JGN-65)
Naumann, K., (DZ&-69)
Navech, J., (ME&-71)
Neimysheva, A.A., (LM&-66), (NEK-70), (NP&-67), (NS&-68)
Nelson, R.M., (NH-76)
Nenner, M., (EPN-67)
Neumann, H., (YK&-76)
Neuveitev, N.P., (N-66)
Newton, M.G., (NCB-66)
Ng, L.K., (FN-76)
Nibler, J.W., (RNW-75)
Niecke, E., (MK&-67), (MNG-66), (MNG-67)
Niemeyer, H.M., (SH&-71a)
Nimrod, D.M., (NFV-68)
Nishimoto, K., (OSN-76)
Nomura, R., (TT&-68)
Noth, H., (VN-68)

207

Nour, T.A., (BQ&-74)
Novick, S., (RH&-67)
Nowakowski, P., (NW-76), (WNB-76)
Nyburg, S.C., (LN-71)
Nyholm, R.S., (CM&-54)

Oae, S., (MU&-69), (OTO-64), (OTO-64a), (OY&-68), (PO-62), (TFO-70), (TT&-68), (YO-69), (YO-70)
Oakenfull, D.G., (URO-70)
Oakland, R.L., (OD-67)
Oberlander, J.E., (MO-73)
O'Brien, D.H., (HO-71)
O'Brien, J.T., (KO-70)
O'Connell, A.M., (OM-67)
O'Donnell, K., (OB&-72)
Odubela, C.T., (DD&-74)
O'Dwyer, J.B., (BO-74)
Oediger, H., (HO-59)
Oehling, H., (OS-70)
Ogura, F., (MOM-75)
Ohkubo, K., (OK-71), (OK-72), (OK-72a), (OY-70), (OY-71), (OYF-69)
Ohno, A., (OTO-64), (OTO-64a)
Ojima, I., (OK-73)
Okamoto, T., (YK&-76)
Okruszek, A., (SOM-76)
Olah, G.A., (O-72), (OKS-69), (OM-71)
Oliver, W.L., (LPO-71)
Olofson, R.A., (OL-66), (OL&-66)
Olsen, J.F., (HO-76)
Omura, H., (KO-71)
Oommen, T.V., (MGB-72)
Orchin, M., (JO-62)
Orgel, L.E., (CM&-54)
Osaki, T., (OOK-73)
Osamura, Y., (OSN-76)
Osipov, O.A., (KOC-66)
Otera, J., (OOK-73)
Ovenall, D.W., (SO-72)
Owens, P.H., (SH&-71a)

Packard, B.S., (LPO-71)
Paddock, N.L., (CP-72), (P-64)
Paddon-Row, M.N., (BQ&-74)
Padley, J.S., (BMP-67)
Paetsch, J., (DP-68)
Pagani, G., (BM&-71), (BMP-70)
Paige, J.N., (HP-67)
Paldi, E., (TB&-70)
Palm, U.A., (NP&-67)
Palmieri, P., (BGP-74), (PZ-67)
Panar, M., (KPW-63)
Panse, P., (RPE-67)
Panshin, O.A., (BK&-65)
Pant, A.K., (PC-68)

Panteleeva, A.R., (BP&-68)
Paquette, L.A., (P-68), (PFW-71), (PH-69), (PH-71), (PPW-71)
Parasaran, T., (PH&-63)
Parkanyi, C., (ZP-65)
Parker, G.A., (SUP-72)
Pascoe, J.D., (BP-71)
Pasternack, G., (WP-67), (WP-69)
Patwardhan, A.V., (RP-68)
Paul, I.C., (PMP-71), (PMP-74)
Pauling, L., (P-50)
Pazdernik, L., (SP&-71)
Pearson, R.G., (PGB-60)
Pedone, C., (AW&-70)
Pennings, J.F.M., (DPB-75)
Perkins, R.G., (P-67)
Perozzi, E.F., (MP-76), (PM-72), (PMP-71), (PMP-74)
Pessini, F.B.T., (KPB-72)
Petersen, J.R., (DP-72)
Peterson, D.J., (P-66)
Peterson, L.K., (AP-65)
Peynircioglu, N.B., (AA&-74)
Pfohl, S., (GH&-71), (RP&-71)
Phillips, J.C., (HPM-70), (PPW-71)
Phillips, W.D., (MP-59)
Phillips, W.G., (PR-70)
Pichal, K., (EPN-67)
Piekos, A., (MPM-75)
Pillai, M.G.K., (PP-68)
Pillai, R.P., (PP-68)
Pillot, J.P., (CD&-75), (PDC-74), (PDC-75)
Pilot, J.F., (HC&-71), (RP&-71), (SC&-71)
Pinnaraia, T.J., (PCH-70)
Pisciotti, F., (CD&-75)
Pitt, C.G., (P-70), (P-71), (PCT-72)
Plachky, M., (RKP-74)
Plattner, G., (GPB-70)
Pletka, H., (HPH-70)
Podszun, W., (EP&-76)
Pohl, W., (RP&-71a)
Polekhin, A.M., (LM&-66)
Polk, M., (PH&-63)
Pople, J.A., (CS&-76)
Porschke, K.R., (SP&-74)
Porter, N.A., (PI-74)
Powell, L., (CC&-71)
Pratt, R.E., (CPH-66)
Prebble, K.A., (MP-72)
Price, C.C., (PH&-63), (PO-62)
Prince, R.H., (GP-69)
Pritchard, D.E., (MP-59a)
Pudjaatmaka, H.H., (SH&-71a)
Pudovic, A.N., (BB&-70), (BP&-68)
Pullman, B., (CP-64)
Pustinger, J.V., (DG&-64)

Queen, A., (BQ&-74)
Quin, L.D., (QBM-71)
Quinn, C.B., (QW-73)

Raban, M., (RC&-72), (RJ-71), (RKJ-69)
Rabinowitz, B.S., (SR-63)
Radeglia, R., (ER&-73)
Rakita, R.E., (DR-70)
Rakshys, J.W., (RTS-68)
Ramirez, F., (GH&-71), (GR&-73), (HC&-71),
    (R-68), (R-70), (RBS-68), (RGS-68),
    (RP&-68), (RP&-71), (RU-74), (SC&-67),
    (SC&-71), (SR&-76), (SRM-76), (UR-72)
Ramos de Carvalho, M.C., (ER-73)
Ramsey, B.G., (R-69), (RB&-74)
Randall, E.W., (RYZ-67), (RZ-66), (RZ-68)
Rankin, D.W.H., (AG&-70), (GR&-70),
    (R-69a), (RR&-69)
Ratts, K.W., (PR-70), (SR-65)
Rauk, A., (RAM-72), (RB&-65), (RMA-70),
    (RWC-69), (WRC-69)
Rayner, D.R., (CD&-70a), (RGM-68),
    (YR&-73)
Razubaev, G.A., (VRD-63)
Rebane, T.K., (R-59)
Reddy, G.S., (GR-62)
Reese, C.B., (KRS-70)
Reetz, M.T., (R-73), (RKP-74)
Reffy, J., (R-75)
Relles, H.M., (DR-64)
Rerat, B., (RR&-67)
Rerat, C., (RR&-67)
Rhee, S.G., (ER-75)
Richardson, Jr., D.I., (UR-70), (URO-70)
Ridd, J.H., (GM&-68)
Rieber, M., (WR-49)
Rieke, R.D., (RWM-76)
Rigau, J.J., (CG&-71)
Robert, J.B., (HW&-69)
Roberts, J.D., (RC-51), (RM-51)
Robiette, A.G., (AG&-70), (AG&-71),
    (GR&-70), (RR&-69), (RS&-68)
Robinson, E.A., (GR-63)
Rojhantalab, H., (RNW-75)
Ross, J.A., (BRL-77), (RSL-76)
Rossi, A.R., (HVR-74)
Rothuis, R., (FR&-74), (RFB-73), (RLB-72)
Rubenstein, P.A., (SH&-71a)
Rüchardt, C., (RPE-67)
Ruidisch, I., (SR-64)
Rudman, R., (RH&-67)
Rünchle, R., (RP&-71a)
Rundle, R.E., (HR-51)
Ruoff, A., (MK&-67)
Rusek, P.E., (BH&-76)

Rushton, B.M., (BEM-65)
Russell, D.R., (HRT-73)
Rynbrandt, R.H., (CA&-71), (CM&-71)

Saenger, W., (SE-72)
Safari, H., (BB&-71)
Saferstein, S.L., (DS-66)
Saito, S., (SM-72)
Sakorai, H., (TS-69)
Salomon, M., (S-75a)
Sands, D.E., (S-63)
Santry, D.P., (S-68), (SS-67)
Sarma, R., (SR&-76), (SRM-76)
Sato, S., (TSK-68)
Sato, T., (SSK-68)
Sauchik, Y.I., (NS&-68)
Saunders, M., (SG-66)
Saunders, V.R., (HS-69), (HS-70), (HS-70a)
Sawodny, W., (BB&-69a), (BGS-68), (BGS-70),
    (HSH-70), (S-69)
Sayanagi, O., (OSN-76)
Schaefer, H.F., (SS-75)
Schäfer, W., (SS-72), (SS&-76)
Schary, K.J., (SB-71)
Schiemenz, G.P., (S-71)
Schinski, W.L., (TSM-69)
Schipper, P., (SJB-75)
Schlegel, H.B., (BC&-75)
Schleyer, P. v. R., (CS&-76)
Schlosberg, R.H., (OKS-69)
Schlosser, M., (WWS-61)
Schmidbaur, H., (S-63a), (SM-70), (SR-64),
    (ST-68), (ST-68a)
Schmidpeter, A., (SBG-67), (SSH-75)
Schmidt, A., (S-68a)
Schmidt, M., (S-65a)
Schmieder, K.R., (SS-76)
Schmutzler, R., (ESS-73), (WM&-74)
Schneider, F.W., (SR-63)
Schoeller, W.W., (EP&-76)
Schomaker, V., (SS-41)
Schotte, L., (BCS-62)
Schowen, R.L., (OB&-72), (SP&-74)
Schriltz, D.M. v., (CD&-70a)
Schuder, F.J., (FS-69), (FS-69a), (FS-70),
    (FSW-72)
Schutt, J.R., (CST-69)
Schwarz, W., (CD&-73a)
Schweig, A., (MSV-75), (OS-70), (SS-72),
    (SS&-76), (WS-72), (WS-72a), (WS-72b)
Schweiger, J.R., (CS-70)
Schweizer, E.E., (AFS-75), (AFS-75a),
    (AS-76)
Schwenzer, G.M., (SS-75)
Scorrano, G., (LM&-70)

Index of Authors and References

Seckar, J.A., (ST-76)
Seebach, D., (GC&-76), (YK&-76)
Segal, G.A., (SS-67)
Seibl, J., (TF&-70)
Seiders, R.P., (RSL-76)
Seidl, H., (BAS-69), (BS-68)
Selgestad, J.M., (WS&-70)
Selve, C., (CS-74)
Semin, G.K., (NP&-67)
Senatore, L., (CSI-72)
Senkler, G.H., (MS&-74), (MS&-75), (SM&-74), (SS&-74)
Senning, A., (JS-67)
Seno, M., (ST&-75)
Seufert, W.G., (EP&-76)
Seyferth, D., (DW&-76), (S-75), (SAV-76), (SFH-66)
Shatenstein, A.I., (SG-69)
Sheldrick, G.M., (AG&-70), (AG&-71), (GR&-70), (RR&-69), (RS&-68)
Sheldrick, W.S., (ESS-73), (RR&-69), (RS&-68), (SSH-75)
Shelochenko, V.V., (LSD-70)
Sheppard, W.A., (RTS-68), (S-62), (S-71a), (SO-72), (ST-72), (SW-69)
Shih, L.S., (DS-74)
Shinski, W.L., (TS&-71)
Shiro, M., (SSK-68)
Siebert, H., (S-53)
Silhite, D.L., (SS-73)
Sim, G.A., (TS-68)
Simes, J.G., (SA-60)
Simm, G.A., (MMS-66)
Simmons, T., (MS&-64)
Simon, Z., (VBS-68)
Simonnin, M.P., (GLS-73)
Simpson, T.D., (BB&-69)
Slater, J.C., (S-30)
Sletten, E., (HSS-66)
Sletten, J., (HSS-66)
Slutsky, J., (KS-72), (SK-73), (SK-73a)
Smith, A.L., (SA-59)
Smith, C.P., (HC&-71), (RBS-68), (RGS-68), (RP&-68), (RP&-71), (SC&-67)
Smith, H.W., (JS&-63)
Smith, J.H., (SIK-72)
Shobleka, J., (VS-73)
Snyder, T.M., (JKS-62)
Sobolev, E.V., (MSE-63)
Soleiman, M., (CCS-76)
Solly, R.K., (FS-68)
Sommer, L.H., (GKS-65), (HS-73), (S-65), (SB-69), (SF-68), (SH&-71), (SKF-72), (SMG-72), (SUP-72), (WSG-47)
Songstad, J., (AS-65)
Sorenson, H.C., (SI-69)

Speziale, A.J., (SR-65)
Spialter, L., (ES&-74), (FS-71), (SP&-71), (SS-73)
Spiewak, J.W., (CA&-71), (CM&-71)
Spitzer, K., (MG&-72), (MS-72)
Spratley, R.D., (SHL-67)
Spratt, R., (JS-70)
Springer, J.P., (CM&-76), (MS&-76)
Stackhouse, J., (MS&-74), (MS&-75), (SM&-74), (SS&-74)
Stade, W., (DS-68)
Stanley, E., (DS-62)
Stanulonis, J.J., (KS-76), (KSB-76)
Stark, F.O., (SH&-71)
Stec, W.J., (MS&-71), (SOM-76)
Steele, W.V., (MMS-69)
Steitz, T.N., (SL-65)
Stepanyants, A.V., (BK&-65)
Stephens, F.S., (S-65b)
Steward, O.W., (SDJ-71)
Stewart, H.F., (SKW-70)
Stewart, J.C.M., (KRS-70)
Stevenson, D.P., (SS-41)
St. Janiak, P., (CTS-66)
Stohrer, W.D., (EP&-76), (S-74), (SS-76)
Streitwieser, A., (S-62a), (SH&-71a)
Strich, A., (SV-73)
Stroyer-Hansen, T., (MG&-72)
Stucky, G.D., (DW&-76)
Sturtz, G., (MMS-68)
Sutton, L.E., (CM&-54), (ES-63)
Swain, C.G., (SP&-74)
Swank, D., (SC&-67), (SC&-71)
Swansinger, W.A., (SP&-71)
Swern, D., (EK&-76)
Swingle, R.B., (DF&-70)
Symons, M.C.R., (LS-72), (MS-74), (S-72)
Szele, I., (SKW-76)

Tada, T., (TT&-68)
Taft, R.W., (GT-65), (RTS-68), (ST-72)
Tagaki, W., (MU&-69), (OTO-64), (OTO-64a), (TT&-68)
Takamizawa, A., (TH&-67)
Tamao, K., (TK-71)
Tamres, M., (KT-75)
Tamura, C., (TS-68), (TSK-68)
Tan, H.W., (BH&-76)
Tang, R., (TM-69)
Tarasenko, N.A., (NT-69)
Tavares, D.F., (TEB-70)
Tebby, J.C., (AT-70), (T-70)
Tenud, L., (TF&-70)
Terauchi, K., (TS-69)
Ternay, Jr., A.L., (MS&-64)
Thayer, J.S., (ST-76)

Thirunamachandran, T., (CT-66)
Thompson, H.W., (T-60)
Thompson, J., (JHT-70)
Thuong, N.T., (TC-75)
Thyagarajan, B.S., (T-69)
Tillett, J.G., (BTW-68), (DTV-62), (T-76a)
Todesco, P.E., (BT-75)
Toren, E.C., (PCT-72)
Torok, F., (TP&-70)
Tourigny, G., (DT-66)
Traetteberg, M., (AB&-63)
Traficante, D.D., (EM&-74), (KK&-75)
Trepka, R.D., (ATC-67), (CTS-66)
Trippett, S., (CDT-70), (CST-69), (HRT-73)
Trivedi, B.C., (CCT-69), (CTW-71)
Tronich, W., (ST-68), (ST-68a)
Trost, B.M., (LT-71), (T-76), (TA-73),
 (TAH-73), (TB-73), (TH-73), (TL-68)
 (TLA-69), (TM-75), (TS&-71), (TSM-69),
 (TZ-69), (TZ-71), (TZ-73)
Tsai, M.R., (ET-73)
Tsang, F.Y., (ATM-68)
Tsolis, E.A., (GH&-71), (RP&-71)
Tsuchiya, S., (ST&-75)
Tsujihara, K., (TFO-70)
Tsvetkov, E.N., (TL&-69)
Tuleen, D.L., (TBM-63)
Tullock, C.W., (TCM-64)
Tun, H.W., (BK&-74), (BT-72)
Turley, J.W., (BTF-68), (TB-68)
Turley, P.C., (CD&-73a)
Turnblom, E.W., (TK-71a)
Tyssee, D.A., (HMT-64)

Ugi, I., (ELU-75), (GH&-71), (GR&-73),
 (GU-71), (RP&-71), (RU-74), (UM&-70),
 (UR-72)
Ulbricht, K., (UJC-68)
Ulland, L.A., (SUP-72)
Ulrich, S.E., (HUZ-71)
Una, M., (BU-63)
Uneyama, K., (MU&-69)
Usher, D.A., (FU-67), (HU&-69), (U-69),
 (UEE-72), (UR-70), (URO-70)
Utley, J.H.P., (GM&-68)

Vahrenkamp, H., (VN-68)
VanCleave, W.C., (GDV-71)
VanderWerf, C.A., (MA&-65)
Vandorffy, M.T., (NV-71)
Van Wazer, J.R., (AV-72), (AV-72a),
 (HV-74), (HVR-74), (MS&-71)
VanWoerden, H.F., (DTV-62)
Vear, C.J., (HJ&-70)
Vedejs, E., (VS-73)
Veillard, A., (SV-73)

Veith, M., (BW&-76)
Vencl, J., (VH&-73)
Vergnani, T., (GC&-76)
Verkade, J.G., (CM&-76), (MS&-76), (NFV-68)
Vermeer, H., (MSV-75)
Viau, R., (DVM-71)
Vick, S.C., (SAV-76)
Vilceanu, R., (VBS-68)
Vilkov, L.V., (VT-69)
Vincent, G.A., (FVF-71)
Vogel, A.L., (CM&-66)
Vol'pin, M.E., (VK&-62)
Voncken, W.G., (VB-74)
Voronkov, M.G., (V-66)
Vorsanger, J.J., (V-68), (V-68a)
Vyazankin, N.S., (EV&-70), (EVK-72)
Vycudilik, M., (CV-70)

Wadsworth, W.S., (WT-74), (WW-76)
Wagner, G., (BWK-72)
Walker, N.S., (MD&-76)
Walsh, E.J., (AW-72), (AW-72a)
Walters, C.A., (KW-72)
Walton, D.R.M., (CEW-70), (EE&-75)
Wang, Y., (DW&-76)
Wanzlick, H.W., (W-62)
Warner, C.M., (BWL-67)
Watts, Jr., P.H., (AWW-73)
Webster, B.C., (CWM-64)
Weidner, U., (WS-72), (WS-72a), (WS-72b)
Weigmann, H.D., (WWS-61)
Weinberger, M.A., (GW-67)
Weiner, M.A., (WP-67), (WP-69)
Wells, P.R., (W-62a)
Weitl, F.L., (CTW-71)
West, C.T., (DD&-74), (DMW-76), (DW-75a),
 (DW-75b), (DW-75c)
West, R., (BW-71a), (CWG-69), (NW-76),
 (SKW-70), (W-69), (WB-59), (WB-63),
 (WB-72), (WB-73), (WB-73a), (WC-63),
 (WL&-71), (WNB-76)
Westheimer, F.H., (AW-73), (CW-63), (DW-66),
 (DW-66a), (HW-61), (KC&-69), (KPW-63),
 (KW-69), (LW-76), (SKW-76), (W-68)
Weston, J., (HW&-72)
Westwood, N.P.C., (FH&-71)
Wetzel, J.C., (AWC-75)
Whangbo, M.H., (BC&-75), (WWB-75)
Whatley, L.S., (WWL-61)
Wheeler, G.L., (AWW-73)
White, C.K., (RWM-76)
White, D.W., (CC&-71), (DWD-71)
Whitehead, M.A., (CW-72), (DLW-60)
Whitesides, G.M., (EM&-74), (KK&-75),
 (WEB-74), (WS&-70)
Whitmore, F.C., (WSG-47)

Index of Authors and References

Wiberg, N., (BW&-76)
Wieber, M., (WM&-74)
Wiebinga, E.H., (HW-59)
Wiegman, A.M. (BB&-59)
Wielesek, R.A., (BH&-76)
Wigfield, Y.Y., (DF-70), (FSW-72)
Wiggins, D.E., (BTW-68)
Wilde, R.L., (WW-76)
Wilhelm, K., (LW-65)
Wilhite, D.L., (ES&-74)
Wilkins, C.J., (RNW-75)
Wilkinson, G., (CW-66)
Williams, D.R., (CC&-71a), (CMW-70)
Williams, F., (GW-74)
Williams, J.M., (BWH-68)
Williams, L.D., (KCP-69)
Willson, M., (WBM-75)
Wilson, Jr., G.E., (AW&-70)
Wingard, R.E., (PPW-71)
Wipff, G., (LW-76a)
Wiseman, J.R., (QW-73)
Wittel, K., (BW&-76)
Wittig, G., (WF-64), (WK-64), (WR-49), (WWS-61)
Wolf, R., (HW&-69)
Wolfe, S., (BC&-75), (EY&-76), (RB&-65), (RWC-69), (WWB-75)
Wolfinger, M.D., (BW-71), (BW-74)
Wong, P.S., (DWJ-72), (JD&-72)

Wudl, F., (SW-69)
Wulfers, T.F., (KKW-65)
Wunderlich, H., (WM&-74)
Wyvratt, M.J., (PFW-71)

Yamabe, T., (OY-70), (OY-71), (OYF-69)
Yamagishi, I.G., (YR&-73)
Yamaguchi, H., (YK&-76)
Yamasaki, R.S., (CY-57)
Yano, Y., (YO-69), (YO-70)
Yates, K., (AYC-76)
Yates, R.L., (EY&-76)
Yeager, S.A., (AA&-74)
Yoder, O.H., (KY-72), (RYZ-67)
Yokayama, M., (OY&-68)
Yokoi, M., (Y-57)
Yoneda, S., (IYY-71)
Yosida, Z., (IYY-71)

Zahradnik, R., (Z-65), (ZP-65)
Zanger, M., (MA&-65)
Zauli, C., (BZ-65), (CZ-62), (LZ-69), (PZ-67)
Zeichner, A., (KZ-75)
Ziman, S.D., (TZ-69), (TZ-71), (TZ-73)
Zon, G., (DZ&-69)
Zoonebelt, S.M., (DD&-74)
Zuckerman, J.J., (HUZ-71), (RYZ-67), (RZ-66), (RZ-68)
Zwicker, E.T., (YR&-73)

IX. Index of Reference-Page Citations

A-60	119	B-74	155, 156	BM-70a	41
A-68	29	B-75	43	BM-70b	16, 66
A-70	22, 131, 132	B-75a	61	BM-71	41
A-75	173	B-76	110	BM-71a	155, 160
AAB-66	16	BA-70	57, 74	BM-72	48
AA&-74	167	BA-70a	56, 74	BM-76	8
AB-69	56, 70	BA-71	147	BM&-71	77
AB-70	22	BAS-69	61	BMB-75	142, 156
AB-73	95	BB-66	116, 117	BMP-67	27
AB-76	142, 143	BB&-59	88	BMP-70	44
AB&-63	15, 16	BB&-69	88, 146, 147	BO-74	78
AFB-69	57	BB&-69a	16	BP-71	155, 160
AFS-75	47	BB&-70	120	BP&-68	118
AFS-75a	49	BB&-71	30	BQ&-74	66
AG&-70	16, 22	BB&-74	111, 155	BRL-77	143
AG&-71	16	BB&-75	157	BS-68	57
AH-72	132	BB&-75a	157	BT-72	44
AHM-72	132	BB&-77	81, 82	BT-75	134
AK-67	28	BB&-77a	81	BTF-68	107, 108
AK-70	48	BBL-73	171	BTW-68	138
AKK-63	28	BC-69	93	BU-63	53, 66, 67
AM-72	171, 172	BC&-71	75, 92	BV&-75	95
AM-76	109, 142	BC&-73	42	BVV-76	82
AP-65	73	BC&-75	82, 88	BW-71	78
AS-65	118, 119	BCS-62	53	BW-71a	160, 161
AS-76	37, 164	BD-73	62	BW-74	78
AT-70	38	BD-74	155, 159	BW&-76	62
ATC-67	40	BD-74a	79	BWH-68	78
ATM-68	39	BD-74b	79	BWK-72	67
AV-72	20	BDC-70	91	BWL-67	154
AV-72a	20	BE-70	119, 120	BZ-65	4, 6
AW-72	132	BE-71	63		
AW-72a	132	BE&-76	63	C-60	88
AW-73	131, 133	BEM-65	158	C-61	14, 15, 24, 29, 72
AW&-70	146, 147	BF-68	127	C-66	29
AWC-75	144, 145	BF-69	18	C-66a	38
AWW-73	60	BF-70	154	C-68	52
AYC-76	102, 103	BG-63	53, 66	C-69	12
B-58	84, 87	BGM-75	46	CA&-71	79
B-60	98, 142, 167	BGP-74	66	CB-54	88
B-62	88	BGS-68	16	CC-63	27
B-64	94	BGS-70	16, 25, 26, 28, 29	CC-71	80, 92
B-64a	29, 31, 33	BH-65	101, 109	CC&-71	111
B-64b	31, 32, 33	BH-71	22	CC&-71a	53
B-65	31, 32, 33	BH&-76	8	CCG-73	174
B-68	40	BJB-74	9	CCL-70	153
B-68a	76, 80	BJK-72	133	CCS-76	128
B-68b	77, 80	BK&-65	40	CCT-69	124
B-68c	16, 30	BK&-74	136	CD-68	21, 165
B-69	20, 27, 60	BLM-67	63, 154, 160	CD&-70	42
B-69a	26	BM-62	145	CD&-70a	139
B-70	121	BM-70	41	CD&-73	41

213

Index of Reference-Page Citations

CD&-73a	122	DH-69	30	EV&-70	33
CD&-75	177	DK-76	41	EVK-72	33
CDD-71	39	DKF-75	16	EY&-76	82
CDT-70	122	DLW-60	13, 35, 39, 81	F-64	108
CE-71	16, 67, 69	DM-75	76	F-72	164
CEW-70	153, 159, 176	DMW-76	178	FCI-61	117, 120
CF-68	139	DP-68	111	FDC-70	100
CF-72	133	DP-72	127	FG-70	83
CFM-75	152	DPB-75	9	FH&-71	70
CG-69	176	DR-64	101	FH&-72	52
CG&-71	22, 25	DR-70	154	FK&-73	34, 46, 51
CH-69	151, 158, 159	DS-62	18	FM-61	165
CH-76	9	DS-66	101	FN-76	76
CL-65	74	DS-68	36	FS-68	156
CL-71	153	DS-74	169	FS-69	75, 91
CLM-68	153	DT-66	140	FS-69a	91
CM-56	4, 81	DTV-62	138	FS-70	91
CM-71	67	DU-69	91	FS-71	50
CM&-54	5	DVM-71	78, 91, 92	FSW-72	41, 50
CM&-66	32	DW-66	18, 134	FU-67	126
CM&-71	92	DW-66a	121, 133	FVF-71	108
CM&-76	108	DW-75a	177	G-59	38, 159
CMW-70	82, 93	DW-75b	177	G-60	7, 96
CP-64	84	DW-75c	177	G-63	7, 8
CP-72	39	DW&-76	64	G-67	17
CPH-66	36, 44, 88	DWD-71	122, 134	G-69	74
CS-70	44, 45	DWJ-72	57	G-70	17, 24
CS-74	174	DZ&-69	124	G-71	51
CS&-76	12	E-60	3, 12	GB-66	117
CST-69	124	E-63	16, 25, 29	GBC-74	100
CST-72	147	E-68	130	GC&-76	106, 107
CT-66	4	E-69	15, 16, 25, 32	GCA-64	65
CTS-66	41, 74, 90	E-70	130	GCF-70	140
CTW-71	125	E-74	174	GDV-71	90
CV-70	61	EA-72	174	GD-76	25
CW-63	121	EB-69	175	GH-68	11, 19
CW-66	5, 8	EBB-74	57	GH-70	122
CW-72	21	EC&-70	124	GH&-68	12, 19, 57
CWG-69	63	EE&-75	89	GH&-71	98, 105, 126
CWM-64	4, 6	EF-63	32	GHB-70	68
CY-57	11	EG-76	157	GI-73	59
CZ-62	5, 6, 81	EHA-74	83	GJC-73	140
D-64	30	EJ-74	151, 177	GK-68	173
D-70	23	EK&-76	61	GK-69	61
D-75	89	ELU-75	105	GKB-69	58, 59
DA-76	126	EM-67	67	GKB-69a	59, 62
DB-73	61, 63, 155	EM&-74	115	GKJ-70	61
DD&-69	114	EP&-76	105	GKS-65	52
DD&-72	169	EPN-67	116, 120	GL-66	52
DD&-74	177	ER-73	68	GL-71	26, 27
DD&-76	51	ER-75	88, 89	GLS-73	50
DDH-73	147	ER&-73	51	GM&-68	27, 90
DF&-70	75, 79, 90, 91, 92, 95	ES-63	40	GM&-68a	74
		ES&-74	49	GP-69	153
DG&-64	37	ESS-73	102	GPB-70	58
DG&-73	154	ET-73	155	GR-62	33, 34

GR-63	14, 17, 27	HU&-69	131	KPW-63	138
GR&-70	15, 24	HUZ-71	50	KRS-70	175
GR&-73	131	HV-74	69	KS-72	154, 155
GT-65	52	HVR-74	113	KS-76	141, 142, 143, 145
GU-71	105, 111	HW-59	10, 11		
GW-67	119	HW-61	121, 137	KSB-76	142, 143
GW-74	9	HW&-69	102	KT-75	90
H-55	16	HW&-72	11, 165	KW-69	136
H-56	117	IYY-71	60	KW-72	148
H-60	6	J-54	5	KY-72	41
H-64	85	J-66	164	KZ-75	42
H-65	111	JA-68	73	L-49	52, 85
H-65a	3	JA-69	27	L-57	118
H-66	111	JC-74	139	L-69	144
H-66a	112	JD&-72	57	L-70	40
H-66b	112	JGN-63	65	L-72	45
H-71	116, 119	JHT-70	160	LC-69	74
H-74	102, 104	JJ-71	147	LCM-76	76
H-75	97	JJ-75	58	LD-61	85
HA-70	126	JJ-75a	58	LF-75	156
HAK-69	75, 79, 80, 91	JKS-62	23	LG-76	105
HB-72	138	JM-72	88	LH-71	110
HBG-70	20, 24	JO-62	52, 55	LM-67	61
HBL-72	169	JS-67	45	LM-70	41
HBM-69	84, 85	JS-70	41	LM-72	51
HBM-71	87, 94, 95	JS&-63	17, 18	LM-75	70
HC&-71	112	K-57	28	LM-76	145
HD-68	112	K-59	25	LM&-66	117, 118, 120
HF-68	3, 11	K-61	175	LM&-70	148
HH&-69	68	K-68	140	LN-71	23
HHH-74	22	K-69	19, 20	LPO-71	45
HHM-72	96, 122, 169	KA-69	55	LS-72	59
HHM-74	34	KB-66	87, 95	LSD-70	117, 120
HJ&-70	30	KB-77	142, 145, 156, 157	LT-71	166, 167, 168
HK-56	117			LW-65	27
HK-60	117	KC&-69	121, 131, 136	LW-75	19
HK-70	26	KD-73	95	LW-76	131
HL-64	173	KD&-69	28	LW-76a	83, 88
HL-72	120	KH-73	122	LZ-69	52
HLI-63	85	KJ-70	141, 143, 145	M-55	38
HLI-64	86	KJ-77	142	M-57	26
HMT-64	33, 34	KK-71	59	M-63	36
HO-59	27	KK-72	59	M-63a	8
HO-71	159	KK-75	97	M-64	45, 46
HO-76	8	KK&-75	110	M-68	5, 6
HP-67	92	KKF-73	34, 46, 51	M-69	13, 21, 28, 39, 77
HPH-70	30	KKK-67	23	M-69a	11, 28, 81, 96, 104, 109
HPM-70	91	KKW-65	18, 137		
HR-51	10	KM-51	17, 24, 27, 55	M-70	109, 124, 134
HRT-73	102	KM-68	110	M-70a	122
HS-69	13	KM-73	171	M-71	124
HS-70	13	KMK-72	9	M-72	96, 113
HS-70a	13	KO-70	84	M-73	56
HS-73	163	KO-71	140, 141	M-74	134
HSH-70	31	KOC-66	90	M-75	124
HSS-66	20	KPB-72	68	M-75a	15

Index of Reference-Page Citations

MA-71	104, 171	NW-76	157	R-63a	97
MA&-65	125	O-72	105	R-68	96, 100, 101, 171
MB&-72	57	OB&-72	151, 177	R-68a	26
MB&-75	81	OD-67	12	R-69	52, 57, 58, 61, 62, 63, 88, 89
MBC-72	123, 124	OK-71	54		
MBK-70	46	OK-72	54, 55	R-69a	16
MC-67	15, 24	OK-72a	54	R-70	96, 100, 171
MC-70	124	OK-73	46	R-73	157
MC-74	145	OKS-69	104	R-75	57, 89
MD-66	105, 113	OL-66	85	RAM-72	96, 163
MD&-76	100	OL&-66	85	RB&-65	91
MDL-73	100, 114	OM-67	17, 18	RB&-74	61
ME&-71	31	OM-71	49	RBS-68	101
MF-75	171	OOK-73	62	RC-51	106
MFA-74	171	OS-70	36	RC&-72	47
MG&-72	54	OSN-76	52	RF&-74	9
MH-68	119	OTO-64	53, 81	RFB-73	9
MH-75	63	OTO-64a	81, 83	RGM-68	145
MK&-67	26	OY-70	55	RGS-68	170, 171
MM-65	100	OY-71	54	RH&-67	113
MM-68	36	OY&-68	142, 148	RJ-71	42, 45, 47
MM-72	88	OYF-69	54	RKJ-69	42
MM&-68	51	P-50	11	RKP-74	157
MMK-75	173	P-64	13, 21, 27, 28, 39, 81	RLB-72	9
MMS-66	17			RM-51	104
MMS-68	49	P-66	88	RMA-70	40
MMS-69	91, 93	P-67	16, 25, 34, 41, 75	RNW-75	15
MNG-66	26	P-68	78, 79, 170	RP&-68	113
MNG-67	26, 27	P-70	56, 61, 69, 89, 159	RP&-71	114
MO-73	135			RP&-71a	31
MOM-75	105	P-71	56, 57, 89	RPE-67	164
MP-59	109, 110	PC-68	15	RR&-67	17, 18
MP-59a	33	PCH-70	161	RR&-69	16
MP-72	54	PCT-72	63	RS&-68	16
MP-76	104, 138	PDC-74	176	RSL-76	143, 144
MPM-75	90	PDC-75	176	RTS-68	39
MS-72	54	PFW-71	74	RU-74	98, 123, 175
MS-74	8	PGB-60	143	RWC-69	40, 42, 74, 75, 76, 80, 92, 95
MS&-64	140	PH-69	78		
MS&-71	49	PH-71	92	RWM-76	70
MS&-74	35	PH&-63	34, 35	RYZ-67	48, 72
MS&-75	35	PI-74	155	RZ-66	34
MS&-76	19	PM-72	113	RZ-68	34, 50
MSE-63	154	PMP-71	104, 138, 148, 170, 171	S-30	3
MSV-75	79			S-53	15, 25
MU&-69	81	PMP-74	148	S-62	109
N-66	78	PO-62	52, 53, 54	S-62a	54
NCB-66	18, 27, 121	PP-68	10	S-63	24, 32
NEK-70	120	PPW-71	93	S-63a	33
NFV-68	26	PR-70	90	S-65	149, 150, 151, 152, 153
NH-70	68	PZ-67	6		
NH-72	70	QBM-71	46	S-65a	54
NH-76	85, 86	QW-73	43	S-65b	20, 21
NP&-67	116, 117, 118	R-59	54	S-68	5, 12
NS&-68	117, 118	R-63	14, 15, 18, 19, 25, 28	S-68a	27
NV-71	57			S-69	27

Index of Reference-Page Citations

S-71	56, 89	ST-68	36	VH&-73	153, 176
S-71a	98, 109	ST-68a	36, 37	VK&-62	65
S-72	59	ST-72	39, 171	VN-68	32
S-74	106	ST-76	154	VRD-63	176, 177
S-75	63, 64, 65	ST&-75	38	VS-73	164
S-75a	73	SUP-72	162	VT-69	16
SA-59	29, 151	SV-73	96	W-62	86
SA-60	17	SW-69	139, 140	W-62a	120
SAV-76	65	T-60	29	W-68	96, 121, 122
SB-69	152	T-69	74	W-69	155, 159, 161, 162
SB-71	135	T-70	37	W-72	22, 42, 83
SBG-67	48	T-76	33	W-73	131, 135, 136, 137
SC&-67	18, 19, 26	T-76a	141	WA-69	85
SC&-71	104	TA-73	167	WB-59	71
SDJ-71	69	TAH-73	168	WB-63	65
SE-70	130	TB-68	107	WB-67	104
SF-68	150	TB-73	107	WB-72	155
SFH-66	168	TBM-63	145	WB-73	155, 163
SG-66	77	TC-75	175	WB-73a	162
SG-69	80	TCM-64	109	WB&-70	28
SH&-71	153	TEB-70	93	WBM-75	171
SH&-71a	76	TFO-70	27, 61, 66	WC-63	12
SHL-67	21	TF&-70	105, 114	WEB-74	115
SI-69	94	TH-73	172	WF-52	11
SIK-72	148	TH&-67	86	WH-69	78, 93, 154, 156, 169
SJB-75	8	TK-71	159	WH-70	140, 149
SK-73	154, 156, 159	TK-71a	97	WH&-75	15
SK-73a	156	TL-68	166	WK-64	97
SKF-72	150	TL&-69	88	WK-71	53
SKW-70	161	TLA-69	5	WK-74	98, 136
SKW-76	114	TM-69	138	WL&-71	155, 163
SL-65	121	TM-75	20, 107, 164, 172	WM&-74	102
SM-70	38	TN-60	16	WNB-76	158
SM-72	24	TP&-70	22, 27, 31	WP-67	88
SM&-74	35	TS-68	17	WP-69	88
SMG-72	150	TS-69	62	WR-49	97
SO-72	109	TS&-71	170	WRC-69	40, 50, 74, 75, 79, 80, 91
SOM-76	136	TSK-68	34		
SP&-71	161	TSM-69	170	WS-72	57
SP&-74	162	TT&-68	81	WS-72a	69
SR-63	154	TZ-69	93	WS-72b	69
SR-64	32	TZ-71	169	WS&-70	33
SR-65	21, 27	TZ-73	169	WSG-47	158
SR&-76	102	U-69	130	WT-74	136, 137
SRM-76	102	UEE-72	130	WW-76	137
SS-41	14, 15, 101	UJC-68	72	WWB-75	20, 164
SS-67	12	UM&-70	98	WWL-61	32, 72
SS-72	58, 88	UR-70	130	WWS-61	164, 165
SS-73	163	UR-72	96, 99	Y-57	16
SS-75	9	URO-70	128	YK&-76	106
SS-76	106, 107, 151	V-66	107	YO-69	80, 81
SS&-74	35	V-68	85	YO-70	94
SS&-76	36	V-68a	94	YR&-73	139, 143
SSH-75	104	VB-74	129	Z-65	52
SSK-68	17, 18	VBS-68	36	ZP-65	52

X. Subject Index

acidity-basicity: 32, 71–73, 76, 81–82, 90, 121–122, 131, 136, 148
activation energy: 83, 99, 103, 110, 112, 114, 117, 138, 141, 142, 143, 144, 156, 157, 161, 167, 170
addition reaction: 88–89
adenosine triphosphate: 129
allylic rearrangements: 141–145, 153–154, 155–156
anisotropy: 32, 33
apical-basal preference: 8, 9, 96–98, 100, 105, 108, 109, 111, 122, 124, 125, 126, 135, 151
Arbuzov reaction: 173–174
aromaticity: 34
axial-equatorial preference in phosphoranes: see apical-basal preference

barrier: see energy barrier
bathochromic shift: 53, 54, 55, 56, 61, 62, 63, 69, 88
Berry pseudorotation: see pseudorotation
betaine: 21, 164–165
biaryls: 166–167
bond lengths and angles: 14, 16, 18, 20, 101, 102, 165
bond order: 15, 18, 20, 27, 31, 107–108
bonding
    $d\pi$: 2–3, 13, 15, 16, 19, 33, 34, 43, 45, 46, 47, 48, 49, 52, 55, 56, 57, 58, 60, 68, 70, 71, 72, 73, 76, 80, 81–82, 88, 89, 90, 93, 96–97, 101, 102, 112, 115, 116, 119, 120, 122, 126, 139, 162, 163, 171, 175
    $d\sigma$: 2–3
    $p\pi$: 2–3, 43, 44, 46, 81, 88, 90, 93, 128, 156
    three center: see hypervalency
bridgehead effects: 43

carbanion: 42, 50, 74–83, 85, 91–94, 123–126, 155, 165–166
carbene: 87, 94
carbonium ion: 104–105, 153, 158, 159, 171, 175–177
charge transfer complex: 56, 57, 72–73, 87, 135
chlorination: 73–74
cholinesterase inhibitors: 116–120
configuration
    pyramidal: 25, 74–76
    square pyramidal: 7, 8–9, 102–104
    tetrahedral: 7, 105, 123, 139
    trigonal bipyramidal: 7, 8–9, 107, 108, 121–137, 146, 148, 154
conformation: 42, 44, 46, 47, 51, 75, 83, 91, 95
conjugation: 17, 34, 67, 70, 81–82, 87, 88, 89, 91, 92
coupling reaction: 165–167
cyanosilanes: 154

decarboxylation: 82–85, 94
dehydration: 171–172
delocalization: 13, 32, 57, 77, 101
    "island": 13, 67, 71
dipole moment: 66

effective nuclear charge: 3
electron affinity: 6, 117
electron density: 54, 60
electron diffraction: 16, 22, 25
electronegativity: 4, 26, 32, 46, 71, 80, 119, 155
energy barrier
    inversion: 40, 45, 144
    rotation: 40–42, 45, 46
    polytopal rearrangement: 98–100, 102, 109, 142
entropy of activation: 45, 117–119, 141, 142, 144, 156, 157
episulfides: 93, 169
episulfones: 77–80, 92
epoxides: 171–172
equilibrium: 123–124, 153, 161, 171
exchange reactions: 34, 36, 75–76, 79, 85, 86, 95, 115, 121, 136, 140, 148, 153

force constant: 25, 27, 30
fragmentation: 168

gauche effect: 42, 43, 83

half-wave reduction potential: 56, 59, 70
homoconjugation: see overlap
hybridization: 12, 22, 45, 96
hydridosilanes: 29, 58, 150, 161, 177–178
hydrolysis: 18, 116–120, 121–136, 137, 147, 162
hyperconjugation: 41–42, 46, 47, 57–58, 59, 62, 69, 79, 81–82, 126, 159, 177
hypervalency: 10, 11, 12, 19–20, 81, 96, 104, 105, 109, 111, 143, 154

Subject Index

inductive effect: 29, 32, 33, 41, 47, 49, 56, 57, 61, 62, 65, 68, 70, 74, 79, 88, 116
inversion or retention of configuration: 41, 74, 122–123, 134, 135, 136, 137, 139, 145, 147, 148, 155, 162
ionization potential: 61, 67, 68, 69
isotope effect: 141, 151, 157, 160
isotope exchange: see exchange reaction

linear free energy relationships: 29, 34, 39, 49, 50, 120, 150, 151, 153, 156, 161
lithium, organic: 75–76, 78, 165, 166, 167
lone pair: 8, 9, 42, 44, 54, 62, 71, 75, 104, 138, 143, 144, 148

MO calculations: 4–6, 8–9, 11, 12, 19, 22, 36, 41, 55, 59, 69, 70, 79, 81–82, 112, 113, 126–127, 151, 163, 164

nitration: 74, 84

octet expansion: see hypervalency
optical activity: 41, 78, 111, 149, 150, 160
orbitals
    in free atom: 2, 4
    size: 3
    p: 154, 156, 159
    d: 2, 12, 54, 55, 57, 79, 81, 82, 84, 94, 126, 143, 148, 155
    HOMO: 56, 58, 61–62, 63, 69
    LUMO: 56, 61–62, 63, 67
    degenerate: 56
    σ^*: 57–58, 79, 87
    π^*: 58, 79, 126
overlap: 20, 53, 57, 62, 63, 81–82, 83, 87, 94, 96, 143, 154, 157, 166, 167, 173
oxidation: 54
oxygen: 54
ozone: 161

pentacoordinate silicon: 107–109, 149–163, 175
peroxides: 54, 101
phosphabenzenes: 36
phosphate esters: 18, 19, 26, 116–120, 121–122, 128–131, 174–175
phosphazenes: 13, 21, 27–28, 39, 132
phosphonium salts: 123–126, 127, 131–132, 173–174
phosphoranes: 4, 96–104, 164–165, 168–170, 171
    pentaaryl: 97, 111–112
    oxy: 100–104, 111, 121–137
phosphoranyl radicals: 8–9

phosphorylation: 174–175
point charge model: 5
polarizability: 81–82
polysilanes: 63
pre-activation factor: see entropy of activation
promotion energy: 6, 10
pseudorotation: 98–100, 102, 109, 110, 111, 112, 113, 121–137, 140, 142, 144, 147, 161–163, 167, 170, 177–178
pyramidal inversion: see energy barrier

racemization: 140–141, 152–153
radical: 8, 59, 145, 163
radical anion: 58, 63, 68
radical contraction: 4, 10, 13, 81
Ramberg-Bäcklund reaction: 78–80, 92
reduction potential: see half-wave reduction potential
resonance: 19, 21, 38, 53, 60, 81–82
retention: see inversion or retention of configuration
ribonucleic acid: 130–131
ring strain: 18–19, 98, 121, 124, 135, 138, 144, 152, 170

screening: 3
Siebert formula: 25
sigmatropic rearrangement: 154, 156
silacyclopropanes: 63–65
silatranes: 107–108
siliconium ion: 152, 158–159
silylcarbinols: 152, 154–155
silyl ketones: 61–62
silylamines: 16, 26, 70, 73, 157, 158, 161
siloxanes: 15, 16, 25, 72
Slater's rules: 3, 4, 60
solvation: 73, 75, 78, 91, 145, 149, 150, 153, 156
spectroscopy
    IR: 25–31, 50, 71–72, 151, 162
    UV: 52–58, 60–63, 66, 80
    microwave: 23, 24
    Raman: 25–31
    NMR: 32–51, 64, 68, 72, 75–76, 98, 113, 114, 122, 131, 138, 143, 147, 164,
    ESR: 57, 59, 68, 70
    photoelectron: 57, 61, 67, 70
steric hindrance: 11, 91, 93, 113, 115, 117, 125, 158, 178
substitution
    aromatic: 73–74
    S_N2: 106–107
    at silicon: 150–163

Subject Index

sulfides, organic: 52, 53, 80–82, 93
sulfonium salts: 54, 171–172
sulfur, elemental: 54
sulfur hexafluoride: 6
sulfuranes: 11, 98, 104, 109, 137–149, 165–169, 171–173
superdelocalizability: 117, 120
symmetry: 42, 103, 126

tetrathioethylenes: 53
thiabenzenes: 35
thiamine: 84–87
thioallylic rearrangement: see allylic rearrangements
thiophenes: 52
thiothiophthenes: 19–20

transition state: 89, 103, 104, 117, 122, 141–142, 149, 157, 158, 159, 160, 173
turnstile rotation: 98, 102–104, 105, 109–110, 113, 115, 122, 149, 150, 151, 152, 178

Wittig reaction: 21, 164–165
Woodward-Hoffmann rules: 78, 93, 169

X-ray diffraction: 10, 18, 20, 22, 24, 101–102, 104, 106, 108, 112, 113, 143, 146, 164

ylides: 20–21, 24, 37, 60–61, 73, 107, 142–143, 157, 164–165
    thiazolium: 84–87, 94–95

Reactivity and Structure

Concepts in Organic Chemistry
Editors: K. Hafner, J.-M. Lehn, C. W. Rees,
P. von Ragué Schleyer, B. M. Trost, R. Zahradník

Volume 1: J. TSUJI

Organic Synthesis by Means of Transition Metal Complexes

A Systematic Approach
4 tables. IX, 199 pages. 1975
ISBN 3-540-07227-6

This book is the first in a new series, *Reactivity and Structure: Concepts in Organic Chemistry,* designed to treat topical themes in organic chemistry in a critical manner. A high standard is assured by the composition of the editorial board, which consists of scientists of international repute. This volume deals with the currently fashionable theme of complexes of transition-metal compounds. Not only are these intermediates becoming increasingly important in the synthesis of substances of scientific appeal, but they have already acquired great significance in large-scale chemical manufacturing. The new potentialities for synthesis are discusses with examples. The 618 references bear witness to the author's extensive coverage of the literature. This book is intended to stimulate organic chemists to undertake further research and to make coordination chemists aware of the unforeseen development of this research field.

Contents: Comparison of synthetic reactions by transition metal complexes with those by Grignard reagents. Formation of σ-bond involving transition metals. – Reactivities of σ-bonds involving transition metals. – Insertion reactions. – Liberation of organic compounds from the σ-bonded complexes. – Cyclization reactions, and related reactions. – Concluding remarks.

Volume 2: K. FUKUI

Theory of Orientation and Stereoselection

72 figures, 2 tables. VII, 134 pages. 1975
ISBN 3-540-07426-0

The "electronic theory" has long been insufficient to interpret various modern organic chemical facts, in particular those of reactivity. The time has come for a book which shows clearly what is within, and what is beyond, the reach of quantum-chemical methods. Graduate students and young researchers in chemistry, both theoretical and experimental, will find this book an invaluable aid in helping them to become accustomed to the quantum-chemical way of thinking. Theory produces new experimental ideas, and, conversely, a host of experimental data opens new theoretical fields. A book such as the present one will constantly maintain its value, although the quantum-chemical approach to the theory of reactivity is, of course, still in the developmental stage.

Contents: Molecular Orbitals. – Chemical Reactivity Theory. – Interaction of Two Reacting Species. – Principles Governing the Reaction Pathway. – General Orientation Rule. – Reactivity Indices. – Various Examples. – Singlet-Triplet Selectivity. – Pseudoexcitation. – Three-species Interaction. – Orbital Catalysis. – Thermolytic Generation of Excited States. – Reaction Coordinate Formalism. – Correlation Diagram Approach. – The Nature of Chemical Reactions.

Springer-Verlag
Berlin
Heidelberg
New York

Lecture Notes in Chemistry

Editors:
G. Berthier, M.J.S. Dewar, H. Fischer, K. Fukui,
H. Hartmann, H.H. Jaffé, J. Jortner, W. Kutzelnigg,
K. Ruedenberg, E. Scrocco, W. Zeil

Volume 1:
G.H. WAGNIÈRE

Introduction to Elementary Molecular Orbital Theory and to Semiempirical Methods

33 figures. V, 109 pages. 1976
ISBN 3-540-07865-7

The aim of these notes is to provide a summary and concise introduction to elementary molecular orbital theory, with an emphasis on semiempirical methods. Within the last decade the development and refinement of *ab initio* computations has tended to overshadow the usefulness of semiempirical methods. However, both approaches have their justification. *Ab initio* methods are designed for accurate predictions, at the expense of greater computational labor. The aim of semiempirical methods mainly lies in a semiquantitative classification of electronic properties and in the search for regularities within given classes of larger molecules. Applications to optical activity, concerted reactions and to polymers are included. (34 references)

Contents: The hierarchy of approximations. − Simple Hückel theory of π electrons. − Many-electron theory of π electrons. − Self-consistent-field (SCF) methods. − All-valence MO procedures. − Special topics. − References. − Subject index.

Volume 2:
E. CLEMENTI

Determination of Liquid Water Structure, Coordination Numbers for Ions and Solvation for Biological Molecules

32 figures, 18 tables.
VI, 107 pages. 1976
ISBN 3-540-07870-3

The structure of liquid water and the solvation of ions and molecules represent an active field of past and current research. The authors have stressed in particular the new quantum mechanical developments that constitute the conceptual base for the recent advancements in this field. They have pointed out, with a variety of examples, the large amount of information embodied in quantum mechanics. The study of solvation represents the ideal field to pass from small chemical systems to large ones, to pass from quantum to statistical mechanics with a first step towards thermodynamics. Particular attention is given to present a unified picture that coherently retains technique and assumption in passing from one type of the description of matter to another. This work is dedicated to Prof. Per-Olov Löwdin, on the occasion of his 60th birthday. (27 references)

Contents: Description of a Chemical System as a Set of m Fixed Nuclei and n Electrons. − Structure of Liquid Water as a Test Case. − Coordination Numbers and Solvation Shells.

Springer-Verlag
Berlin
Heidelberg
New York

STRY LIBRARY

642-3753